Marine Disease Ecology

Marine Disease Ecology

EDITED BY

Donald C. Behringer
University of Florida, USA

Brian R. Silliman
Duke University, USA

Kevin D. Lafferty
U.S. Geological Survey at the University of California Santa Barbara, USA

OXFORD
UNIVERSITY PRESS

Great Clarendon Street, Oxford, OX2 6DP,
United Kingdom

Oxford University Press is a department of the University of Oxford.
It furthers the University's objective of excellence in research, scholarship,
and education by publishing worldwide. Oxford is a registered trade mark of
Oxford University Press in the UK and in certain other countries

First Edition published in 2020

Impression: 1

Published in the United States of America by Oxford University Press
198 Madison Avenue, New York, NY 10016, United States of America

British Library Cataloguing in Publication Data
Data available

Library of Congress Control Number: 2019945710

ISBN 978–0–19–882163–2 (hbk.)
ISBN 978–0–19–882164–9 (pbk.)

DOI: 10.1093/oso/9780198821632.003.0001

Printed and bound by
CPI Group (UK) Ltd, Croydon, CR0 4YY

Preface

As the human population grows ever larger—predicted to reach 9 billion by 2050—there will be increasing pressure on marine ecosystems and the resources they provide. These ecosystems are being subjected to stressors from ocean warming, ocean acidification, coastal eutrophication, overfishing, and habitat degradation. Rising from among these threats is an awareness that infectious disease can emerge as a consequence of environmental and biological stress and can sometimes drive ecosystem change. The rising profile of infectious disease is also concomitant with a broader realization that the parasites responsible for disease are themselves important members of communities. Thus, critical for understanding any marine population, community, or ecosystem is a comprehensive understanding of the relationships between hosts, pathogens, and abiotic and biotic stress in those systems. Infectious diseases are in the ocean whether we like it or not.

One overarching theme that pervades this book is that the ecological and epidemiological patterns and processes on land and in the air do not necessarily apply in the ocean. For example, seawater is 800 times denser and 50 times more viscous than air. These physical properties keep particles, such as infectious microbes, floating on the surface, or

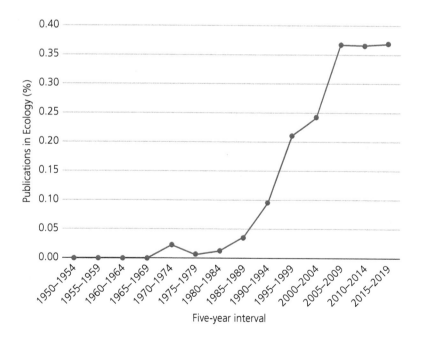

Figure P.1 Percentage of publications focused on marine disease in the field of ecology from 1950 to 2019. Data were gathered from the Web of Science using search terms "marine AND parasite OR disease OR pathogen OR infect" in the subject category of "ecology". Note: zero publications were returned for this search query from 1900 to 1949.

suspended in the water column where they can be transported over thousands of kilometers to new hosts much more efficiently than a terrestrial vector, such as an insect, could possibly achieve. Moreover, the sea is arguably a more stable and hospitable environment to infectious stages than the air, with a higher heat capacity and lower levels of ultraviolet radiation. These characteristics help explain the frequency of motile infective stages, the lack of disease vectors, and rapid pathogen spread in the marine environment. However, this "connectivity" works both ways. Just as a pathogen can more rapidly move between susceptible host populations in the sea, greater connectivity between host populations keeps them supplied with new recruits. Thus, we could fill a book with reasons why we cannot fully understand marine disease ecology from lessons learned on land.

Despite their significant impacts, studies of marine diseases have historically lagged behind their terrestrial counterparts, particularly with regards to their ecological effects. This is most likely due to the comparative inaccessibility of the sea to most scientists until the past century. However, starting in the early 1980s, global research focused on marine disease ecology accelerated dramatically, leveling out at its peak a decade ago (Figure P.1). We have now reached a critical mass of information regarding the role of infectious diseases in marine ecosystems, and it was clearly time to draw on that collective information to determine where our understanding is strong, where gaps lie, and where our future focus belongs. *Marine Disease Ecology* was written by a global team of experts on marine disease from disciplines that run the gamut from pathologists and ecologists to microbiologists and modelers. The culmination of this combined expertise is a book that will serve as a reference and learning tool for researchers, resource managers, and students across the fields of ecology, parasitology, microbiology, virology, aquaculture, fisheries, veterinary science, evolution, conservation, restoration, and others. A marine ecologist's education is incomplete without it.

Donald C. Behringer, Brian R. Silliman,
and Kevin D. Lafferty
July 2019
Gainesville, FL

Contents

Section 4 Working with Infectious Diseases

List of Contributors

David Bass Centre for Environment, Fisheries and Aquaculture Science, Weymouth Laboratory, UK

Kelly S. Bateman Centre for Environment, Fisheries and Aquaculture Science, Weymouth Laboratory, UK

Donald C. Behringer University of Florida, USA

Tal Ben-Horin North Carolina State University, USA

Gorka Bidegain University of the Basque Country

John P. Bignell Centre for Environment, Fisheries and Aquaculture Science, Weymouth Laboratory, UK

Andrea L. Bogomolni Woods Hole Oceanographic Institution, USA

Jamie Bojko University of Florida, USA

Colleen A. Burge University of Maryland Baltimore County, USA

David Bushek The State University of New Jersey, USA

Stephen W. Feist Centre for Environment, Fisheries and Aquaculture Science, Weymouth Laboratory, UK

Alex T. Ford University of Portsmouth, UK

Salvatore Frasca Jr University of Florida, USA

Rebecca J. Gast Woods Hole Oceanographic Institution, USA

Maya L. Groner Prince William Sound Science Center, USA

C. Drew Harvell Cornell University, USA

Paul K. Hershberger U.S. Geological Survey, USA

Eileen Hofmann Old Dominion University, USA

Martin Krkošek University of Toronto, Canada

Kevin D. Lafferty U.S. Geological Survey at the University of California Santa Barbara, USA

Joleah B. Lamb University of California Irvine, USA

Giulio de Leo Stanford University, USA

Erin K. Lipp University of Georgia, USA

Mark Little San Diego State University Research Foundation, USA

Hamish McCallum Griffith University, Australia

John P. McLaughlin University of California Santa Barbara, USA

Dana N. Morton University of California Santa Barbara, USA

Joseph P. Morton Duke University, USA

Katrina M. Pagenkopp Lohan Smithsonian Environmental Research Center, USA

Eric Powell University of Southern Mississippi, USA

Laurie J. Raymundo University of Guam, Guam

Forest Rohwer San Diego State University Research Foundation, USA

Maria Isabel Rojas San Diego State University Research Foundation, USA

Gregory M. Ruiz Smithsonian Environmental Research Center, USA

Brian R. Silliman Duke University, USA

Grant D. Stentiford Centre for Environment, Fisheries and Aquaculture Science, Weymouth Laboratory, UK

Steven M. Szczepanek University of Connecticut, USA

Rebecca Vega Thurber Oregon State University, USA

Mark E. Torchin Smithsonian Tropical Research Institute, USA

Chelsea L. Wood University of Washington, USA

Marine Infectious Diseases and their Ecological Roles

CHAPTER 1

Marine pathogen diversity and disease outcomes

Kelly S. Bateman, Stephen W. Feist, John P. Bignell, David Bass, and Grant D. Stentiford

1.1 Introduction

Disease refers to any condition that impairs the normal functioning of the body. Pathology is the study of disease. In the context of human health, disease may be associated with specific symptoms (a subjective departure from normal, noticed by a patient), and potentially a set of signs (objective measurements of that departure). Pathologists support clinicians by providing further evidence relating to health and disease, which assist the appropriate application of treatment. Medical and veterinary pathologists may specialize in a wide range of disciplines, from cellular pathology through hematology, clinical biochemistry, toxicology, microbiology, parasitology to immunology, or genetics and epigenetics. For this reason, pathology draws upon a wide array of technical specializations to provide behavioral, genetic, physiological, metabolic, and pathological profiles of healthy and diseased tissues and organs.

The aquatic animal pathologist should be similarly as resourceful. However, one major difference exists—animals cannot describe their symptoms. Instead, the aquatic animal pathologist is reliant on the objective observation of signs (in either the whole animal or its tissues and organs) to provide evidence to support diagnosis of a disease state. While the modern aquatic animal pathology laboratory has access to a wide range of tools that can support diagnosis of specific diseases, it is the physical observation of tissues and organs via a range of microscopy techniques that remains central to the study of disease in aquatic hosts (Frasca et al. Chapter 11, this volume). With appropriate adaptation of approach according to host taxonomy and size, cytopathology and histopathology can be readily applied to investigate departures from health in aquatic hosts, providing a snapshot of current health status and an unbiased view on likely prognosis for the host or the population from which it was obtained.

While histopathology provides an essential tool to describe toxicologic pathology in the tissues and organs of aquatic animals, it is often applied to the study of disease caused by infection with pathogens. In a few cases, the clinical signs observed in the whole animal or in histological preparations of tissues and organs may be considered pathognomonic (their collective presence leads to confident diagnosis of the disease). Usually, the aquatic animal pathologist may be faced with hosts of widely differing taxonomic status (e.g., over 500 species may be farmed in the global aquaculture sector alone) and carrying infection by an array of currently undescribed pathogens. In many cases, very little baseline health and disease data may exist for such hosts, making it difficult to definitively associate

Bateman, K.S., Feist, S.W., Bignell, J.P., Bass, D., and Stentiford, G.D., *Marine pathogen diversity and disease outcomes*
In: *Marine Disease Ecology*. Edited by: Donald C. Behringer, Kevin D. Lafferty, and Brian R. Silliman, Oxford University Press (2020).
© Oxford University Press. DOI: 10.1093/oso/9780198821632.003.0001

the presence of an infectious agent with a specific disease outbreak event in a farmed or wild setting.

Ideally, the aquatic animal pathologist would be equipped with a good understanding of the "normal" variation and physiological status of tissues and organs and the presence of symbionts associated with a given host: that is, an appreciation of the appearance of host tissues and organs throughout the reproductive and seasonal cycle, and of the potential symbionts associated with them. In the case of the latter, histology provides an ability to discriminate infection (the presence of an invasive, multiplying agent) from disease (where normal function is lost), which is not an inherent feature of other diagnostic modalities. Manifestation of the disease state may require an environmental catalyst, or the presence of factors otherwise independent of the proposed offending agent (Bojko et al. Chapter 6, this volume; Burge and Hershberger Chapter 5, this volume). Histology and its associated techniques afford the aquatic animal pathologist the potential to understand this complex interaction between infection and health status.

Given the wide array of infectious agents potentially encountered within the organs and tissues of aquatic animals, we base our approach on describing how specific groups of pathogens interact with the cells and tissues of the host. Unicellular and multicellular pathogens are generally visible using light microscopy, and their observation within tissue sections can provide a presumptive diagnosis. For RNA and DNA virus infections, cellular and nuclear changes may be observed, but the pathologist's eyes are drawn to the indirect effects of infection rather than to the agent itself (e.g., enlargement of cell organelles, necrosis, and apoptosis of infected cells). In many cases, inflammatory responses may occur, but where pathogens are solely contained within the cytoplasm or even nucleoplasm of cells, a visible host response is often lacking.

1.2 Diversity of pathogens

In this chapter, we outline the nature of interactions between aquatic hosts and a diverse array of pathogen groups. Pathogens listed here are certainly not an exhaustive list of those agents known to infect aquatic hosts, but rather provide the aspiring

aquatic animal pathologist with a core understanding of the pathogen types. The specific fact sheets associated with the subsections below highlight key diagnostic features and indicate how to recognize these pathogen groups in the tissues and organs of their chosen target host.

The main role of the World Organisation for Animal Health (Office International des Epizooties (OIE)) is to facilitate the international trade in animals and animal products (Bruckner, 2009). The Aquatic Animal Health Code provides standards for improving aquatic animal health worldwide and as such lists diseases that pose the greatest economic and ecological risks to prevent the transboundary spread of these pathogens (Morton et al. Chapter 3, this volume). There are currently ten fish diseases, seven mollusc diseases, nine crustacean diseases, and three amphibian diseases listed as notifiable by the OIE (Chapter 1.3, Aquatic Animal Health Code). Some of the diseases are highlighted within this chapter.

1.2.1 Viruses

Viruses infect all cellular life forms and those from the marine environment are no exception. There are numerous viral infections described in marine organisms, some of which are known to cause disease and some of which appear quiescent. The first invertebrate virus to be described in the marine environment was in the 1960s, from a crab (Vago, 1966). Since this time there have been numerous descriptions of DNA and RNA viruses infecting wide-ranging taxa within marine invertebrates (Arzul et al. 2017; Bateman and Stentiford, 2017; Hewson et al. 2014) and fish (Woo and Bruno, 2011). DNA viruses contain DNA as the genetic material and, likewise, RNA viruses have RNA as the genetic material. DNA viruses are frequently (but not always) found to replicate within the nucleus of the cell, and RNA viruses generally replicate within the cytoplasm of the cell. In some cases, the lack of suitable cell lines (particularly for invertebrates) and the reliance on *in vivo* propagation has hampered formal description. More recently, the development of high-throughput sequencing technologies has aided viral taxonomy, using bioinformatic tools to discriminate host and viral genomic data and underpinning the description of novel pathogens in a range of hosts (Simmonds et al. 2017). Here, we focus on some significant viral pathogens of invertebrates and fish.

White spot syndrome virus (WSSV) was first described in the 1990s in Japanese kuruma shrimp (*Penaeus japonicus*) in Taiwan (Takahashi et al. 1994). The infection rapidly spread between shrimp farming regions around the globe and has been responsible for significant losses in the global industry. This viral infection is somewhat unique in the wide diversity of hosts—all decapod crustacea are listed as being potentially susceptible to this disease (see fact sheet 1).

Despite being an OIE listed disease and one of the most studied viral infections from the marine environment, this pathogen continues to have negative consequences where it emerges, most recently in Australia (Oakey and Smith, 2018). There are currently four other OIE-notifiable viral diseases of marine crustaceans: a DNA virus causing infectious hypodermal and hematopoietic necrosis (Lightner et al. 1983) and RNA viruses causing Taura syndrome (Bonami et al. 1997), yellow head disease (Chantanachookin et al. 1993) (see fact sheet 2), and infectious myonecrosis (Lightner et al. 2004). Unlike DNA virus infections, where histology can be used to identify direct signs of infection (e.g., the presence of hypertrophied nuclei containing viroplasms which are often differentially stained), detection of RNA virus infection can be more difficult. Here, the pathologist is reliant on observation of the indirect effects of virus infection, such as nuclear dissociation, cell necrosis, and apoptosis.

For molluscs there are currently two OIE listed viral diseases, ostreid herpesvirus 1 microvariants (OsHV-1µVar) (Segarra et al. 2010), which infects oyster species (*Crassostrea gigas* and *C. angulata*), and abalone herpesvirus (AbHV) (Hooper et al. 2007), which infects the nerve tissues of abalone (*Haliotis laevigata* and *H. rubra*) in the southern hemisphere. Ostreid herpesvirus 1 (OsHV-1) was initially described infecting hatchery-reared *C. gigas* (Hine et al. 1992), with mortality ranging from 80 to 100 percent reported in farmed oysters in Europe; the causative agent determined was a microvariant of OsHV-1 (Segarra et al. 2010). A viral infection was identified in association with mass mortality in sea star populations off the West Coast of the USA (Hewson et al. 2014). Sea star wasting disease (SSWD) has been described within populations of several sea star species, with mass mortality events described sporadically over the last few decades. A second viral-like pathogen has also been identified (Fahsbender et al. 2015) in sea stars displaying signs of SSWD, suggesting that the cause of the massive mortalities reported in 2014 may have been multifactorial (Burge and Hershberger, Chapter 5, this volume) and providing evidence for a growing diversity of viruses within echinoderms.

Lymphocystis disease causes lesions mainly on the surface of many marine fish species and was originally described by Sandeman (1893). Initially thought to be the egg cells of an unknown invertebrate parasite, it was eventually attributed to a viral pathogen (Weissenberg, 1951). Numerous DNA and RNA viral pathogens, including those from the *Herpesviridae*, *Iridoviridae*, *Rhabdoviridae*, and *Reoviridae*, have since been reported to infect fish (Woo and Bruno, 2011). There are currently six OIE-notifiable viral diseases in marine fish: infectious hematopoietic necrosis (IHN) (Morzunov et al. 1995), infectious salmon anemia (ISA) (Thorud and Djupvik, 1988), Red Sea bream iridoviral disease (RSIV) (Inouye et al. 1992), viral hemorrhagic septicemia (VHS) (Oshima et al. 1993), viral encephalopathy and retinopathy (VER) (Chi et al. 2001), and salmonid alphavirus (SAV) (McLoughlin and Graham, 2007) which affects smolted salmon.

SAV is the causative agent of pancreas disease (PD) or sleeping disease (SD) in Atlantic salmon, rainbow trout, and brown trout (McLoughlin and Graham, 2007). The disease caused significant mortalities in cultured salmonids, but vaccination has been successful in reducing losses since its introduction in 2007. PD was first recognized in marine-reared Atlantic salmon in 1976 (Munro et al. 1984) and in rainbow trout in freshwater where the disease has been called SD (Boucher et al. 1994). The pathogen is transmitted horizontally and is systemic, with clinically affected fish showing necrosis of the exocrine pancreas without pancreatitis, cardiomyocyte necrosis, and inflammation of the somatic musculature. Fish surviving the acute phase of the infection may develop severe fibrosis of the pancreas (preventing full recovery of the tissue).

Fact Sheet 1

Pathogen	White spot syndrome virus
Abbreviation	WSSV
Agent	Family: *Nimaviridae* Genus: *Whispovirus* Species: white spot syndrome virus (WSSV) Currently the only member of this family and genus.
Agent features	Enveloped virions: 120–150 nm diameter, 270–290 nm length Nucleocapsids: 65–70 nm diameter, 300–320 nm length; in some cases a tail-like projection can be seen extending from one end. Nucleocapsid has distinct structure: it is composed of subunits in a stacked series, providing a cross-hatched appearance.
Target hosts	Typically, a disease of cultured penaeids; however, all decapod and non-decapod crustaceans are listed as being susceptible to the virus. It is known that susceptibility varies between host species.
Mortalities	All penaeid species are known to be highly susceptible, resulting in high mortality. Morbidity and mortality in other susceptible species such as crayfish, crabs, and lobsters vary under experimental conditions, and effects upon wild populations are unknown. High-level infection can be seen without clinical signs of disease.
Gross clinical signs	Typically, in farmed penaeid shrimp clinical signs of infection include anorexia, lethargy, loose cuticle, and a reddish discoloration of the whole shrimp. White spots on the cuticle can be seen in some but not all species; these spots can also be caused by other conditions such as bacterial infections. Clinical signs of infection in other species are not well documented; however, cessation of feeding, lethargy, and a marked reduction in clotting time of crustacean hemolymph may be indicative of infection, but these symptoms also occur with many other viral and bacterial diseases of crustacea.
Histopathology	Main target for infection are tissues of ectodermal and mesodermal origin, especially cuticular epithelium and subcuticular connective tissues. Infected nuclei appear hypertrophied, and chromatin is marginalized and contains an eosinophilic inclusion body. Staining characteristics vary as disease progresses, with nuclei staining progressively basophilic as infection develops.

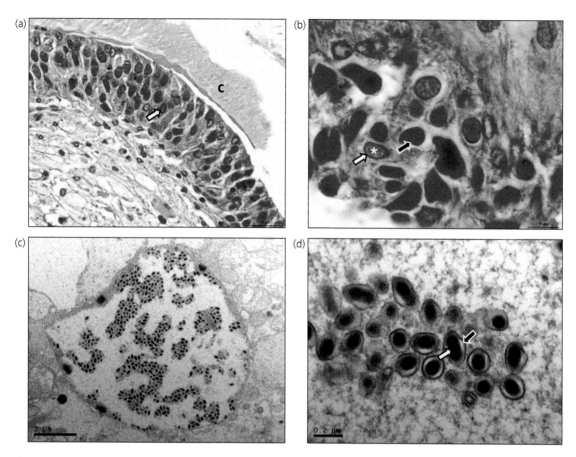

Figure 1.1 **(a)** Infected nuclei (arrows) can be seen within the cuticular epithelium which lies underneath the carapace (c). Infected nuclei appearing hypertrophied with marginalized chromatin. H&E stain. Scale bar = 25 μm. **(b)** Hypertrophied nuclei with marginalized chromatin (white arrow) within cuticular epithelium. Note the granular eosinophilic staining (*) within some nuclei and dense hematoxylin staining in other nuclei (black arrow). H&E stain. Scale bar = 10 μm. **(c)** Transmission electron micrograph of an infected nucleus; ovoid virions can be seen within the nucleus. TEM. Scale bar = 2 μm. **(d)** Virions consist of an electron-dense core, the nucleocapsid (white arrow) surrounded by a clear double membrane, and the envelope (black arrow). TEM. Scale bar = 0.2 μm.

Fact Sheet 2

Pathogen	Yellow head virus 1
Abbreviation	YHV1
Agent	Order: *Nidovirales* Family: *Roniviridae* Genus: *Okavirus* Species: yellow head virus 1 (YHV1)
Agent features	Eight known genotypes in the yellow head complex; YHV1 is the genotype that causes the OIE-notifiable disease yellow head disease. YHV2 is commonly called gill-associated virus (GAV). YHV3–6 occur commonly in healthy *P. monodon* in East Africa, Asia, and Australia, YHV7 in diseased *P. monodon* in Australia, and YHV8 in *P. chinensis* suffering from acute hepatopancreatic necrosis disease (AHPND). YHV1 forms enveloped, rod-shaped particles 40–50 nm × 150–180 nm. Envelopes are studded with prominent peplomers projecting approximately 11 nm from the surface. Nucleocapsids appear as rods (diameter 20–30 nm) and possess a helical symmetry with a periodicity of 5–7 nm.
Target hosts	*P. monodon*, *P. vannamei*, *P. stylirostris*, *P. pugio*, and *M. affinis* are listed as susceptible species. Incomplete evidence is available for the susceptibility of *M. sintangense*, *M. brevicornis*, *P. serrifer*, *P. styliferus*, *P. aztecus*, *P. duorarum*, *P. japonicus*, *P. merguiensis*, *P. setiferus*, and *C. quadricarinatus*.
Mortalities	YHV1 can cause up to 100 percent mortality in affected ponds.
Gross clinical signs	Very high feeding rates followed by sudden cessation. Moribund shrimp at pond edge, with bleached or yellow appearance of cephalothorax.
Histopathology	Ectodermal and mesodermal tissues including lymphoid organ, hemocytes, hemopoietic tissue, gills, and connective tissues. Loss of cuboidal structure to epithelial cells—infected cells are rounded with pyknotic nuclei. Loss of tissue structure within lymphoid organ, stromal matrix cells that comprise tubules become infected, leading to loss of tubular structure, and tubules appear degenerate. Lymphoid organ spheroids (LOS) develop during infection. Ectopic spheroids can be found in other tissues such as heart and gill.

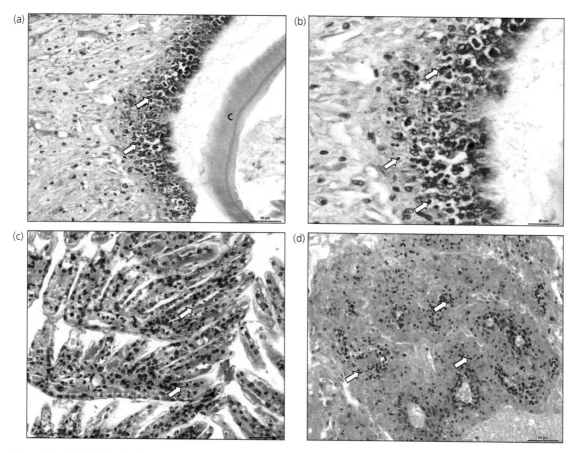

Figure 1.2 (a) Pyknotic nuclei (arrows) can be seen within the cuticular epithelium which lies underneath the carapace (c). H&E stain. Scale bar = 50 μm. **(b)** YHV1 infected cells are rounded with pyknotic nuclei (arrows). H&E stain. Scale bar = 25 μm. **(c)** Pyknotic nuclei (arrows) can be seen within the cuticular epithelium which lies underneath the carapace that lines the gills. H&E stain. Scale bar = 50 μm. **(d)** Lymphoid organ tubules become infected, and stromal matrix cells which comprise the tubules are rounded with densely stained pyknotic nuclei (arrows). Scale bar = 50 μm.

The virus has been detected in wild marine flatfish (McCleary et al. 2014; Snow et al. 2010) and Ballan wrasse (Ruane et al. 2018) not showing clinical signs. However, the possible role of these species and other marine organisms as reservoirs or vectors of infection remains unknown (Morton et al. Chapter 3, this volume).

Infection with infectious salmon anemia virus (ISAV) (family Orthomyxoviridae; genus *Isavirus*) relates to infections with a pathogenic highly polymorphic region (HPR)—deleted ISAV and the nonpathogenic HPR0 ISAV (without the deletion). First recorded in Norway during the early 1980s, the disease remains a major concern for salmonid aquaculture, with cumulative mortalities of up to 90 percent having been reported. Affected fish show severe anemia resulting from systemic infection of endothelial cells leading to collapse of the circulatory system. Hemorrhaging and necrosis of internal organs is prominent with petechial hemorrhaging of the gills and musculature (Rimstad et al. 2011). Most occurrences of ISA have been in Atlantic salmon, but the disease has also been reported in cultured coho salmon (*Oncorhynchus kisutch*) in Chile (Kibenge et al. 2001). Experimental replication of the virus has been demonstrated in several fish species, but evidence that these act as reservoirs of infection to salmonids is lacking.

1.2.2 Bacteria

Bacteria are responsible for a range of infectious diseases in numerous marine species including fish, invertebrates, and corals (Beaz-Hidalgo et al. 2010; Birkbeck et al. 2011; Sheridan et al. 2013; Toranzo et al. 2005). In marine systems, diseases caused by members of the genus *Vibrio* have been widely reported, but other significant pathogenic taxa are also encountered.

Following a mass mortality of king scallops (*Pecten maximus*) in south-west England in 2013 and 2014, the causative agent, an *Endozoicomonas*-like organism (ELO), was identified (Cano et al. 2018). Intracellular microcolonies (IMC) of bacteria were identified in the epithelial cells of the gills, IMC being associated with pathogenicity and mortality in infected hosts. Histologically, two types of bacterial microcolonies were identified within the gill epithelium: type I colonies consisting of tightly packed globular cells, while type II colonies contained small basophilic rod-shaped cells, both colony types consisting of Gram-negative bacteria (see fact sheet 3).

Intracellular microcolonies have until recently remained uncharacterized and were initially thought to be a rickettsia-like organism (RLO). Phylogenetic analysis based on the 16S rRNA gene and six bacterial housekeeping genes demonstrated that the IMC in king scallops are in fact members of the γ subgroup of Proteobacteria. Metagenomic analysis resulted in a draft genome of these bacteria and showed highly significant protein-level similarity with other *Endozoicomonas* species. Members of the *Endozoicomonas* are emerging as symbionts of diverse marine hosts including sponges, corals, molluscs, and fish (Jensen et al. 2010; Neave et al. 2016).

Francisellosis in cod (*Gadua morhua*) caused by the bacterium *Francisella noatunensis* was first described in farmed cod in Norway and was a significant concern with regard to the anticipated increase in the culture of cod (Nylund et al. 2006; Olsen et al. 2006). The pathogen also affects Atlantic salmon in Chile (Birkbeck et al. 2007). The intracellular bacterium, also a member of the γ subgroup of Proteobacteria, gives rise to a chronic systemic granulomatous disease, with macroscopic white nodular lesions primarily affecting visceral organs such as the liver and spleen, and is associated with high mortalities (Birkbeck et al. 2011). Over 30 years prior to these reports in cultured fish, a condition referred to as visceral granulomatosis was reported affecting wild cod in the North Sea at a prevalence of up to 30 percent (Bucke, 1989), often with large fluid-filled lesions in the liver. The etiological agent was not identified at that time, but using archived materials, Zerihun et al. (2011) were able to demonstrate that the disease was caused by the same pathogen. Francisellosis was in part responsible for the reduction in the number of farms rearing cod in Norway between 2007 and 2012 (Hjeltnes, 2014) and provides an example of the need to evaluate disease status and the pathobiome of wild species, particularly where they are being considered for aquaculture purposes.

Acute hepatopancreatic necrosis disease (AHPND) is probably the most important non-viral disease to affect cultured shrimp. The causative agent was identified as an isolate of *Vibrio parahaemolyticus* containing a 70-kbp plasmid that encodes for *pir*AB toxins (Tran et al. 2013). The *pir*AB toxins are the homologs of the *Photorhabdus* insect-related (Pir) binary toxins PirA and PirB (Lee et al. 2015). Presumptive gross signs of infection include an empty stomach and midgut and a pale and shrunken hepatopancreas; no clear causative agent was identified via histological analysis. Extensive sloughing of the tubular epithelium sloughing of the cells of the tubular epithelium present in acute infections is caused by the toxins rather than the bacteria itself (Tran et al. 2013). The disease was first reported in the People's Republic of China around 2009 and was initially called covert mortality disease; the condition was also referred to as early mortality syndrome (EMS) due to mass mortalities occurring within the first 35 days of culture. The syndrome describes a range of shrimp health problems that lead to early mortalities, so a more descriptive term (AHPND) was adopted. The disease has been reported in numerous shrimp-producing nations in Asia and the Americas and is an OIE listed disease. It most commonly affects whiteleg shrimp (*Penaeus vannamei*) but has also been reported within tiger shrimp (*P. monodon*) and fleshy prawn (*P. chinensis*). Recent studies have demonstrated the presence of the AHPND-causing plasmid in other *Vibrio* spp. (Kondo et al. 2015; Xiao et al. 2017), suggesting that the pathogenicity factor originally associated with *V. parahaemolyticus* is mobile.

A second OIE-notifiable disease in cultured marine shrimp is necrotizing hepatopancreatitis (NHP), which describes the infection and disease due to *Candidatus Hepatobacter penaei* (Nunan et al. 2013). *Hepatobacter penaei* is a member of the proteobacteria, is Gram negative and pleomorphic (possessing both rod-shaped and helical forms), and occurs within the host cell cytoplasm. Infection is often associated with high mortalities, lethargy, atrophied hepatopancreas, anorexia, and an empty gut. Histologically, bacterial cells are associated with atrophy of hepatopancreatic tubular epithelia, forming large edematous spaces and attracting hemocytes which result in multifocal encapsulations of one or more of the tubules. The infection has been reported in both wild and cultured stocks of penaeid shrimp (*Peneaus setiferus, P. duorarum, P. stylirostris, P. merguiensis, P. marginatus, P. azetcus,* and *P. monodon*) in the western hemisphere. Following experimental exposure, positive PCR results were also reported in the American lobster, *Homarus americanus,* but an active infection was not demonstrated (Avila-Villa et al. 2012).

Gaffkemia or red tail disease is most commonly associated with lobsters (*H. americanus, H. gammarus*), but the pathogen has also been shown capable of infecting other crab and shrimp species during exposure studies (Stewart, 1980). The causative agent is a bacteria, *Aerococcus viridans* var. *homari,* which forms a characteristic Gram-positive coccus tetrad. Lobsters usually acquire the infection via damaged cuticle, allowing entry of the bacteria and subsequent colonization of the hemolymph. The condition was described by Snieszko and Taylor (1947) following a mortality event among lobsters held in pounds and tanks. The cause of the mortalities was initially thought to be the use of DDT at local fish canneries; however, bacteria were later identified in blood smears and cultures from diseased lobsters. The disease is associated with higher summer temperatures and diseased lobsters showing a pink discoloration, lethargy, and reduced hemocyte counts. Despite the high economic significance of the lobster fisheries, relatively little is known about the prevalence of Gaffkemia in wild lobster stocks.

1.2.3 Microeukaryotes

To avoid any confusion arising from use of the words "protist" and "protozoan," we use "micro-eukaryote" in this chapter to refer to all microbial eukaryotes excluding the macroscopic/multicellular lineages that arise from within them. It also includes fungi—microsporidia (if considered fungi (see section below on "Microsporidia"), chytrids, and other zoosporic fungi being as "protist-like." Our definition would exclude, for example, the stramenopile brown algae arising within stramenopiles, plants and plant-like algae within Archaeplastida, and most metazoans.

Fact Sheet 3

Pathogen	*Endozoicomonas* sp.
Abbreviation	ELO
Agent	Phylum: Proteobacteria Class: Gammaproteobacteria Order: Oceanospirillales Family: Hahellaceae Genus: *Endozoicomonas*
Agent features	Gram-negative short rods.
Target hosts	King scallops (*Pecten maximus*) and likely to infect diverse bivalve species.
Mortalities	The *Endozoicomonas*-like organism (ELO) infecting king scallops has been associated with mass mortality events on several occasions in European waters. Intracellular microcolonies of similar ELO bacteria infecting other bivalve species, including Pacific oyster (*Crassostrea gigas*), Ezo giant scallop (*Patinopecten yessoensis*), and the Pacific razor clam (*Siliqua patula*) have also shown pathogenicity with mortalities.
Gross clinical signs	Gross clinical signs are limited to small pale lesions on the surface of severely affected gills. Evidence of infection in other organs or tissues is not discernible by the naked eye.
Histopathology	Histological features of infection are most apparent in gill tissue where pathological changes are closely linked to intensity of infection. Initially, small intracellular microcolonies are present, followed by bacterial replication forming larger colonies causing disruption of gill tissue and distribution to other tissues and organs, where small microcolonies can be discerned by light microscopy (i.e., digestive gland tissue).

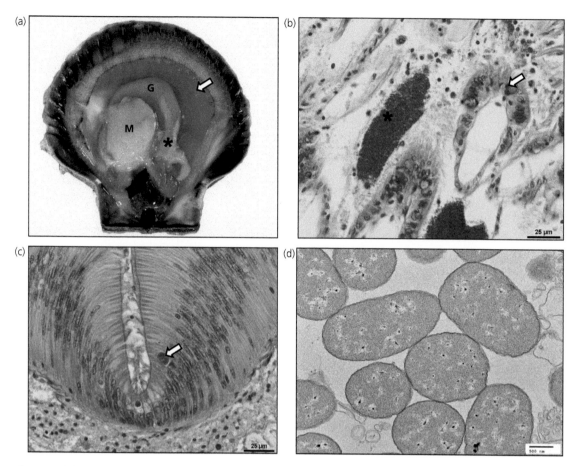

Figure 1.3 **(a)** Anatomy of king scallop, showing the gill (arrow), gonad (G), adductor muscle (M), and digestive gland (asterisk). **(b)** Section through a gill, with two conspicuous colonies of ELO (asterisk) showing pronounced basophilia. Note the normal gill tissue adjacent with ciliated epithelium containing mucous cells (arrow). H&E stain. Scale bar = 25 μm. **(c)** Detail from the intestinal epithelium containing small microcolonies of ELO (arrow). Note lack of host response. H&E stain. Scale bar = 25 μm. **(d)** Ultrastructure of ELO, with a double membrane cell wall and faint granular cytoplasm and electron-dense granules. Scale bar = 500 nm.

Fungi

There is no consensus about which biological characters "define" fungi. Various authorities recognize different sets of taxonomic groups as belonging to fungi. Holomycota (Liu et al. 2009) is a term used to classify a large set of candidate taxa (Dikarya, Blastocladiales, zygomycete fungi, Opisthosporidia, to name a few). Richards et al. (2017) report a range of definitions that have been proposed and used to circumscribe fungi. They make the point that insight into and opinions about the classification boundary between fungi and other eukaryotes will change as new information (derived from additional sampling and genomic-level evolutionary analyses) is available. Within the Holomycota the boundary between fungi and non-fungi is currently undecided, but wherever it is placed the underlying phylogeny is unaffected. Fungal analogs (Richards et al. 2012) comprise fungal-like organisms and those previously considered as fungi, including many parasites relevant to aquatic animal health, which branch in very different regions of the eukaryotic tree. These are not directly related to fungi, and no classification system could reasonably include them within the taxon fungi. Fungal analogs include oomycetes (*Lagenidium*, *Halioticida*, *Haliphthoros*, *Phytophthora*, etc.), hyphochytriomycetes, labyrinthulids, thraustochytrids (all stramenopiles), ascetosporeans and phytomyxids (Rhizaria), ichthysporeans, and non-parasitic slime molds.

An oomycete-like infection has recently been reported in European lobsters *Homarus gammarus*. *Halioticida noduliformans* was isolated from discolored eggs sampled from *H. gammarus* in the UK (Holt et al. 2018), infected eggs appearing white, pink, or gray in comparison to uninfected eggs. Eggs showed a reduction or complete lack of egg yolk protein and were filled with large hyphal structures. A similar infection had been previously reported in abalone (*Haliotis rubra*) from Australia, mantis shrimp (*Oratosquilla oratoria*) from Japan, and cultured abalone (*H. midae*) from South Africa (Atami et al. 2009; Macey et al. 2011; Muraosa et al. 2009; Sekimoto et al. 2007), phylogenetic analysis highlighting a 99–100 percent similarity between the infection in lobsters and abalone (Holt et al. 2018). The consequences *Halioticida noduliformonas* and other similar

oomycete infections may have on wild populations are unknown (Holt et al. 2018).

There have been few reports of infection of marine fish by true fungi as opposed to fungal-like organisms such as oomycetes and *Ichthyophonus* (Mesomycetozoa). Of these, *Exophiala* spp. have been reported as a systemic infection in marine-reared salmonids (Blazer and Wolke, 1979; Johnson et al. 2018; Langvad et al. 1985; Richards et al. 1978), in other freshwater fish species (Alderman and Feist, 1985), and, rarely, in cod (see fact sheet 4).

During the 1980s, a fungal disease caused by *Exophiala salmonis* was identified as a cause of mortalities of up to 50 percent in Atlantic salmon smolts and post-smolts (Langvad et al. 1985). Originally described from cutthroat trout with a cerebral lesion (Carmichael, 1966), *E. salmonis* has now been recognized as a pathogen of low host specificity affecting several fish genera (Blazer and Wolke, 1979) from Europe, Canada, and Australia. Clinical signs of infected salmonids include erratic swimming behavior, exophthalmia (abnormal bulging of the eye from the socket), and abdominal distension (Bruno, 2016). Internally, infections can be systemic, although renal lesions are frequently present associated with a granulomatous response with giant cell formation engulfing the fungal hyphae. However, systemic spread of the agent is not prevented and in severe cases lesions in other organs and tissues, including the musculature, are affected. *Ichthyophonus* is another important pathogen of wild cold-water fish species and was responsible for a significant population decline in Atlantic herring (*Clupea herengus*) (Tibbo and Graham, 1963). Granulomatous lesions primarily affect liver, kidney, spleen, and heart tissues and may elicit an acute or chronic disease condition depending on the susceptibility of the host. While chronic infections may result in mortalities several months post-infection, acute infections often result in significant invasion of organs, followed by tissue necrosis and death (Noga, 1993).

Microsporidia

Microsporidia are obligate intracellular parasites that can infect a wide range of hosts and tissues. They were shown to have evolved from fungi and were placed within that kingdom (Cavalier-Smith, 1998; James et al. 2013). However, the latest evidence suggests that they are more closely related to rozellids (James et al. 2013) and that both rozellids and microsporidians should be classed within the microeukaryotes (Ruggiero et al. 2015). Microsporidia are a diverse parasite phylum infecting hosts from terrestrial and aquatic environments; almost half of the described microsporidia infect aquatic organisms, approximately twenty genera infect fish, fifty genera infect aquatic arthropods, and at least twenty-one genera infect aquatic non-arthropod invertebrates, protists, and hyperparasites of aquatic hosts (Stentiford et al. 2013b). Historically structural characteristics, such as size, shape, and number of turns of the polar filament, have been used to classify the microsporidia within different taxa. Recent studies have highlighted the potential for plasticity in development of spores and suggest that phylogeny should be based upon the informed use of molecular sequence data (Bateman et al. 2016; Stentiford et al. 2013a). Microsporidia are thought to be an emerging issue due to their opportunistic nature and propensity to infect immune-suppressed organisms (Stentiford et al. 2016).

Microsporidia are especially common in crustacean species, with muscle and hepatopancreas tissues being frequent sites of infection. Within these tissues, infections are associated with disruption of normal tissue structure, high parasite numbers, and in the case of muscle tissue the appearance of externally visible signs such as whitening and increased opacity (Stentiford et al. 2013b; Stentiford et al. 2017a). Multiple taxonomic descriptions are reported in the literature, so here we highlight a few pertinent examples.

Microsporidian infection of crustacean hosts by *Enterocytozoon hepatopenaei* (Tourtip et al. 2009) and *Enterospora canceri* (Stentiford et al. 2007) and fish hosts by *Paranucleospora theridion* (Nylund et al. 2011) have revealed close relationship between these pathogens and the human pathogen *Enterocytozoon bienusi*. In humans, *E. bienusi* is a common pathogen of immunocompromised patients (such as those with AIDS). The potential role of aquatic tissues as a source of zoonotic infections in humans requires further investigation (Stentiford et al. 2016). *Enterocytozoon*

hepatopenaei (EHP) is also the causative agent in a significant emerging disease within the global penaeid shrimp industry. The agent was initially discovered in 2004 (Chayaburakul et al. 2004) and characterized in 2009 (Tourtip et al. 2009). The infection was found within the hepatopancreas of *Penaeus monodon* affected by so-called Monodon slow growth syndrome (MSGS) and concomitantly infected with viruses. The disease has more recently been described in *P. vannamei* and *P. japonicus* (Hudson et al. 2001). When initially described, the microsporidian was rarely seen (Tourtip et al. 2009); however, in 2018 it was highly prevalent among shrimp farms in Asia and is associated with slow growth.

To date there have been two microsporidians described infecting bivalve molluscs: an unidentified infection of queen scallops (Lohrmann et al. 2000) and, *Steinhausia* sp. The species *S. mytilovum* (see fact sheet 5) and *S. ovicola* affect bivalve molluscs belonging to the families *Mytilidae* and *Ostreidae* respectively, with a *Steinhausia*-like parasite being described affecting the cockle *Cerastoderma edule* (Comtet et al. 2003). Microsporidians have also been reported within hosts from the class Osteichthys (bony fish) and rarely the Chondrichthyes (cartilaginous fish) (Diamant et al. 2010), and at least twenty genera have been described. The focus of these descriptions appears to be commercially exploited species, so it is likely that microsporidian infections are vastly underreported within fish hosts (Appeltans et al. 2012; Stentiford et al. 2013b).

Amoebozoa

Amoebae constitute an abundant fauna in marine ecosystems, but, to date, only a few taxa have been described formally in fish (Dykova et al. 1999; Zillberg and Munday, 2006). In most cases, their capacity to form disease remains unclear, but examples such as *Grellamoeba* (Amoebozoa, class Variosea) from pikeperch (*Sander lucioperca*) cause active infections in kidney tissue (Dykova et al. 2010). Amoebic gill disease (AGD) of salmonids, caused by the distantly related *Paramoeba perurans* (class Discosea), was described in cage-reared coho salmon in the USA (Kent et al. 1988) and Atlantic salmon and rainbow trout in Tasmania (Munday et al. 1993).

Fact Sheet 4

Pathogen	*Exophiala* sp.
Abbreviation	None
Agent	Kingdom: Fungi Division: Ascomycota Class: Chaetothyriomycetes Order: Chaetothyriales Family: Herpotrichiellaceae Genus: *Exophiala*
Agent features	Dematiaceous fungus producing melanin pigments. Septate and branching.
Target hosts	Cod (*Gadus morhua*), marine-reared salmonids, and several other marine fish species.
Mortalities	Not recorded for cod. Incidental findings during surveys or reported by anglers when filleting. High levels of mortalities have been reported in cultured striped jack (*Pseudocaranax dentex*).
Gross clinical signs	In salmonids, abnormal swimming behavior, poor feeding, and distension of the abdomen have been reported. Ulceration of the skin and exophthalmia may occur. Internally, renomegaly may be present, and in systemic infections, pigmented lesions in the somatic musculature can occur.
Histopathology	The primary host response is an acute multifocal inflammation leading to chronic granulomatous lesions, frequently with giant cell formation in affected organs and tissues.

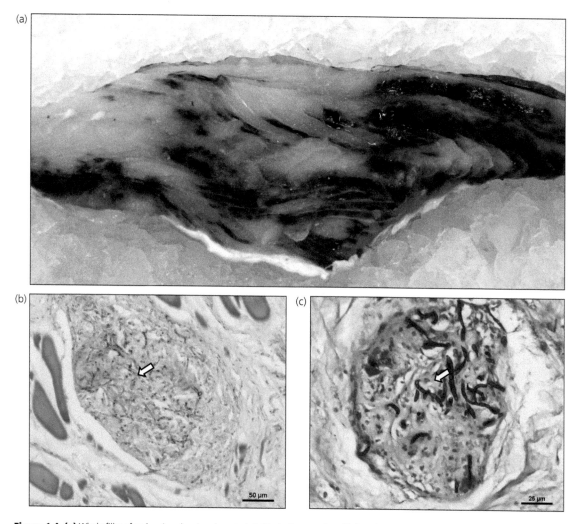

Figure 1.4 (a) Whole fillet of cod on ice, showing characteristic black pigmentation of infected musculature. **(b)** Multi-focal and disseminated granulomata in the musculature with numerous hyphae present (arrow). Note the relative absence of hyphae in surrounding tissues. H&E stain. Scale bar = 50 μm. **(c)** High-power view of a granulomatous lesion. Numerous red pigmented branching and septate fungal hyphae (arrow) are largely contained within the lesion. PAS stain. Scale bar = 25 μm.

Fact Sheet 5

Pathogen	*Steinhausia mytilovum*
Abbreviation	None
Agent	Phylum: Microsporidia Genus: *Steinhausia* Species: *Steinhausia mytilovum*
Agent features	*Steinhausia* presents developmental stages and spores exhibiting diplokarya. Spores form within a sporopherous vesicle surrounded by a single membrane within the oocyte cytoplasm or nucleus. A merogonial plasmodia located within this vacuole undergoes nuclear division and cellular segmentation to form individual sporonts. Sporonts develop into sporoblasts, pre-spores, and spores.
Target hosts	Bivalve molluscs within the family *Mytilidae*.
Mortalities	The effect of *S. mytilovum* on the host is poorly understood.
Gross clinical signs	The parasite can be observed in healthy mussels and those in poor condition, in relation to gonad and storage cells. Condition factor cannot be used as a clinical sign of infection since other endogenous and exogenous factors also affect this.
Histopathology	Target for infection are oocytes of all developmental stages. Sporopherous vesicles containing numerous basophilic spores are usually located within cytoplasmic vacuoles of affected oocytes, although they can also be seen within the nucleus. The parasite can elicit a pronounced host response, with seemingly few parasites resulting in significant inflammation of gonadal follicles and surrounding vesicular connective tissue. An accompanying and marked increase in atretic oocytes affecting a large proportion of the gonad can result.

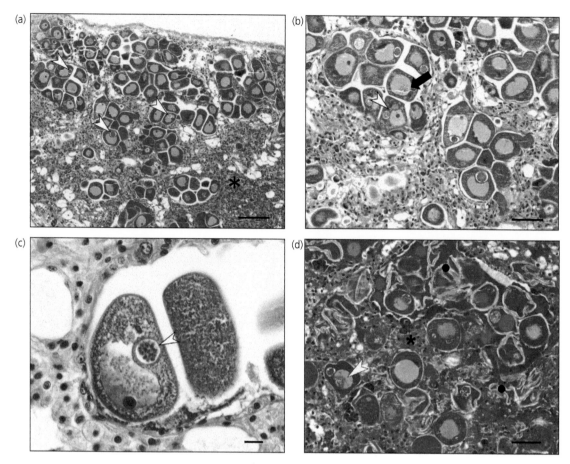

Figure 1.5 (a) Oocytes of *Mytilus edulis* infected with *S. mytilovum* (arrowheads). Infection with parasite can be discrete, although it is often accompanied by pronounced hemocytic infiltration within gonadal follicles and vesicular connective tissue (asterisk). H&E stain. Scale bar = 100 μm. **(b)** Detail of (a) showing sporopherous vesicles in vacuole located within oocyte cytoplasm. Note presence of a single vacuole within nucleus of oocyte (arrow) and multiple vacuoles within cytoplasm (arrowhead). H&E stain. Scale bar = 50 μm. **(c)** Sporopherous vesicle situated within oocyte cytoplasm containing many individual maturing spores. Note presence of basophilic membrane (arrowhead) surrounding sporopherous vesicle. Scale bar = 10 μm. **(d)** *S. mytilovum* infection accompanied by hemocytic infiltration (asterisk) and significant atresia (black circles). Soporiferous vesicle can be seen occupying both the cytoplasm and nucleus of an individual oocyte (arrowhead). H&E stain. Scale bar = 50 μm.

The disease is now a major issue for marine salmon aquaculture globally (Mouton et al. 2014; Palmer et al. 1997; Steinum et al. 2008; Behringer et al. Chapter 10, this volume). Associated with sea urchin mortalities (Feehan et al. 2013), *P. perurans* causes an intense proliferative response of the gill epithelium, resulting in compromised respiratory efficiency and significant mortalities. The disease has also been reported in turbot (*Psetta maxima*), sea bass (*Dicentrarchus labrax*), and more recently Ballan wrasse (*Labrus bergylta*), and it seems likely that other marine hosts will be recognized in due course (Bojko et al. 2018). Affected fish become lethargic and display flared opercula and excess mucous production. The marked host response to the parasite significantly changes the cellular composition of the gill, with epithelial hyperplasia, lamellar fusion, and formation of lacunae which may contain cellular debris or occasionally amoebae. In cultured stocks, treatment involves repeated freshwater bathing (Clark et al. 2003), which reduces the parasite burden and initiates healing of the gills. The diversity and health impacts of amoebic infections in wild marine fish are largely unknown.

Haplosporida

Haplosporidians are rhizarian parasites of marine invertebrates belonging to the class Ascetosporea. Haplosporida contains the causative agents of many commercially significant oyster diseases, important pathogens with global impacts upon aquaculture. There are currently four recognized genera within this family: *Bonamia*, *Minchinia*, *Urosporidium*, and *Haplosporidium*.

Bonamia currently contains four described species (Berthe and Hine, 2003; Carnegie et al. 2006; Cochennec-Laureau et al. 2003; Pichot et al. 1980; Behringer et al. Chapter 10, this volume) and all of these are economically important oyster diseases, with two infections (*B. ostreae* and *B. exitiosa*) listed by the OIE as notifiable diseases. *Bonamia* species are also known as microcell parasites due to their characteristic form, with intracellular, uninucleate cells, 2–3 μm, with a prominent nucleus. Infection with this pathogen is often lethal to the host.

Minchinia currently contains eleven species (Arvy, 1949; Bearham et al. 2008; Ford et al. 2009; Haskin et al. 1966; Hillman et al. 1990; Labbe, 1896;

Marchand and Sprague, 1979; Ormieres and de Puytorac, 1968; Sprague, 1963; van Banning, 1977; Ward et al. 2019), all of which are parasitic infections of molluscs and crustaceans. Infected oysters often appear thin and watery and have a characteristic brownish coloration caused by the masses of spores which form in the connective tissues. The disease is fatal to the host; however, prevalence of infection is usually low (< 1 percent) within cultured oysters.

Urosporidium currently contains five species (Caullery and Mesnil, 1905; Howell, 1967; Ormieres et al. 1973; Perkins, 1971; Perkins et al. 1975). Members of this genus occur exclusively within the marine environment and are hyperparasitic infections of trematode worms.

Haplosporidium is the largest of the four genera and contains approximately forty species (Hartikainen et al. 2014a). Most known members of this genus infect molluscs, but a few have been described infecting crustaceans and annelids. The discovery and descriptions of haplosporidian parasites have been largely based upon histological and molecular characterization of infected host tissues. As expected, most of the descriptions have been identified in commercially significant species, and it is likely that there are many more species that are yet to be discovered and described.

Paramyxida

Paramyxida are also rhizarian parasites which belong to the class Ascetosporea, mainly infecting marine invertebrates, molluscs, crustaceans, and annelids. The order contains the causative agents of many commercially significant oyster diseases. Paramyxids can be distinguished from haplosporidians as they produce characteristic spores which are formed by the production of multiple daughter cells within a primary cell, a distinctive cell-within-cell arrangement. The order has undergone multiple revisions regarding its phylogeny and classification (Carrasco et al. 2015; Desportes and Perkins, 1990; Feist et al. 2009; Itoh et al. 2014) and it is currently accepted that there are five genera within this family—*Marteilia*, *Paramarteilia*, *Paramyxa*, *Marteilioides*, and *Eomarteilia* (Ward et al. 2016).

Marteilia contains four species, for which gene sequence data are available (Grizel et al. 1974; Perkins and Wolf, 1976); other species have been described but their phylogenetic affinities remain unknown (Comps, 1976; Comps, 1983). All *Marteilia* spp. are known to infect bivalve molluscs. *Marteilia refringens* is one of the most significant pathogens of bivalve molluscs, an OIE-notifiable disease, and initially described as a pathogen of oysters. Another species, *M. maurini* (Comps et al. 1982), was described as a pathogen in mussels. Le Roux et al. (2001) compared *Marteilia* infections in mussels and oysters. They suggested that the infections were caused by the same pathogen and that there were two types of *M. refringens*: O type (oyster) and M type (mussel) (= *M. maurini*). In 2007 *M. maurini* and *M. refringens* were synonymized by a working party of the European Food Safety Authority (Algers et al. 2007), conflating the two types into the single OIE-notifiable parasite *M. refringens*. The M-type lineage has recently been resurrected as a distinct species from *M. refringens*—*M. pararefringens* (Kerr et al. 2018) (see fact sheet 6). The full life cycle of these parasites has not been determined; however, copepods have been shown to be vectors in the life cycle of *M. refringens* (Arzul et al. 2013) and the polychaete *Nephtys australis* has been shown to host *M. sydneyi* (Adlard and Nolan, 2015) in part of its life cycle.

Paramarteilia currently contains two species, both of which infect crustaceans (Feist et al. 2009; Ginsburger-Vogel and Desportes, 1979). A similar parasite, *Paramarteilia* sp., was isolated in spider crab (*Maja squinado*); morphology was comparable to that described for *P. canceri* in *C. pagurus* and in-situ hybridization (ISH) confirmed that this pathogen belonged to *Paramarteilia* (Ward et al. 2016).

Paramyxa currently contains two species (Chatton, 1911; Ward et al. 2016), both of which affect polychaete worms. *Marteilioides* currently contains two species (Anderson and Lester, 1992; Comps et al. 1986), both of which affect oyster species. *Eomarteilia* is a newly erected genus and currently has a single species (Ward et al. 2016).

Mikrocytida

Mikrocytida is the last of the Ascetosporea known to contain pathogens of marine invertebrates and is putatively a sister group to the Haplosporida (Hartikainen et al. 2014b). There are currently two genera within this family, *Mikrocytos* and *Paramikrocytos*, infecting crustaceans and molluscs.

Mikrocytos currently has three species (Farley et al. 1988; Hartikainen et al. 2014b). *Mikrocytos mackini* causes disease and mortalities in several economically important oyster species along the West Coast of North America and is an OIE-notifiable disease. Initially called Denman Island disease due to the location of the initial outbreak in the 1960s, the disease is characterized by the appearance of focal green lesions within the mantle, body wall, labial palps, or adductor muscle. These lesions render the oysters unmarketable and in severe infections the disease causes high mortalities (Behringer et al. Chapter 10, this volume). A second species was described infecting Pacific oysters in Norfolk in the UK. *Mikrocytos mimicus* (Hartikainen et al. 2014b) showed similar gross pathology. It was the first incidence of a *Mikrocytos* infection in Europe and was described as a new species. The third species, *M. boweri*, closely related to *M. mackini* (Abbott et al. 2011), has caused pathology in Olympia oysters (*Ostrea lurida*) (Hartikainen et al. 2014b).

Paramikrocytos is a newly erected genus and currently contains a single species, *Paramikrocytos canceri* (Hartikainen et al. 2014b). This pathogen was described infecting the antennal gland of the juvenile edible crab (*Cancer pagurus*) in the UK (see fact sheet 7). Initially thought to be a Haplosporidian-like infection due to its microcell-like appearance, uninuclear stages, and plasmodia within the epithelial cells, phylogenetic analysis showed it to be a sister group to *Mikrocytos mackini*. This pathogen is highly prevalent within pre-recruit populations of edible crab (i.e., non-fished juvenile crabs below minimum landing size) within the UK but has not been described infecting crabs landed by the fishery, suggesting it may be specific to juvenile crabs (Bateman et al. 2011). How this pathogen may affect the recruitment of crabs into the fishery is unknown, as is the mortality rate; further work is needed to investigate and highlight the true pathogenesis of this organism.

Dinoflagellata

Parasitic dinoflagellates have been reported to infect algae, protozoans, annelids, crustaceans, molluscs, ascidians, rotifers, and fishes (Cachon and Cachon, 1987; Coats, 1999; Shields, 1994). The genus *Hematodinium* belongs to the family Syndiniceae,

Fact Sheet 6

Pathogen	*Marteilia pararefringens*		
Abbreviation	None		
Agent	Phylum: Ascetosporea Order: Paramyxida Family: Marteiliidae Genus: *Marteilia* Species: *Marteilia pararefringens* *M. pararefringens* was previously described as *M. refringens* (M-type).		
Agent features	Primary cells contain a single uninucleate secondary cell that divides to produce eight sporangia. These mature to form sporonts containing four spore primordia (tertiary cells). These each cleave internally, developing into mature spores containing three nucleated sporoplasms. Each sporo-plasm is located inside the other, forming tricellular spores. Refringent structures appear within sporonts during spore maturation.		
Target hosts	*M. pararefringens* preferentially infects *Mytilus* sp. However, detection was achieved using polymerase chain reaction (PCR) and not confirmed histologically.		
Mortalities	Pathogenicity of *M. pararefringens* in mussels is unclear; however, *M. refringens* in oysters is known to cause mass mortalities.		
Gross clinical signs	The digestive gland of affected individuals commonly appears pale brown in color. This is particularly evident in heavily affected individuals, although is not necessarily specific to infection with *Marteilia*. Light infections may not elicit immediate gross signs of disease.		
Histopathology	Pathogenesis of *Marteiliosis* by *M. pararefringens* is not fully elucidated. Main targets of infection are tissues of the digestive system. Basophilic primary cells are seen affecting the stomach epithelium, although these can also be observed affecting epithelia of the primary and secondary digestive tubules, and smaller digestive tubules. Following sporulation, mature refringent granule-containing acidophilic sporangia can be identified within the digestive tubules. Sporulation predominantly occurs within the digestive tubules, although acidophilic structures, corresponding to plate-like inclusions, can occasionally be seen within primary cells affecting the stomach and secondary digestive tubule epithelium.		

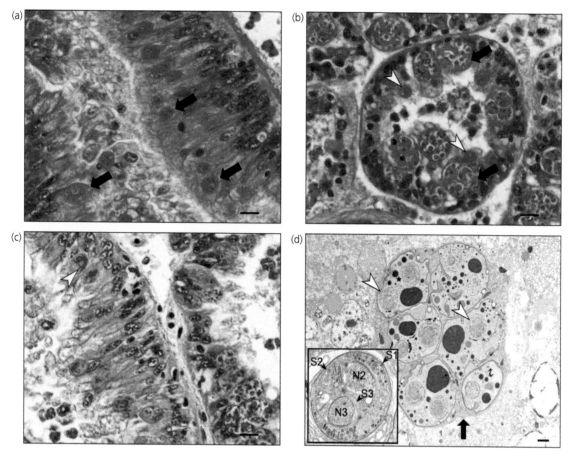

Figure 1.6 (a) Basophilic immature primary cells (arrows) containing a secondary cell (presporangia), within epithelial cells of a secondary digestive tubule. Several primary cells contain multiple secondary cells. H&E stain. Scale bar = 10 μm. **(b)** Enlarged "sporangia-containing" primary cell with acidophilic refringent granules (arrows), situated within the epithelium of the digestive tubule. Presence of immature primary cells (arrowheads) can also be seen. H&E stain. Scale bar = 10 μm. **(c)** Immature primary cells situated within the epithelium of secondary digestive tubule. Note presence of acidophilic plate-like inclusions (arrowhead). Gomori one-step trichrome stain. Scale bar = 10 μm. **(d)** Electron micrograph of primary cell (arrow) containing sporangia (asterisk) of maturing spores. Internal cleavage of spore primordia elicits mature spores (arrowheads) containing three nucleated sporoplasms. Scale bar = 2 μm. Inset demonstrates mature spore outermost (S1), intermediate (S2), and innermost (S3) sporoplasm. Nuclei of corresponding sporoplasms are denoted N2 and N3.

Fact Sheet 7

Pathogen	*Paramikrocytos canceri*
Abbreviation	None

Agent	Phylum: Cercozoa
	  Class: Ascetosporea
	    Order: Mikrocytida
	      Family: Mikrocytidae
	        Genus: *Paramikrocytos*
	          Species: *Paramikrocytos canceri*

Agent features	Spherical microcell parasites, approximately 3 μm in diameter within the cytoplasm of labyrinth and coleomosac epithelial cells of the antennal gland and bladder. Pathogen is associated with juvenile life stages, i.e., crabs below minimum landing size.
Target hosts	Juvenile life stages of edible crabs (*Cancer pagurus*).
Mortalities	Pathogenicity is unknown.
Gross clinical signs	Hypertrophy of the antennal gland and bladder. Crabs do not appear to display any external signs of infection, but upon dissection a layer of yellow, gelatinous tissue can be identified at the periphery of the shell.
Histopathology	Uninucleate and multinucleate plasmodial life stages can be identified within the cytoplasm of labyrinth and coleomosac epithelial cells of the antennal gland and bladder. Parasites are ovoid in shape. Uninucleate stages undergo nuclear fission without cell division to form binucleate and multinucleate plasmodia. Plasmodia undergo plasmotomy to form uninucleate life stages. All life stages can be liberated into the lumen of the bladder.

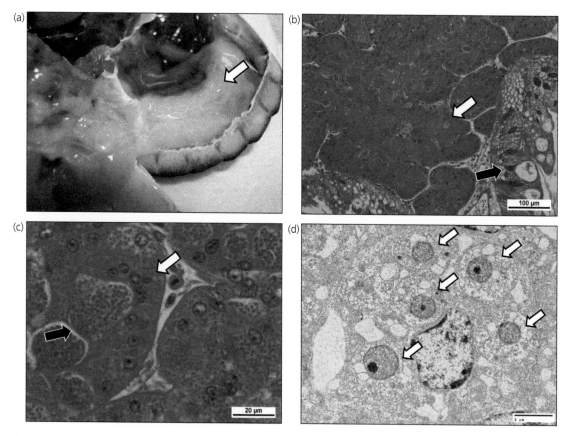

Figure 1.7 **(a)** Upon dissection, infected crabs display a yellow, gelatinous tissue (arrow) at periphery of shell. **(b)** Sections of hypertrophied antennal gland (white arrow) possess a distinct eosinophilic appearance. Tubules appear to contain granular packets and can be distinguished from adjacent hepatopancreas (black arrow). H&E stain. Scale bar = 100 μm. **(c)** Antennal gland contains discrete granular packets (arrows) within cytoplasm of bladder epithelial cells. Epithelial cells contain multinucleate inclusions (black arrow) or uninucleate forms in direct contact with cytoplasm (white arrow). H&E stain. Scale bar = 20 μm. **(d)** Uninucleate stages of the parasite (arrows) within cytoplasm of host epithelial cell. TEM. Scale bar = 2 μm.

order Syndinida, members of which have complex life cycles with at least three phases of development, although the complete life cycle of the type species (*H. perezi*) is unknown (Stentiford and Shields, 2005). *Hematodinium* species have caused major mortality among crustacean fishery and aquaculture industries and have been reported in more than forty crustacean species across several continents (Small, 2012; Wang et al. 2017). Bacteria-like endosymbionts have been reported within the dinospore life stage, suggesting a potential role of endosymbiosis in the survival of free-living stages of the parasite (Stentiford et al. 2012). One of the most common features of this parasitic infection is the hyperpigmentation or "cooked" appearance of the host carapace (Stentiford et al. 2002; Stentiford and Shields, 2005). *Hematodinium* has been shown to cause pathologic changes within the tissues of crabs and lobsters (see fact sheet 8), causing a bitter taste to the meat and the common term "bitter crab disease" (Meyers et al. 1987; Taylor and Khan, 1995).

There are two described species: the type species *Hematodinium perezi* (Chatton and Poisson, 1931; Messick and Shields, 2000) (infecting *Carcinus maenas*, *Liocarcinus depurator*, and *C. sapidus*) and *H. australis* (Hudson and Shields, 1994) (infecting *Portunas pelagicus*), with infections being described from other host species needing further characterization (Small, 2012). Peak prevalence has been reported in winter and spring (Field et al. 1992; Field et al. 1998; Stentiford et al. 2001) and infections appear more significantly abundant in juveniles (Messick, 1994; Messick and Shields, 2000). Infections are fatal and have the potential to significantly impact wild crustacean populations and associated fisheries (Small, 2012; Behringer et al. Chapter 10, this volume). In recent years, *Hematodinium* sp. infections have been associated with losses in cultured crabs *Portunas trituberculatus* and *Scylla serrata* (Li et al. 2013; Li et al. 2008) and the shrimp *Exopalaemon carinicauda* (Xu et al. 2010).

Perkinsea

The pathogen *Perkinsus marinus*, also known as "dermo" disease, was initially isolated and described by Mackin et al. (1950). The disease has been responsible for mass mortalities (Behringer et al. Chapter 10,

this volume), but the condition can be controlled with good husbandry practices. The disease is associated with warm summer temperatures when pathogenicity and mortalities are at their highest. There are currently seven species of *Perkinus* (Blackbourn et al. 1998; Dungan and Reece, 2006; Mackin et al. 1950; McLaughlin et al. 2000; Moss et al. 2008; Sandra et al. 2004), all of which infect bivalve molluscs around the globe. The classification of the genus *Perkinsus* has been reviewed over the years, initially being placed in the phylum Apicomplexa; sequence analysis indicated that the genus was more closely related to the Dinoflagellida (Saldarriaga et al. 2003), Zhang et al. (2011) suggesting that the genus *Perkinsus* be positioned as an independent lineage (Perkinsozoa) between the phyla of Apicomplexa and Dinoflagellata. Recently, tumor-forming X-cell parasites of fish have been shown to form a sister clade to *Perkinsus*—the new family Xcellidae (Freeman et al. 2017).

1.2.4 Metazoans

Metazoan parasites of marine fish are extremely diverse and numerous, with many taxa requiring several hosts to complete their life cycle. Helminth infections caused by trematodes, cestodes, and acanthocephalans do not generally cause severe disease conditions, although individuals may suffer intense infections with associated pathological responses. Copepod crustacean parasites (e.g., *Lepeoptheirius salmonis*) are notorious pathogens of cultured salmonids (Behringer et al. Chapter 10, this volume) but are a less significant threat to wild fish living in open waters. Monogeneans are very common on the gills and skin of wild marine fish, but their presence usually causes little harm to the host. Acanthocephalans (thorny headed worms) inhabit the intestine of fish and other vertebrates and their attachment may cause localized inflammation, but they are not associated with mortalities. The next section highlights myxozoa as agents of significant disease in wild and cultured fish, and nematodes that are ubiquitous in marine environments and cause significant infections in marine fish and mammals as well as having zoonotic potential.

Myxozoa

The Myxozoa comprise a subphylum of the phylum Cnidaria. They are a highly specialized and extremely diverse group of parasites infecting predominately fish and invertebrates (annelids and bryozoa) in the aquatic environment (Okamura et al. 2015). Most species affecting marine fish are coelozoic, inhabiting body cavities such as the urinary bladder and gall bladder, and do not cause serious disease. However, certain species associated with infections of organs and tissues cause significant pathological changes and can result in impaired physiology and function, leading to mortality.

In wild fish, most infections are detected in adult or subadult fish and are identified by the presence of spores in affected tissues. Coelozoic infections generally cause little harm, but a few species can result in significant tissue damage and altered function. Gallbladders infected with *Myxidium* species may become hypertrophied or hardened or have a cheese-like consistency (Lom, 1984). Saithe (*Pollachius virens*) infected with *M. gadi* were found to harbor intense infections in the gallbladder which resulted in marked proliferation of the lining epithelium into papillomatous folds covered in sporulating plasmodia completely filling the bladder (Feist and Bucke, 1992). Enteromyxosis caused by histozoic infections of the intestine by *Enteromyxon scophthalmi* in turbot (Branson et al. 1999) and *E. leei* in cultured gilthead seabream (Diamant, 1992) causes severe enteritis-associated infection and destruction of the intestinal epithelium, respectively, followed by mortalities. Transmission is direct via cohabitation with infected fish, so that high-density populations such as those encountered in farm or hatchery situations are at particular risk. Information on the impact of this disease in wild populations is lacking.

Muscle infections of marine fish by myxosporeans of the genus *Kudoa* are well known as the cause of "milky flesh" following enzymatic myoliquefaction post-mortem (Moran et al. 1999), while less severe infections can produce large melanized cysts which still render the fish unmarketable. High-value fish species such as tuna are also susceptible to infection and *Kudoa* spp. are a threat to tuna aquaculture. However, the life cycle and nature of the infective stage for *Kudoa* spp. are unknown and remain a constraint to strategies for avoiding infection. Similar muscle infections in salmonid species caused by *Henneguya salmonicola* affect marketability, where the quality of fillets is reduced (Awakura and Kimura, 1977). Cranial cartilage infections caused by *Myxobolus aeglifini* (see fact sheet 9) are found in a wide range of host fish species, particularly gadoids and flatfish species where deformities in both the cartilage and otoliths can result, potential affecting balance. As with most myxosporean infections in wild fish, knowledge on pathogenicity is largely based on examination of adult or subadult survivors. Evidence from the freshwater environment for infections in juvenile fish in their first year suggest that this life phase is particularly at risk from infections with myxozoan parasites (Longshaw et al. 2010) and this may also be the case for marine fish.

Nematoda

Nematodes are extremely common, globally dispersed parasites of a diverse range of marine fish and have attracted significant attention due to their zoonotic potential. The main species of interest in that regard are the herring worms (*Anisakis* spp.) (see fact sheet 10) (Mattiucci and Nascetti, 2006; Smith and Wootten, 1978), which cause anisakiasis, and the cod worm (*Pseudoterranova* spp.), whose larvae have low host specificity.

Hosts for *Anisakis* include herring (*Clupea harengus*), mackerel (*Scomber scombrus*), several gadoid species, and Atlantic salmon (*Salmo salar*). The life cycle of *Anisakis* involves several hosts, with adult stages occurring in marine mammals. Marine mammals and humans become the final hosts upon ingestion of infected fish or crustaceans, where maturation of the parasites occurs. Infections in fish are frequently innocuous and cause limited pathology, but larvae can occur in high numbers in individuals and have been associated with significant inflammatory responses, such as "red vent syndrome" in Atlantic salmon (Noguera et al. 2009; Twigg et al. 2008). The impacts of fish nematode infections at the population level are unknown. Infections in humans produce similar pathology, and allergic hypersensitivity to the presence of anisakid proteins can occur (Del Rey Moreno et al. 2006).

Fact Sheet 8

Pathogen	*Hematodinium* spp.

Agent	Phylum: Dinoflagellata
	  Class: Syndiniophyceae
	    Order: Syndiniales
	      Family: Syndiniaceae
	        Genus: *Hematodinium*

Agent features	Identified as dinoflagellates due to their typical dinokaryon, alveolate pellicle, presence of naked, athecate gymnodinoid dinospores (or zoospores), and classic form of mitosis known as dinomitosis. Due to lack of distinct characteristics and poorly understood life cycles there are only two described species, *H. perezi* and *H. australis*.
Target hosts	*Nephrops norvegicus, Callinectes* spp., *Cancer* spp., *Carcinus maenas, Chionoecetes* spp., *Trapezia* spp., *Scylla serrata, Portunus pelagicus, Portunus latipes, Ovalipes ocellatus, Panopeus herbstii, Neopanope sayi, Necora puber, Menippe mercenaria, Maja squinado, Liocarcinus depurator, Libinia emarginata, Hexapanopeus angustifrons*
Mortalities	Mortalities to pre-recruit and adult populations can reach 100 percent.
Gross clinical signs	Hyperpigmentation of carapace often accompanied by "chalky" or "cooked" appearance; discoloration of arthrodial membranes and genital pores in females. Opaque to creamy coloration of hemolymph and in heavily infected crabs coagulopathy (lack of clotting ability). Crabs and lobsters exhibit hemocytopenia (decline in hemocytes).
Histopathology	**Muscle**—lysis of peripheral fiber regions and separation of sarcolemma from myofibrils, degeneration of claw muscle, disorganization of filaments in region of the Z-line. **Hepatopancreas (HP)**—parasites have been observed in close association, possibly even attached to basal lamina of HP. In patently infected hosts, HP exhibits loss of structure, and hemal arterioles are grossly dilated and filled with large numbers of parasitic plasmodial cells. Parasites can be seen within lumen of intact tubules in heavy infections. **Gonad**—arterioles of ovary and testis can be heavily infiltrated with the parasite during patent infections. Infected females do not develop mature ovaries. Plasmodial forms of the parasite have also been described in hemal spaces of the gill, heart, eye-stalk, connective tissue of gut, sinuses of the antennal gland, and brain.

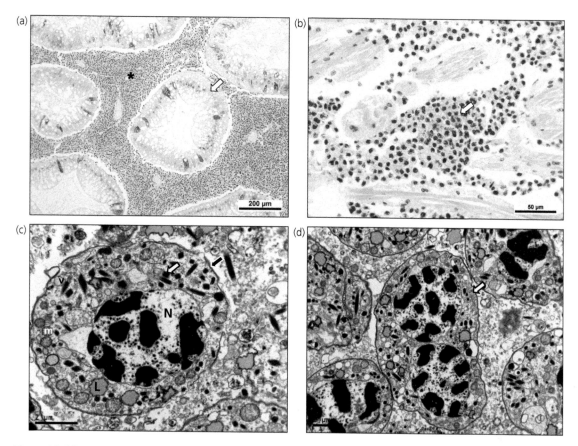

Figure 1.8 (a) *Hematodinium* spp. infection within the hepatopancreas. Hepatopancreas tubules (arrow) themselves are not affected; hemal sinuses (asterisk) between tubules are filled with parasite cells. H&E stain. Scale bar = 200 μm. **(b)** *Hematodinium* spp. infection within the heart; parasite cells are present between islands of muscle cells (arrow). Parasite cells are ovoid and possess a distinctive nucleus. H&E stain. Scale bar = 50 μm. **(c)** A uninucleate *Hematodinium* parasite. An alveolar membrane (black arrow) encases a single nucleus (N), trichocysts (white arrow), mitochondria (m), lipid droplets (L), and vacuoles (v). TEM. Scale bar = 2 μm. **(d)** A binucleate parasite cell (arrow); parasites typically contain between one and four nuclei. TEM. Scale bar = 0.2 μm.

Fact Sheet 9

Pathogen	*Myxobolus aeglifini*
Abbreviation	None
Agent	Phylum: Cnidaria Class: Myxozoa Order: Bivalvulida Family: Myxobilidae Genus/species: *Myxobolus aeglifini*
Agent features	Formation of large macroscopic cysts in the cranial cartilage of marine fish. Large plasmodium containing numerous sporogonic stages. Mature spores are bivalvular, elliptical, subspherical in shape (approximately 11 μm long × 9 μm wide), with a prominent sutural ridge joining the two spore valves and containing two pyriform polar capsules.
Target hosts	Gadoid species including blue whiting *Micromesistius poutassou*, whiting *Merlangius merlangus*, poor cod *Trispoterus minutus*, cod *Gadus morhua*, haddock *Melanogrammus aeglifinus*, and hake *Merluccius merluccius*. Other reported hosts are lumpfish *Cyclopterus lumpus* and flatfish species including dab *Limanda limanda* and plaice *Pleuronectes platessa*.
Mortalities	Mortalities directly associated with *M. aeglifini* have not been reported.
Gross clinical signs	Pale cysts in cranial cartilage. Ocular lesions are prominent. Deformation of otoliths has been reported.
Histopathology	Infections are histozoic, infecting cartilage, including cranial cartilage and bones, scleral cartilage of the eye, and gill cartilage. Lysis of the cartilage associated with growth of the parasite plasmodium.

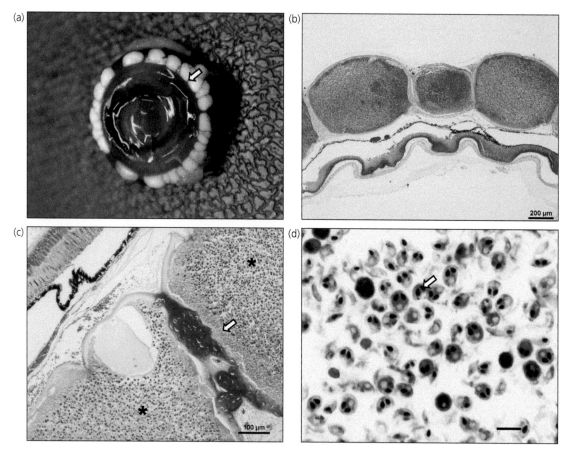

Figure 1.9 (a) Ocular cysts (arrow) of *Myxobolus aeglifini* surrounding the eyeball of a blue whiting. **(b)** Low power view of the infected region of the eyeball showing the location of the parasite cysts in the scleral cartilage. H&E Stain. Scale bar = 200µm. **(c)** Detail from the infected cartilage showing cysts containing numerous spores. Note the peripheral region of the cyst comprised of late sporogonic stages (arrow) and the central regions filled with mature spores (asterisks). Giemsa Stain. Scale bar = 100µm. **(d)** Sections of mature spores showing the characteristic myxobolid subspherical form with two deeply stained pyriform polar capsules (arrow). Giemsa stain. Scale Bar = 10µm.

Fact Sheet 10

Pathogen	*Anisakis* sp.
Abbreviation	None
Agent	Phylum: Nematoda Class: Ascaridida Order: Secernentea Family: Anisakidae Genus: *Anisakis*
Agent features	Anisakid nematodes are unsegmented and have a vermiform body, circular in cross section. The outer surface is covered by a cuticle secreted by the underlying epidermis. Fish infecting stages are typically coiled.
Target hosts	Very wide host range, including gadoids (cod *Gadus morhua* and haddock *Melanogrammus aeglifinus*), Atlantic salmon (*Salmo salar*), and flatfishes. Adult stages found in marine mammals (cetaceans and pinnipeds).
Mortalities	Not associated with mortalities in fish.
Gross clinical signs	Infection with large numbers of nematodes in the visceral cavity may result in emaciation of the host, but this does not appear to be a consistent feature. With superficial infections on the surface of the liver, several nematodes may be encapsulated by a fibrous tissue response and appear opaque compared to healthy individuals. Histozoic infections as with red vent syndrome (RVS) in salmonids are associated with severe localized swelling and hemorrhaging around the vent, with skin and scale loss.
Histopathology	Main histologic features in anisakid infections are dependent on the severity of infection. Typically, the primary response is inflammation with dilation of blood vessels and infiltration of lymphocytes and macrophages, leading to the formation of granulomatous inflammatory reaction and often associated with increased numbers of eosinophil granule cells. Host attempts to isolate the nematodes in a fibrous capsule lead to more extensive fibrosis in chronic infections, with some nematodes becoming necrotic.

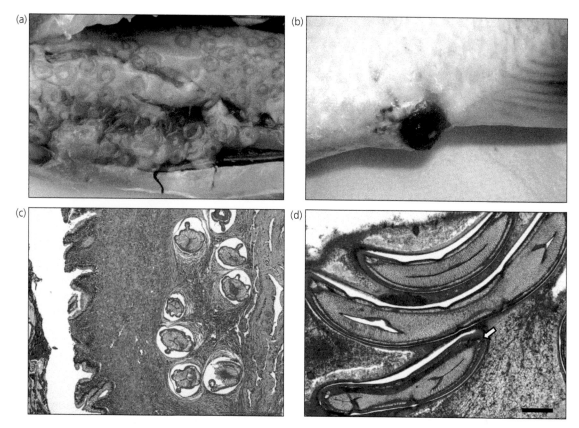

Figure 1.10 **(a)** Typical "coiled spring" appearance of anisakid nematodes, several of which are embedded in the surface of the liver of a gadoid fish. Fibrous encapsulation is not evident. **(b)** Anisakiasis in Atlantic salmon with RVS, showing a swollen and hemorrhagic vent area. Image used with permission from *Diseases of Aquatic Organisms* (Figure 2 in Beck et al. (2008) DAO 82:61–65). **(c)** Longitudinal section through the vent area of a salmon with RVS. Several transverse sections of anisakid nematodes embedded in inflamed connective tissue. H&E stain. Scale bar = 0.5 mm. **(d)** Profiles of anisakids embedded in the liver of hake; the cuticle of the nematode is prominent (arrow) with the digestive tract clearly visible. Note lack of host response. Mallory trichrome. Scale bar = 50 μm.

The European eel (*Anguilla anguilla*) and the North American eel (*A. rostrata*) have been adversely affected by the nematode *Anguillicoloides crassus* (Wielgoss et al. 2008), since its introduction into Europe from Asia in the 1980s. The parasite infects the swimbladder and results in a severe aerocystitis, fibrosis, and, in severe infections, collapse of the swimbladder. It has been postulated that the loss of swimbladder function could prevent eels reaching their spawning ground in the Sargasso Sea and may be a key factor in eel population declines (Palstra et al. 2007). However, other factors such as habitat loss and pollution have also been suggested as potential factors (Kennedy, 2007).

1.3 Pathology and the "pathobiome": future direction

The advent of high-throughput sequencing technologies has provided a valuable new perspective in pathology (Frasca et al. Chapter 11, this volume). Two fundamentally different types of investigation are now used routinely: PCR amplicon (metabarcoding) and metagenomics (shotgun sequencing of total nucleic acid component of a sample). PCR amplicons target a particular region of the genome, most commonly a taxonomic marker gene such as the small subunit ribosomal RNA gene. PCR primers can be designed so that they are broadly targeted, for example with the intention of amplifying a wide range of bacterial, fungal, or other taxonomic groups, through to very specific, amplifying only a single pathogen strain or clade of parasites (Hartikainen et al. 2014a; Hartikainen et al. 2014b). These environmental DNA (eDNA)-based approaches allow thorough sampling of the diversity of the target genes in a sample, aided by sensitive and exponentially amplifying PCR, but with the limitation that such PCR assays have inherent biases, for example against genetically divergent versions of the gene and those significantly longer or more structurally complex than their relatives. On the other hand, in metagenomics samples all the genetic material in a sample—coding and non-coding regions, organellar and nuclear DNA, from all organisms/material present—contributes to the sequence analysis. This means that a large amount of sequencing and bioinformatics effort is required to taxonomically

identify all organisms present, but this approach is taxonomically much less biased, simultaneously sampling the genomes of viruses, prokaryotes, and eukaryotes within a sample.

This expanding molecular toolkit is increasingly being used synergistically in parallel with histological techniques. For example, metagenomic sequencing has been used to characterize extremely genetically divergent parasites that were detected in histological screens but were refractory to all PCR-based amplification attempts (Hartikainen et al. 2014b). Another virtue of metagenomics is the ability to assemble genomic fragments from novel pathogens, often in the case of viruses and bacteria, complete genomes (Bayliss et al. 2017; van Aerle and Santos, 2017). Environmental DNA is increasingly used as the first line of investigation into parasite–host systems and to investigate pathogen ecology (Bass et al. 2015). For example, Hartikainen et al. (2014a) designed haplosporidian-specific primers to investigate the environmental diversity of this clade of parasites in coastal environmental and invertebrate samples. This revealed a large and previously unsuspected diversity of haplosporidians, providing insight into their evolutionary relationships, life cycles, and potential hosts. Subsequent studies have followed such initial insights to characterize novel parasites and their hosts, and the presence of known and novel parasites associated with new hosts and different habitat types (Ward et al. 2016; Ward et al. 2019).

Finally, new concepts of infection and disease are considering the broader context in which specific pathogens and other host-associated organisms can influence host health (Stentiford et al. 2017b). There has been a move away from the single pathogen–single disease paradigm to one in which assemblages of host-associated organisms, including epi- and endosymbiotic eukaryotes (including metazoan and microbial parasites), bacteria, and viruses (together comprising the "holobiome") contribute not only to the observed "disease" phenotype but also to the susceptibility of the host to establishment of certain prominent pathogens (Mosser et al. 2015; Rebollar et al. 2016). Related to this, disruption in the diversity of microbes associated with a host at certain time points in its life history can cause shifts in susceptibility to infection and disease by a given pathogen later in the life of those hosts (Knutie et al. 2017). High-throughput sequencing technologies, particularly metagenomics, are essential for such investigations into the cumulative effect of microbial and symbiotic diversity on host health and disease states, and can be applied to different components of both the host (gut, blood, tissues, and external surfaces) and the environment (water, sediment, surfaces) (Bass et al. 2015). In this way, the "pathobiome" (expansion of the single pathogen–single disease paradigm (Egan and Gardiner, 2016; Jakuschkin et al. 2016; Vayssier-Taussat et al. 2014)) can be investigated and understood, including information about the interaction between all holobiont members, including the host (Broberg et al. 2018; Pitlik and Koren, 2017). In doing so, we seek not only to identify which other organisms interact with those supposed primary pathogens (e.g., OIE-listed pathogens) but also more broadly how these microbial ecosystems operate within the host. By considering such factors, the definition of susceptibility is likely to be significantly more complex than the system relied upon in international legislation governing trade.

Given all of this, it is important to embrace these emerging technologies within the armoury of the modern aquatic animal pathologist and to ensure their applications are focused on better understanding of disease processes in aquatic systems. The aquatic pathologist needs to employ a variety of methods to explore how pathogens transmit and spread, and use techniques to taxonomically identify organisms that are present within a sample to enable comparisons with known disease conditions and generate the "pathobiome" of an individual (see Frasca et al. Chapter 11, this volume). The development of point-of-need diagnostic testing devices (accurate low-cost devices which can be used in the field) along with development of remote sensing devices will increase the number of tests performed and it is expected that these technologies will limit further impact, via implementation of strategies to prevent and control the spread of disease.

1.4 Summary

- Histopathology is a frontline tool used to discriminate infection (the presence of an invasive, multiplying agent) from disease (where normal function is impaired or lost).
- Understanding "normal" variation and physiological status of host tissues and organs is key when determining host response to potential pathogens, since limited baseline data exist for many marine species.
- Pathogen diversity within the marine environment is high, but not all pathogens cause disease and mortalities.
- Environmental DNA analysis is increasingly used in the study of pathogen ecology and to investigate potential risks for novel infections and disease emergence.
- New concepts of infection and disease are now being considered, with a move away from the single pathogen–single disease paradigm, and it is more likely that the accumulative effects of multiple host-associated organisms affect an organism's susceptibility to infection.
- Pathobiome can be used to identify shifts in the diversity of microbes and highlight how this may affect this susceptibility to infection.
- The aquatic animal pathologist needs to use an array of techniques to understand disease processes in aquatic systems and to identify their etiologies, which may include pathogenic organisms.

Acknowledgments

The authors acknowledge the support of the Department for Environment, Food and Rural Affairs (DEFRA) under contracts FX001 and FB002 to S.W.F. and D.B. We thank Patrick Hooper and Dr Irene Cano for assistance with sections on scallop ELO, and Dr Jaime Martinez-Urtaza for review and comments.

References

Abbott, C. L., Gilmore, S. R., Lowe, G., Meyer, G., Bower, S., 2011. Sequence homogeneity of internal transcribed spacer rDNA in *Mikrocytos mackini* and detection of *Mikrocytos* sp. in a new location. Diseases of Aquatic Organisms. 93, 243–50.

Adlard, R., Nolan, M., 2015. Elucidating the life cycle of *Marteilia sydneyi*, the aetiological agent of QX disease in the Sydney rock oyster (*Saccostrea glomerata*). International Journal of Parasitology. 45, 419–26.

Alderman, D. J., Feist, S. W., 1985. Exophiala infection of kidney of rainbow trout recovering from proliferative kidney disease. Transactions of the British Mycological Society. 84, 157–9.

Algers, B., Blokhuis, H. J., Broom, D. M., Costa, P., Domingo, M., Greiner, M., et al. 2007. Scientific opinion of the panel on animal health and welfare on possible vector species and live stages of susceptible species not transmitting disease as regards certain mollusc diseases. The EFSA Journal. 597, 1–116.

Anderson, T. J., Lester, R. J., 1992. Sporulation of *Marteilioides branchialis* n. sp. (Paramyxea) in the Sydney rock oyster, *Saccostrea commercialis*: an electron microscope study. Journal of Protozoology. 39, 502–8.

Appeltans, W., Ahyong, S. T., Anderson, G., Angel, M. V., Artois, T., Bailly, N., et al. 2012. The magnitude of global marine species diversity. Current Biology. 22, 2189–202.

Arvy, L. 1949. Presentation de documents relatifs à l'ovogenese chez le Dentale et a deux parasites de ce Scaphopode: *Cercaria prenanti* n. sp. et *Haplosporidium dentali*. Bulletin de la Societe Zoologique de France. 74, 292–4.

Arzul, I., Chollet, B., Boyer, S., Bonnet, D., Gaillard, J., Baldi, Y., et al. 2013. Contribution to the understanding of the cycle of the protozoan parasite *Marteilia refringens*. Parasitology. 141, 227–40.

Arzul, I., Corbeil, S., Morga, B., Renault, T., 2017. Viruses infecting marine molluscs. Journal of Invertebrate Pathology. 147, 118–35.

Atami, H., Muraosa, Y., Hatai, K., 2009. Haliotisida infection found in wild mantis shrimp *Oratosquilla oratoria* in Japan. Fish Pathology. 44, 145–50.

Gollas-Galván, T., Martínez-Porchas, M., Hernández-López, F. J., 2012. Experimental infection and detection of necrotizing hepatopancreatitis bacterium in the American lobster *Homarus americanus*. The Scientific World Journal. 2012, 979381.

Awakura, T., Kimura, T., 1977. On the milky condition in smoked coho salmon (*Oncorhynchus kisutch*) caused by myxosporidian parasite. Fish Pathology. 12, 179–84.

Bass, D., Stentiford, G. D., Littlewood, D. T. J., Hartikainen, H., 2015. Diverse applications of environmental DNA methods in parasitology. Trends in Parasitology. 31, 499–513.

Bateman, K. S., Stentiford, G. D., 2017. A taxonomic review of viruses infecting crustaceans with an emphasis on wild hosts. Journal of Invertebrate Pathology. 147, 86–110.

Bateman, K. S., Hicks, R. J., Stentiford, G. D., 2011. Disease profiles differ between non-fished and fished populations of edible crab (*Cancer pagurus*) from a major commercial fishery. ICES Journal of Marine Science. 68, 2044–52.

Bateman, K. S., Wiredu-Boakye, D., Kerr, R., Williams, B. A. P., Stentiford, G. D., 2016. Single and multi-gene phylogeny of Hepatospora (Microsporidia)—a general-

ist pathogen of farmed and wild crustacean hosts. Parasitology. 143, 971–82.

Bayliss, S. C., Verner-Jeffreys, D. W., Bartie, K. L., Aanensen, D. M., Sheppard, S. K., Adams, A., et al. 2017. The promise of whole genome pathogen sequencing for the molecular epidemiology of emerging aquaculture pathogens. Frontiers in Microbiology. 8, 121.

Bearham, D., Spiers, Z., Raidal, S. R., Jones, J. B., Nicholls, P. K., 2008. Spore ornamentation of *Minchinia occulta* n. sp. (Haplosporidia) in rock oysters *Saccostrea cuccullata* (Born, 1778). Parasitology. 135, 1271–80.

Beaz-Hidalgo, R., Balboa, S., Romalde, J. L., Figueras, M. J., 2010. Diversity and pathogenicity of *Vibrio* species in cultured bivalve molluscs. Environmental Microbiology Reports. 2, 34–43.

Berthe, F. C. J., Hine, P. M., 2003. *Bonamia exitiosa* Hine et al. 2001 is proposed instead of *B. exitiosus* as the valid name of *Bonamia* sp. infecting flat oysters *Ostrea chilensis* in New Zealand. Diseases of Aquatic Organisms. 57, 181.

Birkbeck, T. H., Bordevik, M., Froystad, M. K., Baklien, A., 2007. Identification of *Francisella* sp. from Atlantic salmon, *Salmo salar* L., in Chile. Journal of Fish Diseases. 30, 505–7.

Birkbeck, T. H., Feist, S. W., Verner-Jeffreys, D. W., 2011. Francisella infections in fish and shellfish. Journal of Fish Diseases. 34, 173–87.

Blackbourn, J., Bower, S. M., Meyer, G. R., 1998. *Perkinsus qugwadi* sp.nov. (incertae sedis), a pathogenic protozoan parasite of Japanese scallops, *Patinopecten yessoensis*, cultured in British Columbia, Canada. Canadian Journal of Zoology. 76, 942–53.

Blazer, V. S., Wolke, R. E., 1979. An Exophiala-like fungus as the cause of a systemic mycosis of marine fish. Journal of Fish Diseases. 2, 145–52.

Bojko, J., Stebbing, P. D., Dunn, A. M., Bateman, K. S., Clark, F., Kerr, R. C., et al. 2018. Green crab *Carcinus maenas* symbiont profiles along a North Atlantic invasion route. Diseases of Aquatic Organisms. 128, 147–68.

Bonami, J. R., Hasson, K. W., Mari, J., Poulos, B. T., Lightner, D. V., 1997. Taura syndrome of marine peaneid shrimp: characterisation of the viral agent. Journal of General Virology. 78, 313–19.

Boucher, P., Castric, J., Laurencin, F. B., 1994. Observations of virus-like particles in rainbow trout *Oncorhynchus mykiss* infected with sleeping disease virulent material. Bulletin of the European Association of Fish Pathologists. 14, 215–16.

Branson, E., Riaza, A., Alvarez-Pellitero, P., 1999. Myxosporean infection causing intestinal disease in farmed turbot, *Scophthalmus maximus* (L.), (Teleostei: Scophthalmidae). Journal of Fish Diseases. 22, 395–9.

Broberg, M., Doonan, J., Mundt, F., Denman, S., McDonald, J. E., 2018. Integrated multi-omic analysis of host–microbiota interactions in acute oak decline. Microbiome. 6, 21.

Bruckner, G. K., 2009. The role of the World Organisation for Animal Health (OIE) to facilitate the international trade in animals and animal products. Onderstepoort Journal of Veterinary Research. 76, 141–6.

Bruno, D. W., 2016. Infection with *Exophiala salmonis*. ICES Identification Leaflets for Diseases and Parasites of Fish and Shellfish. Leaflet No. 42.

Bucke, D., 1989. Observations on visceral granulomatosis and dermal necrosis in population of North Sea cod (*Gadus morhua*). ICES CM 1989/E:17.

Cachon, J., Cachon, M., 1987. Parasitic dinoflagellates. In: F. J. R. Taylor (ed.), The Biology of Dioflagellates. Blackwell Scientific Publications, Oxford, pp. 571–610.

Cano, I., van Aerle, R., Ross, S., Verner-Jeffreys, D. W., Paley, R. K., Rimmer, G. S. E., et al. 2018. Molecular characterization of an *Endozoicomonas*-like organism causing infection in the king scallop (*Pecten maximus* L.). Applied and Environmental Microbiology. 84.

Carmichael, J. W., 1966. Cerebral mycetoma of trout due to a phialophora-like fungus. Sabouraudia: Journal of Medical and Veterinary Mycology. 5, 120–3.

Carnegie, R. B., Burreson, E. M., Hine, P. M., Stokes, N. A., Audemard, C., Bishop, M. J., et al. 2006. *Bonamia perspora* n. sp. (*Haplosporidia*), a parasite of the oyster *Ostreola equestris*, is the first bonamia species known to produce spores. Journal of Eukaryotic Microbiology. 53, 232–45.

Carrasco, N., Green, T., Itoh, N., 2015. *Marteilia* spp. parasites in bivalves: a revision of recent studies. Journal of Invertebrate Pathology. 131, 43–57.

Caullery, M., Mesnil, F., 1905. Sur le gene *Aplosporidium* (nov) et l'ordre nouveau des Aplosporidies. Comptes Rendus des Seances de la Societe de Biologie Paris. 51, 789–91.

Cavalier-Smith, T., 1998. A revised six-kingdom system of life. Biological Reviews of the Cambridge Philosophical Society. 73, 203–66.

Chantanachookin, C., Boonyaratpalin, S., Kasornchandra, J., Direkbusarakom, S., Ekpanithanpong, U., Supamattaya, K., et al. 1993. Histology and ultrastructure reveal a new granulosis-like virus in *Penaeus monodon* affected by Yellowhead disease. Diseases of Aquatic Organisms. 17, 145–57.

Chatton, E., 1911. Sur une Cnidosporidie sans cnidoblaste (*Paramyxa paradoxa* n. g., n. sp.). Comptes Rendus de l'Academie des Sciences, Paris. 152, 631–3.

Chatton, E., Poisson, R., 1931. Sur l'existence, dans le sang des crabs, de peridiniens parasites: *Hematodinium perezi* n.g., n.sp. (Syndinidae). Compes Rendus des Seances de la Societe de Biologie et de ses Filiales. 105, 553–7.

Chayaburakul, K., Nash, G., Pratanpipat, P., Sriurairatana, S., Withyachumnarnkul, B., 2004. Multiple pathogens found in growth-retarded black tiger shrimp *Penaeus monodon* cultivated in Thailand. Diseases of Aquatic Organisms. 60, 89–96.

Chi, S. C., Lo, B. J., Lin, S. C., 2001. Characterization of grouper nervous necrosis virus (GNNV). Journal of Fish Diseases. 24, 3–13.

Clark, G., Powell, M., Nowak, B., 2003. Effects of commercial freshwater bathing on reinfection of Atlantic salmon, *Salmo salar*, with amoebic gill disease. Aquaculture. 219, 135–42.

Coats, D. W., 1999. Parasitic life styles of marine dinoflagellates. Journal of Eukaryotic Microbiology. 46, 402–9.

Cochennec-Laureau, N., Reece, K. S., Berthe, F. C., Hine, P. M., 2003. *Mikrocytos roughleyi* taxonomic affiliation leads to the genus *Bonamia* (Haplosporidia). Diseases of Aquatic Organisms. 54, 209–17.

Comps, M., 1976. *Marteilia lengehi* n. sp., parasite de l'hutre *Crassostrea cucullata* Born. Revue des Travaux de l'Institut des Peches Maritimes. 40, 347–9.

Comps, M., 1983. Etude morphologique de *Marteilia christenseni* sp. n. parasite du lavignon *Scrobicularis piperata* P. (mollusque pelecypode). Revue des Travaux de l'Institut des Peches Maritimes. 47, 99–104.

Comps, M., Pichot, Y., Papagianni, P., 1982. Recherche sur *Marteilia maurini* n. sp. parasite de la moule *Mytilus galloprovincialis* Lmk. Revue des Travaux de l'Institut des Peches Maritimes. 45, 211–14.

Comps, M., Park, M. S., Desportes, I., 1986. Etude ultrastructurale des *Marteilioides chungmuensis* n.g., n. sp. parasite des ovocytes de l'huître *Crassostrea gigas* Th. Protistologica. XXII, 279–85.

Comtet, T., Garcia, C., Le Coguic, Y., Joly, J. P., 2003. Infection of the cockle *Cerastoderma edule* in the Baie des Veys (France) by the microsporidian parasite *Steinhausia* sp. Diseases of Aquatic Organisms. 57, 135–9.

Del Rey Moreno, A., Valero, A., Mayorga, C., Gomez, B., Torres, M. J., Hernandez, J., et al. 2006. Sensitization to *Anisakis simplex* s.l. in a healthy population. Acta Tropica. 97, 265–9.

Desportes, I., Perkins, F.O., 1990. Phylum paramyxea. In: M. C. J. O. Margulis, M. Melkonian, and D. J. Chapman (eds), Handbook of Protoctista. Jones and Bartlett Publishing, Boston, pp. 30–5.

Diamant, A., 1992. A new pathogenic histozoic *Myxidium* (myxosporea) in cultured gilthead seabream *Sparus aurata*. Bulletin of the European Association of Fish Pathologists. 12, 64–6.

Diamant, A., Goren, M., Baki Yokes, M., Galil, B.S., Klopman, Y., Huchon, D., Szitenberg, A., Karhan, U., 2010. Dasyatispora levantinae gen. et sp. Nov., a new microsporidian parasite from the common stingray Dasyatis pastinaca in the eastern Mediterranean. Diseases of Aquatic Organisms. 91, 137–50.

Dungan, C. F., Reece, K. S., 2006. In vitro propagation of two *Perkinsus* spp. parasites from Japanese Manila clams *Venerupis philippinarum* and description of *Perkinsus honshuensis* n. sp. Journal of Eukaryotic Microbiology. 53, 316–26.

Dykova, I., Figueras, A., Novoa, B., 1999. Epizoic amoebae from the gills of turbot *Scophthalmus maximus*. Diseases of Aquatic Organisms. 38, 33–8.

Dykova, I., Kostka, M., Peckova, H., 2010. *Grellamoeba robusta* gen. n., sp. n., a possible member of the family *Acramoebidae Smirnov*, Nassonova et Cavalier-Smith, 2008. European Journal of Protistology. 46, 77–85.

Egan, S., Gardiner, M., 2016. Microbial dysbiosis: rethinking disease in marine ecosystems. Frontiers in Microbiology. 7, 991–1.

Fahsbender, E., Hewson, I., Rosario, K., Tuttle, A. D., Varsani, A., Breitbart, M., 2015. Discovery of a novel circular DNA virus in the Forbes sea star, *Asterias forbesi*. Archives of Virology. 160, 2349–51.

Farley, C. A., Wolf, P. H., Elston, R. A., 1988. A long term study of "microcell" disease in oysters with a description of a new genus, *Mikrocytos* (g. n.), and two new species, *Mikrocytos mackini* (sp. n.) and *Mikrocytos roughleyi* (sp. n.). Fishery Bulletin United States. 86, 581–93.

Feehan, C. J., Johnson-Mackinnon, J., Scheibling, R. E., Lauzon-Guay, J. S., Simpson, A. G., 2013. Validating the identity of *Paramoeba invadens*, the causative agent of recurrent mass mortality of sea urchins in Nova Scotia, Canada. Diseases of Aquatic Organisms. 103, 209–27.

Feist, S. W., Bucke, D., 1992. *Myxidium gadi* Georgevitch, 1916 infections in *Pollachius virens* L. from the North Sea. Bulletin of the European Association of Fish Pathologists. 12, 211–14.

Feist, S. W., Hine, P. M., Bateman, K. S., Stentiford, G. D., Longshaw, M., 2009. *Paramarteilia canceri* sp. n. (Cercozoa) in the European edible crab (*Cancer pagurus*) with a proposal for the revision of the order *Paramyxida* Chatton, 1911. Folia Parasitologica. 56, 73–85.

Field, R. H., Chapman, C. J., Taylor, A. C., Neil, D. M., Vickerman, K., 1992. Infection of the Norway lobster *Nephrops norvegicus* by a *Hematodinium*-like species of dinoflagellate on the west coast of Scotland. Diseases of Aquatic Organisms. 13, 1–15.

Field, R. H., Hills, J. M., Atkinson, R. J. A., Magill, S., Shanks, A. M., 1998. Distribution and seasonal prevalence of *Hematodinium* sp. infection of the Norway lobster (*Nephrops norvegicus*) around the west coast of Scotland. ICES Journal of Marine Science. 55, 846–58.

Ford, S. E., Stokes, N., Burreson, E. M., McGurk, E., Carnegie, R., Kraeuter, J., et al. 2009. *Minchinia mercenariae* n. sp (Haplosporidia) in the hard clam *Mercenaria mercenaria*: implications of a rare parasite in a commercially important host. Journal of Eukaryotic Microbiology. 56, 542–51.

Freeman, M. A., Fuss, J., Kristmundsson, A., Bjorbaekmo, M. F. M., Mangot, J. F., Del Campo, J., et al. 2017. X-cells are globally distributed, genetically divergent fish parasites related to perkinsids and dinoflagellates. Current Biology. 27, 1645–1651.e3.

Ginsburger-Vogel, T., Desportes, I., 1979. Etude ultrastructurale de la sporulation de *Paramarteilia orchestiae* gen. n., sp. n., parasite de l'amphipode *Orchestia gammarellus* (Pallas). Journal of Protozoology. 26, 390–403.

Grizel, H., Comps, M., Bonami, J. R., Cousserans, F., Duthoit, J. L., Le Pennec, M. A., 1974. Recherche sur l'agent de la maladie de la glande digestive de *Ostrea edulis* Linne. Science des Peche Bulletin des Institute Peches Maritime. 240, 7–29.

Hartikainen, H., Ashford, O. S., Berney, C., Okamura, B., Feist, S. W., Baker-Austin, C., et al. 2014a. Lineage-specific molecular probing reveals novel diversity and ecological partitioning of haplosporidians. ISME Journal. 8, 177–86.

Hartikainen, H., Stentiford, G. D., Bateman, K. S., Berney, C., Feist, S. W., Longshaw, M., et al. 2014b. Mikrocytids are a broadly distributed and divergent radiation of parasites in aquatic invertebrates. Current Biology. 24, 807–12.

Haskin, H. H., Stauber, L. A., Mackin, J. A., 1966. *Minchinia nelsoni* n. sp. (Haplosporida, Haplosporidiidae): causative agent of the Delaware Bay oyster epizootic. Science. 153, 1414–16.

Hewson, I., Button, J. B., Gudenkauf, B. M., Miner, B., Newton, A. L., Gaydos, J. K., et al. 2014. Densovirus associated with sea-star wasting disease and mass mortality. Proceedings of the National Academy of Sciences. 111, 17278–83.

Hillman, R. E., Ford, S. E., Haskin, H. H., 1990. *Minchinia teredinis* n. sp. (Balanosporida, Haplosporidiidae), a parasite of teredinid shipworms. Journal of Protozoology. 37, 364–8.

Hine, P. M., Wesney, B., Hay, B. E., 1992. Herpesviruses associated with mortalities among hatchery-reared larval Pacific oysters *Crassostrea gigas*. Diseases of Aquatic Organisms. 12, 135–42.

Hjeltnes, B., 2014. The Health Situation in Norwegian Aquaculture 2013. Norwegian Veterinary Institute, Oslo.

Holt, C., Foster, R., Daniels, C. L., van der Giezen, M., Feist, S. W., Stentiford, G. D., et al. 2018. *Halioticida noduliformans* infection in eggs of lobster (*Homarus gammarus*) reveals its generalist parasitic strategy in marine invertebrates. Journal of Invertebrate Pathology. 154, 109–16.

Hooper, C., Hardy-Smith, P., Handlinger, J., 2007. Ganglioneuritis causing high mortalities in farmed Australian abalone (*Haliotis laevigata* and *Haliotis rubra*). Australian Veterinary Journal. 85, 188–93.

Howell, M., 1967. The trematode *Bucephalus longicornutus* (Manter, 1954) in the New Zealand mud-oyster *Ostrea lutdaria*. Transactions of the Royal Society of New Zealand (Zoology). 8, 221–37.

Hudson, D. A., Shields, J., 1994. *Hematodinium australis* n. sp., a parasitic dinoflagellate of the sand crab *Portunus pelagicus* from Moreton Bay, Australia. Diseases of Aquatic Organisms. 19, 109–19.

Hudson, D. A., Hudson, N. B., Pyecroft, S. B., 2001. Mortalities of *Penaeus japonicus* prawns associated with microsporidean infection. Australian Veterinary Journal. 79, 504–5.

Inouye, K., Yamano, K., Maeno, Y., Nakajima, K., Matsuoka, M., Wada, Y., et al. 1992. Iridovirus infection of cultured red sea bream, *Pagrus major*. Fish Pathology. 27, 19–27.

Itoh, N., Yamamoto, T., Kang, H.-S., Choi, K.-S., Green, T., Carrasco, N., et al. 2014. A novel paramyxean parasite, *Marteilia granula* sp nov (Cercozoa), from the digestive gland of Manila clam *Ruditapes philippinarum* in Japan. Fish Pathology. 49, 181–93.

Jakuschkin, B., Fievet, V., Schwaller, L., Fort, T., Robin, C., Vacher, C., 2016. Deciphering the pathobiome: intra- and interkingdom interactions involving the pathogen *Erysiphe alphitoides*. Microbial Ecology. 72, 870–80.

James, T. Y., Pelin, A., Bonen, L., Ahrendt, S., Sain, D., Corradi, N., et al. 2013. Shared signatures of parasitism and phylogenomics unite Cryptomycota and microsporidia. Current Biology. 23, 1548–53.

Jensen, S., Duperron, S., Birkeland, N.-K., Hovland, M., 2010. Intracellular Oceanospirillales bacteria inhabit gills of Acesta bivalves. FEMS Microbiology Ecology. 74, 523–33.

Johnson, K. E., Freeman, M. A., Laxdal, B., Kristmundsson, A., 2018. Aetiology and histopathology of a systemic phaeomycosis in farmed lumpfish, *Cyclopterus lumpus*. Bulletin of the European Association of Fish Pathologists. 38, 194–7.

Kennedy, C. R., 2007. The pathogenic helminth parasites of eels. Journal of Fish Diseases. 30, 319–34.

Kent, M. L., Sawyer, T. K., Hedrick, R. P., 1988. *Paramoeba pemaquidensis* (Sarcomastigophora: Paramoebidae) infestation of the gills of coho salmon *Oncorhynchus kisutch* reared in sea water. Diseases of Aquatic Organisms. 5, 163–9.

Kerr, R., Ward, G. M., Stentiford, G. D., Alfjorden, A., Mortensen, S., Bignell, J. P., et al. 2018. *Marteilia refringens* and *Marteilia pararefringens* sp. nov. are distinct parasites of bivalves and have different European distributions. Parasitology. 145, 1483–92.

Kibenge, F. S., Garate, O. N., Johnson, G., Arriagada, R., Kibenge, M. J., Wadowska, D., 2001. Isolation and identification of infectious salmon anaemia virus (ISAV) from coho salmon in Chile. Diseases of Aquatic Organisms. 45, 9–18.

Knutie, S. A., Wilkinson, C. L., Kohl, K. D., Rohr, J. R., 2017. Early-life disruption of amphibian microbiota decreases later-life resistance to parasites. Nature Communications. 8, 86.

Kondo, H., Van, P. T., Dang, L. T., Hirono, I., 2015. Draft genome sequence of non-*Vibrio parahaemolyticus* acute hepatopancreatic necrosis disease strain KC13.17.5, isolated from diseased shrimp in Vietnam. Genome Announcements. 3(5), e00978-15.

Labbe, A., 1896. Recherches zoologiques, cytologiques et biologiques sure les coccides. Archives de Zoologie Experimentale et Generale. 4, 517–654.

Langvad, F., Pederson, O., Engjom, K., 1985. A fungal disease caused by *Exophiala* sp. nov. in farmed Atlantic salmon in western Norway. In: A. E. Ellis (ed.), Fish and Shellfish Pathology. Academic Publishers, Cambridge, MA, pp. 323–8.

Le Roux, F., Lorenzo, G., Peyret, P., Audemard, C., Figueras, A., Vivares, C., et al. 2001. Molecular evidence for the existence of two species of *Marteilia* in Europe. Journal of Eukaryotic Microbiology. 48, 449–54.

Lee, C. T., Chen, I. T., Yang, Y. T., Ko, T. P., Huang, Y. T., Huang, J. Y., et al. 2015. The opportunistic marine pathogen *Vibrio parahaemolyticus* becomes virulent by acquiring a plasmid that expresses a deadly toxin. Proceedings of the National Academy of Sciences of the United States. 112, 10798–803.

Li, C., Song, S., Liu, Y., Chen, T., 2013. *Hematodinium* infections in cultured Chinese swimming crab, *Portunus trituberculatus*, in northern China. Aquaculture. 396–9, 59–65.

Li, Y. Y., Xia, X. A., Wu, Q. Y., Liu, W. H., Lin, Y. S., 2008. Infection with *Hematodinium* sp. in mud crabs *Scylla serrata* cultured in low salinity water in southern China. Diseases of Aquatic Organisms. 82, 145–50.

Lightner, D. V., Redman, R. M., Bell, T. A., 1983. Infectious hypodermal and hematopoietic necrosis, a newly recognized virus disease of penaeid shrimp. Journal of Invertebrate Pathology. 42, 62–70.

Lightner, D. V., Pantoja, C. R., Poulos, B. T., Tang, K. F. J., Redman, R. M., Pasos de Andrade, T., et al. 2004. Infectious myonecrosis: new disease in Pacific white shrimp. Global Aquaculture Advocate. 7, 85.

Liu, Y., Steenkamp, E. T., Brinkmann, H., Forget, L., Philippe, H., Lang, B. F., 2009. Phylogenomic analyses predict sistergroup relationship of nucleariids and fungi and paraphyly of zygomycetes with significant support. BMC Evolutionary Biology. 9, 272.

Lohrmann, K. B., Feist, S. W., Brand, A. R., 2000. Microsporidiosis in queen scallops (*Aequipecten opercularis* L.) from UK waters. Journal of Shellfish Research. 19, 71–5.

Lom, J., 1984. Diseases caused by protistans. In: O. Kinne (ed.), Diseases of Marine Animals. Biologische Anstalt Helgoland, Hamburg.

Longshaw, M., Frear, P. A., Nunn, A. D., Cowx, I. G., Feist, S. W., 2010. The influence of parasitism on fish population success. Fisheries Management and Ecology. 17, 426–34.

Macey, B. M., Christison, K. W., Mouton, A., 2011. *Halioticida noduliformans* isolated from cultured abalone (*Haliotis midae*) in South Africa. Aquaculture. 315, 187–95.

Mackin, J. G., Owen, H. M., Collier, A., 1950. Preliminary note on the occurrence of a new protistan parasite, *Dermocystidium marinum* n. sp. in *Crassostrea virginica* (Gmelin). Science. 111, 328–9.

Marchand, J., Sprague, V., 1979. Ultrastructure de *Minchinia cadomensis* sp. n. (Haplosporida) parasite du décapode *Rhithropanopeus harrisii* tridentatus Maitland dans le Canal de Caen à la Mer (Calvados, France). Journal of Protozoology. 26, 179–85.

Mattiucci, S., Nascetti, G., 2006. Molecular systematics, phylogeny and ecology of anisakid nematodes of the genus *Anisakis* Dujardin, 1845: an update. Parasite. 13, 99–113.

McCleary, S., Giltrap, M., Henshilwood, K., Ruane, N. M., 2014. Detection of salmonid alphavirus RNA in Celtic and Irish Sea flatfish. Diseases of Aquatic Organisms. 109, 1–7.

McLaughlin, S. M., Tall, B. D., Shaheen, A., Elsayed, E. E., Faisal, M., 2000. Zoosporulation of a new *Perkinsus* species isolated from the gills of the softshell clam *Mya arenaria*. Parasite. 7, 115–22.

McLoughlin, M. F., Graham, D. A., 2007. Alphavirus infections in salmonids—a review. Journal of Fish Diseases. 30, 511–31.

Messick, G. A., 1994. *Hematodinium perezi* infections in adult and juvenile blue crabs *Callinectes sapidus* from coastal bays of Maryland and Virginia, USA. Diseases of Aquatic Organisms. 19, 77–82.

Messick, G. A., Shields, J. D., 2000. Epizootiology of the parasitic dinoflagellate *Hematodinium* sp. in the American blue crab *Callinectes sapidus*. Diseases of Aquatic Organisms. 43, 139–52.

Meyers, T. R., Koeneman, T. M., Bothelho, C., Short, S., 1987. Bitter crab disease; a fatal dinoflagellate infection and marketing problem for Alaskan Tanner crabs *Chionoecetes bairdi*. Diseases of Aquatic Organisms. 3, 37–43.

Moran, J. D. W., Whitaker, D. J., Kent, M. L., 1999. A review of the myxosporean genus *Kudoa* Meglitsch, 1947, and its impact on the international aquaculture industry and commercial fisheries. Aquaculture. 172, 163–96.

Morzunov, S. P., Winton, J. R., Nichol, S. T., 1995. The complete genome structure and phylogenetic relationship of infectious hematopoietic necrosis virus. Virus Research. 38, 175–92.

Moss, J. A., Xiao, J., Dungan, C. F., Reece, K. S., 2008. Description of *Perkinsus beihaiensis* n. sp., a new *Perkinsus* sp. parasite in oysters of southern China. Journal of Eukaryotic Microbiology. 55, 117–30.

Mosser, T., Talagrand-Reboul, E., Colston, S. M., Graf, J., Figueras, M. J., Jumas-Bilak, E., et al. 2015. Exposure to pairs of *Aeromonas* strains enhances virulence in the *Caenorhabditis elegans* infection model. Frontiers in Microbiology. 6, 1218.

Mouton, A., Crosbie, P., Cadoret, K., Nowak, B., 2014. First record of amoebic gill disease caused by *Neoparamoeba perurans* in South Africa. Journal of Fish Diseases. 37, 407–9.

Munday, B. L., Lange, K., Foster, C. A., Lester, R. J., Handlinger, J., 1993. Amoebic gill disease in sea-caged salmonids in Tasmanian waters. Tasmanian Fisheries Research. 28, 14–19.

Munro, A. L. S., Ellis, A. E., McVicar, A. H., McLay, A. H., Needham, E. A., 1984. An exocrine pancreas disease of farmed Atlantic salmon in Scotland. Helgolander Meeresuntersuchungen. 37, 571–86.

Muraosa, Y., Morimoto, K., Sano, A., Nishimura, K., Hatai, K., 2009. A new peronosporomycete, *Halioticida noduliformans* gen. et sp. nov., isolated from white nodules in the abalone *Haliotis* spp. from Japan. Mycoscience. 50, 106–15.

Neave, M. J., Apprill, A., Ferrier-Pagès, C., Voolstra, C. R., 2016. Diversity and function of prevalent symbiotic marine bacteria in the genus *Endozoicomonas*. Applied Microbiology and Biotechnology. 100, 8315–24.

Noga, E. J., 1993. Fungal diseases of marine and estuarine fishes. In: J. A. Couch (ed.), Pathobiology of Marine and Estuarine Organisms. CRC Press, Boca Raton, FL, pp. 85–100.

Noguera, P., Collins, C., Bruno, D., Pert, C., Turnbull, A., McIntosh, A., et al. 2009. Red vent syndrome in wild Atlantic salmon *Salmo salar* in Scotland is associated with *Anisakis simplex* sensu stricto (Nematoda: Anisakidae). Diseases of Aquatic Organisms. 87, 199–215.

Nunan, L. M., Pantoja, C. R., Gomez-Jimenez, S., Lightner, D. V., 2013. "*Candidatus* Hepatobacter penaei," an intracellular pathogenic enteric bacterium in the hepatopancreas of the marine shrimp *Penaeus vannamei* (Crustacea: Decapoda). Applied and Environmental Microbiology. 79, 1407–9.

Nylund, A., Ottem, K. F., Watanabe, K., Karlsbakk, E., Krossøy, B., 2006. *Francisella* sp. (family *Francisellaceae*) causing mortality in Norwegian cod (*Gadus morhua*) farming. Archives Microbiology. 185, 383–92.

Nylund, S., Andersen, L., Saevareid, I., Plarre, H., Watanabe, K., Arnesen, C., et al. 2011. Diseases of farmed Atlantic salmon (*Salmo salar*) associated with the microsporidian *Paranucleospora theridion*. Diseases of Aquatic Organisms. 94, 41–57.

Oakey, H. J., Smith, C. S., 2018. Complete genome sequence of a white spot syndrome virus associated with a disease incursion in Australia. Aquaculture. 484, 152–9.

Okamura, B., Gruhl, A., Bartholomew, J. L., 2015. Myxozoan Evolution, Ecology and Development. Springer, Berlin.

Olsen, A. B., Mikalsen, J., Rode, M., Alfjorden, A., Hoel, E., Straum-Lie, K., et al. 2006. A novel systemic granulomatous inflammatory disease in farmed Atlantic cod, *Gadus morhua* L., associated with a bacterium belonging to the genus *Francisella*. Journal of Fish Diseases. 29, 307–11.

Ormieres, R., de Puytorac, P., 1968. Ultrastructure des spores de l'haplosporidiae *Haplosporidium ascidiarum* endoparasite du tunicier *Sydniuim elegans*. Comptes Rendus de l'Academie des Sciences, Paris. 266, 1134–6.

Ormieres, R., Sprague, V., Bartoli, P., 1973. Light and electron microscope study of new species of *Urosporidium* (Haplosporidia), hyperparasite of trematode sporocysts in the clam *Abra ovata*. Journal of Invertebrate Pathology. 21, 71–86.

Oshima, K. H., Higman, K. H., Arakawa, C. K., de Klinkelin, P., Jorgensen, P. E. V., Meyers, T. R., et al. 1993. Genetic comparison of viral hemorrhagic septicemia virus isolates from North America and Europe. Diseases of Aquatic Organisms. 19.

Palmer, R., Carson, J., Riuttledge, M., Drinan, E., Wagner, T., 1997. Gill disease associated with *Paramoeba*, in sea reared Atlantic salmon in Ireland. Bulletin of the European Association of Fish Pathologists. 17, 112–14.

Palstra, A. P., Heppener, D. F. M., van Ginneken, V. J. T., Székely, C., van den Thillart, G. E. E. J. M., 2007. Swimming performance of silver eels is severely impaired by the swim-bladder parasite *Anguillicola crassus*. Journal of Experimental Marine Biology and Ecology. 352, 244–56.

Perkins, F. O., 1971. Sporulation in the trematode hyperparasite *Urosporidium crescens* De Turk, 1940 (Haplosporida: Haplosporidiidae): an electron microscope study. Journal of Parasitology. 57, 9–23.

Perkins, F. O., Wolf, P. H., 1976. Fine structure of *Marteilia sydneyi* sp. n.—Haplosporidian pathogen of Australian oysters. Journal of Parasitology. 62, 528–38.

Perkins, F. O., Zwerner, D. E., Dias, R. K., 1975. The hyperparasite *Urosporidium spisuli* sp. n. (Haplosporea), and its effects on the surf clam industry. Journal of Parasitology. 61, 944–9.

Pichot, Y., Comps, M., Tige, G., Grizel, H., Rabouin, M.A., 1980. Recherches sur *Bonamia ostreae* gen. n., sp., n., parasite nouveau de l'huitre plate *Ostreae edulis*. Revue des Travaux de l'Institut des Peches Maritimes. 43, 131.

Pitlik, S. D., Koren, O., 2017. How holobionts get sick—toward a unifying scheme of disease. Microbiome. 5, 64.

Rebollar, E. A., Hughey, M. C., Medina, D., Harris, R. N., Ibáñez, R., Belden, L. K., 2016. Skin bacterial diversity of Panamanian frogs is associated with host susceptibility and presence of *Batrachochytrium dendrobatidis*. ISME Journal. 10, 1682.

Richards, R. H., Holliman, A., Helgason, S., 1978. *Exophiala salmonis* infection in Atlantic salmon *Salmo salar* L. Journal of Fish Diseases. 1, 357–68.

Richards, T. A., Jones, M. D. M., Leonard, G., Bass, D., 2012. Marine fungi: their ecology and molecular diversity. Annual Review of Marine Science. 4, 495–522.

Richards, T. A., Leonard, G., Wideman, J. G., 2017. What defines the "kingdom" fungi? Microbiology Spectrum. 5(3).

Rimstad, E., Ole, D., Dannevig, H. B., Falk, K., 2011. Infectious salmon anaemia. In: P. T. K. Woo, D. W. Bruno (eds), Fish Diseases and Disorders. Volume 3: Viral, Bacterial and Fungal Infections, 2nd edition. CAB International, Wallingford, UK, pp. 143–65.

Ruane, N. M., Swords, D., Morrissey, T., Geary, M., Hickey, C., Collins, E. M., et al. 2018. Isolation of salmonid alphavirus subtype 6 from wild-caught ballan wrasse, *Labrus bergylta* (Ascanius). Journal of Fish Diseases. 41, 1643–51.

Ruggiero, M. A., Gordon, D. P., Orrell, T. M., Bailly, N., Bourgoin, T., Brusca, R. C., et al. 2015. A higher level classification of all living organisms. PLOS ONE. 10, e0119248.

Saldarriaga, J. F., McEwan, M. L., Fast, N. M., Taylor, F. J. R., Keeling, P. J., 2003. Multiple protein phylogenies show that *Oxyrrhis marina* and *Perkinsus marinus* are early branches of the dinoflagellate lineage. International Journal of Systematic and Evolutionary Microbiology. 53, 355–65.

Sandeman, G., 1893. On the multiple tumours in plaice and flounders. 11th Annual Report of Scottish Fisheries Board for 1892, 391–2.

Sandra, M. C., Amalia, G., Kimberly, S. R., Kathleen, A., Carlos, A., Antonio, V., 2004. *Perkinsus mediterraneus* n. sp., a protistan parasite of the European flat oyster *Ostrea edulis* from the Balearic Islands, Mediterranean Sea. Diseases of Aquatic Organisms. 58, 231–44.

Segarra, A., Pepin, J. F., Arzul, I., Morga, B., Faury, N., Renault, T., 2010. Detection and description of a particular ostreid herpesvirus 1 genotype associated with massive mortality outbreaks of Pacific oysters, *Crassostrea gigas*, in France in 2008. Virus Research. 153, 92–9.

Sekimoto, S., Hatai, K., Honda, D., 2007. Molecular phylogeny of an unidentified *Haliphthoros*-like marine oomycete and *Haliphthoros milfordensis* inferred from nuclear-encoded small- and large-subunit rRNA genes and mitochondrial-encoded cox2 gene. Mycoscience. 48, 212–21.

Sheridan, C., Kramarsky-Winter, E., Sweet, M., Kushmaro, A., Leal, M. C., 2013. Diseases in coral aquaculture: causes, implications and preventions. Aquaculture. 396–399, 124–35.

Shields, J. D., 1994. The parasitic dinoflagellates of marine crustaceans. Annual Review of Fish Diseases. 4, 241–71.

Simmonds, P., Adams, M. J., Benkő, M., Breitbart, M., Brister, J. R., Carstens, E. B., et al. 2017. Virus taxonomy in the age of metagenomics. Nature Reviews Microbiology. 15, 161.

Small, H. J., 2012. Advances in our understanding of the global diversity and distribution of *Hematodinium* spp.—significant pathogens of commercially exploited crustaceans. Journal of Invertebrate Pathology. 110, 234–46.

Smith, J. W., Wootten, R., 1978. Anisakis and anisakiasis. Advances in Parasitology. 16, 93–163.

Snieszko, S. F., Taylor, C. C., 1947. A bacterial disease of the lobster (*Homarus americanus*). Science. 105(2732).

Snow, M., Black, J., Matejusova, I., McIntosh, R., Baretto, E., Wallace, I. S., et al. 2010. Detection of salmonid alphavirus RNA in wild marine fish: implications for the origins of salmon pancreas disease in aquaculture. Diseases of Aquatic Organisms. 91, 177–88.

Sprague, V., 1963. *Minchinia louisiana* n. sp. (Haplosporidia, Haplosporidiidae), a parasite of *Panopeus herbstii*. Journal of Protozoology. 10, 267–74.

Steinum, T., Kvellestad, A., Ronneberg, L. B., Nilsen, H., Asheim, A., Fjell, K., et al. 2008. First cases of amoebic gill disease (AGD) in Norwegian seawater farmed Atlantic salmon, *Salmo salar* L., and phylogeny of the causative amoeba using 18S cDNA sequences. Journal of Fish Diseases. 31, 205–14.

Stentiford, G. D., Shields, J. D., 2005. A review of the parasitic dinoflagellates *Hematodinium* species and *Hematodinium*-like infections in marine crustaceans. Diseases of Aquatic Organisms. 66, 47–70.

Stentiford, G. D., Neil, D. M., Atkinson, R. J. A., 2001. The relationship of *Hematodinium* infection prevalence in a Scottish *Nephrops norvegicus* population to season, moulting and sex. ICES Journal of Marine Science. 58, 814–23.

Stentiford, G. D., Green, M., Bateman, K., Small, H. J., Neil, D. M., Feist, S. W., 2002. Infection by a *Hematodinium*-like parasitic dinoflagellate causes pink crab disease (PCD) in the edible crab *Cancer pagurus*. Journal of Invertebrate Pathology. 79, 179–91.

Stentiford, G. D., Bateman, K. S., Longshaw, M., Feist, S. W., 2007. *Enterospora canceri* n. gen., n. sp., intranuclear within the hepatopancreatocytes of the European edible crab *Cancer pagurus*. Diseases of Aquatic Organisms. 75, 61–72.

Stentiford, G. D., Bateman, K. S., Small, H. J., Pond, M., Ungfors, A., 2012. *Hematodinium* sp. and its bacteria-like endosymbiont in European brown shrimp (*Crangon crangon*). Aquatic Biosystems. 8(1), 24.

Stentiford, G. D., Bateman, K. S., Feist, S. W., Chambers, E., Stone, D. M., 2013a. Plastic parasites: extreme dimorphism creates a taxonomic conundrum in the phylum Microsporidia. International Journal for Parasitology. 43, 339–52.

Stentiford, G. D., Feist, S. W., Stone, D. M., Bateman, K. S., Dunn, A. M., 2013b. Microsporidia: diverse, dynamic, and emergent pathogens in aquatic systems. Trends in Parasitology. 29, 567–78.

Stentiford, G. D., Becnel, J., Weiss, L. M., Keeling, P. J., Didier, E. S., Williams, B. P., et al. 2016. Microsporidia—emergent pathogens in the global food chain. Trends in Parasitology. 32, 336–48.

Stentiford, G. D., Ross, S., Minardi, D., Feist, S. W., Bateman, K. S., Gainey, P. A., et al. 2017a. Evidence for trophic transfer of *Inodosporus octospora* and *Ovipleistophora arlo* n. sp. (Microsporidia) between crustacean and fish hosts. Parasitology. 145(8), 1–13.

Stentiford, G. D., Sritunyalucksana, K., Flegel, T. W., Williams, B. A. P., Withyachumnarnkul, B., Itsathitphaisarn, O., et al. 2017b. New paradigms to help solve the global aquaculture disease crisis. PLOS Pathogens. 13, e1006160.

Stewart, J. E., 1980. Diseases. In: J. S. Cobb, B. F. Phillips (eds), The Biology and Management of Lobsters. Academic Press, New York, pp. 301–42.

Takahashi, Y., Itami, T., Kondo, M., Maeda, M., Fujii, R., Tomonaga, S., et al. 1994. Electron microscopic evidence of bacilliform virus infection in kuruma shrimp (*Penaeus japonicus*). Fish Pathology. 29, 121–5.

Taylor, D. M., Khan, R. A., 1995. Observations on the occurrence of *Hematodinium* sp. (Dinoflagellata: Syndinidae), the causative agent of bitter crab disease in Newfoundland snow crab (*Chionoecetes opilio*). Journal of Invertebrate Pathology. 65, 283–8.

Thorud, K. E., Djupvik, H. O., 1988. Infectious salmon anaemia in Atlantic salmon (*Salmo salar* L.). Bulletin of the European Association of Fish Pathologists. 8, 109–11.

Tibbo, S. N., Graham, T. R., 1963. Biological changes in herring stocks following an epizootic. Journal of the Fisheries Research Board of Canada. 20, 435–49.

Toranzo, A. E., Magariños, B., Romalde, J. L., 2005. A review of the main bacterial fish diseases in mariculture systems. Aquaculture. 246, 37–61.

Tourtip, S., Wongtripop, S., Stentiford, G. D., Bateman, K. S., Sriurairatana, S., Chavadej, J., et al. 2009. *Enterocytozoon hepatopenaei* sp. nov. (Microsporida: Enterocytozoonidae), a parasite of the black tiger shrimp *Penaeus monodon* (Decapoda: Penaeidae): fine structure and phylogenetic relationships. Journal of Invertebrate Pathology. 102, 21–9.

Tran, L., Nunan, L., Redman, R. M., Mohney, L. L., Pantoja, C. R., Fitzsimmons, K., et al. 2013. Determination of the infectious nature of the agent of acute hepatopancreatic necrosis syndrome affecting penaeid shrimp. Diseases of Aquatic Organisms. 105, 45–55.

Twigg, M., Evans, R., Feist, S., Stebbing, P., Longshaw, M., Harris, E., 2008. *Anisakis simplex* sensu lato associated with red vent syndrome in wild adult Atlantic salmon *Salmo salar* in England and Wales. Diseases of Aquatic Organisms. 82, 61–5.

Vago, C., 1966. A virus disease in Crustacea. Nature. 209, 1290.

van Aerle, R., Santos, E. M., 2017. Advances in the application of high-throughput sequencing in invertebrate virology. Journal of Invertebrate Pathology. 147, 145–56.

van Banning, P., 1977. *Minchinia armoricana* sp. nov. (Haplosporida), a parasite of the European flat oyster, *Ostrea edulis*. Journal of Invertebrate Pathology. 30, 199–206.

Vayssier-Taussat, M., Albina, E., Citti, C., Cosson, J. F., Jacques, M. A., Lebrun, M. H., et al. 2014. Shifting the paradigm from pathogens to pathobiome: new concepts in the light of meta-omics. Frontiers in Cellular and Infectious Microbiology. 4, 29.

Wang, J. F., Li, M., Xiao, J., Xu, W. J., Li, C. W., 2017. *Hematodinium* spp. infections in wild and cultured populations of marine crustaceans along the coast of China. Diseases of Aquatic Organisms. 124, 181–91.

Ward, G. M., Bennett, M., Bateman, K., Stentiford, G. D., Kerr, R., Feist, S. W., et al. 2016. A new phylogeny and environmental DNA insight into paramyxids: an increasingly important but enigmatic clade of protistan parasites of marine invertebrates. International Journal for Parasitology. 46, 605–19.

Ward, G. M., Feist, S. W., Noguera, P., Marcos-Lopez, M., Ross, S., Green, M., et al. 2019. Detection and characterisation of *Minchinia mytilii* n. sp., a haplosporidian parasite of the blue mussel *Mytilus edulis*. Diseases of Aquatic Organisms. 133(1), 57–68.

Weissenberg, R., 1951. Positive result of a filtration experiment supporting the view that the agent of the *lympho-cysitis* disease of fish is a true virus. The Anatomical Record. 111, 166–7.

Wielgoss, S., Taraschewski, H., Meyer, A., Wirth, T., 2008. Population structure of the parasitic nematode *Anguillicola crassus*, an invader of declining North Atlantic eel stocks. Molecular Ecology. 17, 3478–95.

Woo, P., K., P. T., Bruno, D., 2011. Fish Diseases and Disorders, Volume 3: Viral, Bacterial and Fungal Infections, 2nd edition. Marine Scotland, Edinburgh.

Xiao, J., Liu, L., Ke, Y., Li, X., Liu, Y., Pan, Y., et al. 2017. Shrimp AHPND-causing plasmids encoding the PirAB toxins as mediated by pirAB-Tn903 are prevalent in various *Vibrio* species. Scientific Reports. 7, 42177.

Xu, W., Xie, J., Shi, H., Li, C., 2010. *Hematodinium* infections in cultured ridgetail white prawns, *Exopalaemon carinicauda*, in eastern China. Aquaculture. 300, 25–31.

Zerihun, M. A., Feist, S. W., Bucke, D., Olsen, A. B., Tandstad, N. M., Colquhoun, D. J., 2011. *Francisella noatunensis* subsp. *noatunensis* is the aetiological agent of visceral granulomatosis in wild Atlantic cod *Gadus morhua*. Diseases of Aquatic Organisms. 95, 65–71.

Zhang, H., Campbell, D. A., Sturm, N. R., Dungan, C. F., Lin, S., 2011. Spliced leader RNAs, mitochondrial gene frameshifts and multi-protein phylogeny expand support for the genus *Perkinsus* as a unique group of alveolates. PLOS ONE. 6, e19933.

Zillberg, D., Munday, B. L., 2006. Phylum Amoebozoa. In: P. T. K. Woo (ed.), Fish Diseases and Disorders. Vol. 1: Protozoan and Metazoan Infections, 2nd edition. CAB International, Wallingford, UK.

CHAPTER 2

Parasites in marine food webs

John P. McLaughlin, Dana N. Morton, and Kevin D. Lafferty

2.1 Introduction

The term "disease" conveys adverse impacts and "marine disease" implies that parasites cause disorder in the ocean. As a result, "marine disease ecology" is often goal-oriented, aimed at improving organismal health. Other approaches, such as marine parasitology, employ a more neutral eye, observing species, whether whales or whale lice, as equivalent participants in, and contributors to, marine systems. A neutral view reveals the role all parasites play in marine ecosystems, better equipping marine disease ecologists, to find and deal with those that cause infectious disease problems (Raymundo et al. Chapter 9, this volume). In this chapter, we take a neutral view and use food webs as a conceptual lens to focus on the question "What can we learn from parasites in marine ecosystems?"

In 1966, Robert Paine changed marine ecology forever when he applied a food-web perspective to the rocky intertidal environment (Lafferty and Suchanek 2016; Paine 1966). Ecologists have since assembled food webs for more than a hundred marine systems (Figure 2.1 shows the subset of marine webs that include at least fifty species). By tracing energy flow through ecosystems, food webs function like ecological maps, illustrating potential indirect effects, bottom-up processes, trophic cascades, and resource competition. Food webs help ecologists represent marine systems as networks that are a common currency for describing ecological

complexity and comparing different systems. Networks have two elements: nodes (sometimes called vertices) and links (sometimes called edges). In food webs, nodes are often species or life stages (e.g., larvae or adults), whereas the links connect who eats whom.

Ecologists can use food webs to explore ecosystem stability and responses to perturbations like fishing or conservation interventions (Yodzis 1996). An early and much-analyzed marine food web describes twenty-nine nodes connected by 203 links in the Benguela current fishery (Yodzis 1998). Researchers used this web to determine that proposed marine mammal culls would not increase fishery yields in the system, and would likely have the opposite effect (Yodzis 1998). Like other marine food webs, nodes in the Benguela food web are close together, just 1.6 links away from each other on average (Dunne et al. 2004). The short distance between nodes (characteristic path length) makes marine systems like Benguela, less modular than terrestrial systems and suggests that perturbations such as overfishing could spread rapidly through the entire system (Dunne et al. 2004; Shannon et al. 2000). On the other hand, the densely connected structure of marine webs makes them robust to secondary extinctions. In Benguela, removing 30 percent of nodes only resulted in 11 percent of the remaining nodes going extinct (Dunne et al. 2004).

Like most marine food webs, parasites were omitted from the Benguela food web (Figure 2.1).

McLaughlin, J.P., Morton, D.N., and Lafferty, K.D., *Parasites in marine food webs* In: *Marine Disease Ecology*. Edited by: Donald C. Behringer, Kevin D. Lafferty, and Brian. R. Silliman, Oxford University Press (2020). © Oxford University Press.
DOI: 10.1093/oso/9780198821632.003.0002

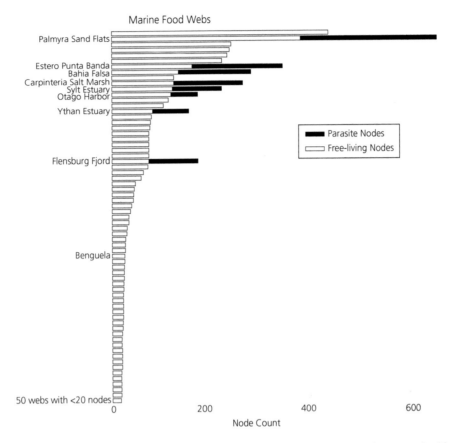

Figure 2.1 Marine food webs. Only 7 percent of marine food webs include parasites. But adding parasites can increase species richness a lot. Benguela is the only labeled food web that does not include parasites.

If parasites affect food-web structure to the same degree as similar free-living consumer groups, then integrating parasites into marine food webs will help us better understand food webs and marine-disease dynamics.

2.2 How parasites affect food webs

Whether or not they receive attention, parasites are the most abundant organisms in the oceans, and parasitism the most common lifestyle. Viruses, described as "a piece of bad news wrapped up in a protein" (Medawar and Medawar 1985), rule the sea at 10 billion per liter (Fuhrman 1999). Most viruses are bacteriophages and about 10^{23} viral infections occur in the ocean every second (Suttle 2007). But all marine species have viruses and

other specialist parasites (Dobson et al. 2008; Théodoridès 1989), suggesting that there could be more parasitic than free-living species (Windsor 1998).

Although we don't know their exact contribution to biodiversity, when researchers have included them, parasites increase insight into taxonomic and functional diversity in marine systems. In tropical and estuarine systems, over one-third of metazoan species are parasites (Hechinger et al. 2011b; McLaughlin 2018) (see Figure 2.2B in Box 2.1). This contribution to taxonomic diversity by parasites extends beyond species counts. Some marine taxa are all (e.g., Orthonectida) or mostly (e.g., Platyhelminthes) parasitic, and including parasites adds six new phyla to estuarine webs. In addition, parasites extend functional diversity by bringing unique consumer strategies (e.g., parasitoidism, parasitic castration) to food webs.

Parasites also balance how consumer–resource body size ratios change with trophic level (Lafferty and Kuris 2002). For instance, giant gray whales (*Eschrichtius robustus*) ingest tiny benthic amphipods from the muddy seafloor. At the same time, the whales host tiny parasitic amphipods (whale lice) that eat the whale's flaking skin. Parasitism is one of the few ways that tiny organisms can eat giant ones (micropredation, like gnathiid isopods feeding on parrotfish, is another), balancing what would otherwise be a lopsided consumer–resource body size ratio landscape. Thus, their uniqueness and ubiquity suggests parasites may affect marine food-web structure and dynamics (Morton et al. Chapter 3, this volume).

Comparing food webs that include parasites with versions that do not illustrates how parasites affect food-web structure (Figure 2.2). Almost all published marine food webs that include parasites are from temperate estuaries (Dunne et al. 2013), though Figure 2.2 in Box 2.1 provides a new example from a tropical marine system (McLaughlin 2018). In these food webs, parasites increase complexity (Dunne et al. 2013) and dominate structure by participating in 75 percent of trophic interactions (Lafferty et al. 2006), links that would not be accounted for if parasites were omitted. Including parasites increases food-web complexity, nestedness, and trophic-level resolution, challenging our current ideas about food-web structure (Dunne et al. 2013; Lafferty et al. 2006) (Figure 2.2). Although we now understand that by omitting parasites, ecologists have underestimated food-web complexity, we are just beginning to investigate the implications for energy flow through marine systems.

As consumers, parasites take energy from hosts for their own maintenance, growth, reproduction, and metabolism. Therefore, a parasite's direct effects should be proportional to its biomass. In estuarine

Box 2.1 Case study: Palmyra Atoll—parasites in a tropical marine food web

Located in the Northern Line Islands, 1,000 miles south of Hawaii, Palmyra is a low-lying coral atoll. An uninhabited National Wildlife Refuge, Palmyra has never supported commercial or subsistence fisheries, and its marine ecosystems are relatively pristine. The waters surrounding Palmyra support a high biomass of large apex-predators (Sandin et al. 2008), making this trophically intact system an ideal place to understand parasite contributions to the structure of tropical marine systems.

Over the course of 6 years (2009–2015), researchers surveyed the free-living and parasitic organisms inhabiting the intertidal sand flats at Palmyra Atoll, quantifying the abundance, body size, and biomass of everything from trematodes to trevally. The sand flats are a series of soft-bottom intertidal habitats, 314 ha in area, supporting over 275 metazoan species (McLaughlin 2018). In scale and species richness, the Palmyra sand flats are comparable to the three temperate estuarine systems where parasites have also been quantified. Despite their small size and high trophic position, parasites at Palmyra are as abundant as similar free-living consumers and their biomass can exceed free-living consumer groups like birds (McLaughlin 2018). This suggests that parasites play important roles in the ecosystem energetics at Palmyra just as they do in temperate estuaries.

Researchers also assembled trophic (Figure 2.2A–D), transmission (Figure 2.2E), and ontogenetic networks (Figure 2.2F) for the organisms inhabiting the Palmyra sand flats (McLaughlin 2018). The Palmyra sand flats support 195 free-living species, which participate in 11,003 trophic interactions when broken into their 389 distinct ontogenetic life stages (Figure 2.2A). Parasites make up 30 percent of the diversity on the sand flats. There are eighty parasite species comprising 290 distinct ontogenetic stages (Figure 2.2B). These parasites also account for 58 percent of all trophic interactions at Palmyra. As consumers, parasites participate in 2,381 interactions (Figure 2.2C). On the other hand, parasites are resources in 12,956 interactions (Figure 2.2D). Palmyra parasites use 347 transmission pathways to navigate this food web and infect their hosts without being eaten (Figure 2.2E). In order for parasites with complex life cycles to persist, each of their 204 developmental links must remain intact (Figure 2.2F). Thus, parasites at Palmyra dominate food-web structure, just as in the other marine systems (McLaughlin 2018).

continued

Box 2.1 *Continued*

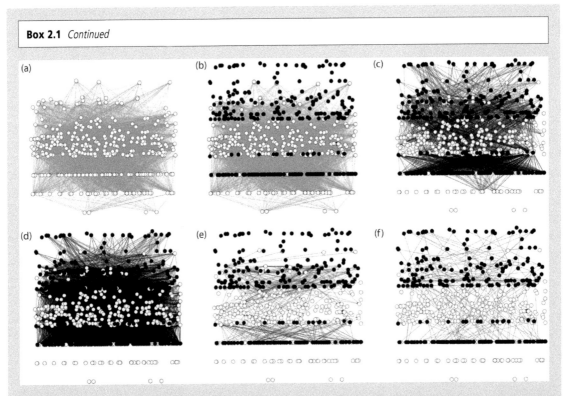

Figure 2.2 Food web for the sand flats at Palmyra Atoll (McLaughlin 2018). Palmyra is a remote tropical atoll and the only non-estuarine system for which we have comprehensive parasite surveys. In all visualizations, nodes represent life stages for free-living (white) and parasite (black) species. Feeding interactions for free-living (gray) and parasite (black) species are indicated by lines. Vertical height indicates consumer short-weighted trophic level (SWTL). The four most basal nodes indicate dissolved nutrients (SWTL = 0) available to primary producers (SWTL = 1). The next coherent trophic level (SWTL = 1.5) indicates free-living and parasitic consumer life stages that are non-feeding (e.g., eggs, miracidia). (A) Palmyra as a typical marine food web including free-living species (white nodes) and interactions (gray links) only. (B) Parasite diversity (black nodes) missing from a typical version of the Palmyra food web. (C) Parasites as consumers at Palmyra. Black lines indicate links where parasites (black nodes) feed on hosts (white nodes) or each other. (D) Parasites as resources at Palmyra. Black lines indicate links where free-living species (white nodes) feed on parasites (black nodes). This includes both concurrent predation and predation on free-living infective stages. (E) Parasite transmission at Palmyra. Black lines indicate transmission routes for parasites at Palmyra. This includes both trophic transmission and typical direct transmission. (F) Parasite complex life cycles at Palmyra. Black lines indicate the ontogenetic development that parasites must undertake to complete their life cycles. Severing any black line will lead to extinction of a parasite. Parasites with direct life cycles do not have black lines.

food webs, parasites have the same biomass density as similar-sized free-living species, after accounting for trophic level (Hechinger et al. 2011a). This suggests that as a group, energetic contributions from parasites are proportional to those from other consumer groups. Indeed, in estuaries, trematode biomass in snails exceeds bird biomass (Kuris et al. 2008). But parasites alter energy flow beyond the host tissue they eat. First, just as they try to avoid being eaten, hosts can try to avoid being infected, often at some cost (Weinstein et al. 2018). Second, infection risk alone forces hosts to invest in immune systems (Moret and Schmid-Hempel 2000). An important innate immune response was discovered in a marine organism, when in 1882, Ilya Mechnikov observed phagocytes attacking a splinter he had introduced into a sea star larva (Tauber 2003). Lastly, hosts must repair and replace tissue damaged by parasites (Allen and Wynn 2011), a cost that does not occur in predator–prey interactions. For these reasons, the impact that parasites have on food webs extends beyond their biomass density, just as some predators affect prey populations as much through the fear they induce as by the individuals they eat.

2.3 Parasites across trophic levels

From primary producers to top predators, parasites infect organisms across all trophic levels (Figure 2.2C). Basal species have parasites that can impair ecosystem production (Harvell & Lamb Chapter 8, this volume). Even top predators don't escape consumption. Orcas, for instance, have at least fifteen parasite species (Gibson et al. 2005). Here, we explore infectious processes across trophic levels and how they can alter both bottom-up and top-down processes in marine food webs.

Most marine ecosystems build on the photosynthesis done by phytoplankton, macroalgae, or algae–coral symbioses (Falkowski et al. 2004), and these primary producers are also subject to infection (Morton et al. Chapter 3, this volume). Viruses, which infect almost all phytoplankton (Fuhrman 1999), can end phytoplankton blooms (Bratbak et al. 1993) and density-dependent dynamics have been demonstrated in the laboratory (Brussaard 2004). Parasites also infect benthic macrophytes, sometimes with dramatic effect. Between 1931 and 1934, a wasting disease (caused by *Labyrinthula zosterae*) reduced eel grass populations by 90 percent in the temperate Atlantic (Muehlstein 1989), resulting in cascading changes to nearshore benthic communities. For example, the waters surrounding the Delmarva peninsula lost upper trophic-level taxa ranging from bay scallops (*Agropecten irradians*) to brant (*Branta bernicla*) (Orth et al. 2006). In warmer waters, Caribbean elkhorn coral (*Acropora palmata*) has been decimated by white pox disease caused by an enterobacterium (*Serratia marscens*) associated with the human gut (Patterson et al. 2002). Elkhorn coral declines have simplified reef structure, increased algal cover, and altered invertebrate communities (Aronson and Precht 2001; Gladfelter 1982). Although these examples show that disease outbreaks can dramatically reduce lower trophic levels, it is less clear if the impacts of marine parasites on primary producers are comparable to those of marine herbivores.

Although parasites can act as herbivores in marine systems, they don't just reduce plant biomass. Viruses can facilitate phytoplankton by releasing iron from lysed bacteria (Poorvin et al. 2004). Further, viral infection can increase nitrogen uptake and diversify the nitrogen sources an infected algal cell can use (Monier et al. 2017). By preventing fast-growing species from displacing slower-growing, but more resistant species, viruses can increase phytoplankton diversity (Suttle 2007). Horizontal gene transfer by viruses can contribute to evolution in bacteria and other marine pathogens (Little et al. Chapter 4, this volume). Thus, frequency-dependent infectious processes can reduce or increase the abundance of marine primary producers, whereas density-dependent infectious processes can and do regulate their populations.

Marine primary producers are also affected by parasites indirectly through trophic cascades (Morton et al. Chapter 3, this volume). For instance, sea urchins can increase to densities that defoliate temperate kelp beds, creating urchin barrens. However, urchins can also reach densities that support disease outbreaks, which in turn decimate urchin populations. Kelp then increases after diseases knock urchin populations back below densities that support disease (Behrens and Lafferty 2004; Scheibling 1986). Although sea urchin epizootics can protect kelp forests, they can harm coral reefs. Perhaps the best example of marine disease altering a food web was in 1983–84 when an unknown pathogen in the Caribbean reduced urchin (*Diadema antillarium*) densities by 94 percent (Lessios 1988). This disease-driven urchin die-off led to a phase shift from coral to macroalgae (Dudgeon et al. 2010) that has yet to recover. Thus, consumer mass mortalities caused by disease can have destabilizing impacts that extend beyond indirect effects on primary producers.

Marine diseases sometimes cause mass mortalities in consumers as well as producers. Echinoderms appear particularly susceptible to disease-induced population fluctuations, which can alter the important roles they play in marine systems (Uthicke et al. 2009). The latest example occurred between 2013 and 2015, when sea star wasting disease (an etiological agent hypothesized to be a virus) devastated sea star populations along the North American West Coast, from Baja California to Alaska (Montecino-Latorre et al. 2016) (Harvell & Lamb Chapter 8, this volume). The dominant sea star in rocky intertidal habitats, *Pisaster ochraceus*, experienced population declines between 59 and 84 percent (Menge et al. 2016), depressing predation rates on the foundational mussel species, *Mytilus californianus* (Menge et al. 2016). A similar mass mortality occurred in 1978, when an unknown pathogen devastated *Heliaster kubiniji* populations in the Gulf of California (Dungan et al. 1982). This keystone predator was the "most common, obvious and widely distributed shore starfish in the Gulf" (Steinbeck and Ricketts 2009), but as late as 2008 the sun star population had not recovered at most sites (Herrero-Pérezrul 2008). What led to these various echinoderm mass mortalities remains a mystery, though warming or pathogen introductions are speculated to be involved (Harvell et al. 1999).

Other invertebrate mass mortalities are better understood. In 1994, the largest remaining black abalone populations experienced mass mortalities in southern California (Lafferty and Kuris 1993) due to a novel rickettsial pathogen that increases in lethality with temperature and increases infectivity with temperature variation (Ben-Horin et al. 2013). Black abalone are grazers, and because their population has failed to recover, the intertidal community has undergone a phase shift from a bare-rock substrate populated by crustose coralline algae to one dominated by sessile invertebrates and sea urchins (Miner et al. 2006). Marine mass mortalities illustrate how parasite-driven trophic cascades (Buck and Ripple 2017) can shift marine ecosystems in directions that people may or may not find desirable.

Although less studied, marine mammal parasites likely also alter food webs through trophic cascades. For instance, *Toxoplasma gondii* is an apicomplexan parasite that can only reproduce in cats, but can infect and kill marine mammals. *Toxoplasma gondii* infection may increase sea otter (*Enhydra lutris*) mortality. Sea otters are exposed to this terrestrial parasite by freshwater runoff. About half of all sea otters test positive for *T. gondii*, and dead sea otters are twice as likely to test positive (Miller et al. 2002). Otters are keystone predators in kelp forests (Estes and Palmisano 1974); therefore, such otter diseases could destabilize kelp forest food webs. *Toxoplasma* seems to infect just about any marine mammal. Morbilliviruses also infect marine mammals worldwide (Van Bressem et al. 2001), and epizootics are triggered when naive populations are exposed to new viruses. Such outbreaks have caused mass die-offs in pinniped (Härkönen et al. 2006) and cetacean populations around the world (Guardo et al. 2005). These marine mammal mass mortalities could have indirect effects. Many marine mammals feed at high trophic levels and have high caloric requirements (Hammill and Stenson 2000), and reductions in their populations could have cascading consequences for the lower trophic levels that support them. As illustrated by the Benguela food-web models, fluctuations in marine mammal populations can have surprising effects on fisheries (Yodzis 1998), and understanding marine mammal diseases in a food-web context is an area for future research sparked by conservation efforts to recover depleted marine mammal stocks.

2.4 How parasites navigate food webs

Incorporating parasites into food webs requires considering them as resources, and leads to an unusual question: What eats parasites? (Figure 2.2D). The answer is many things; parasites are prey in over half the interactions in estuarine food webs. Some predators eat parasites on purpose, and cleaning symbioses are common in marine ecosystems (Grutter 1999). For example, topsmelt (*Atherinops affinis*) pick lice off gray whales in the whales' estuarine breeding grounds (Swartz 1981). Many parasites have free-living infective stages that might be food for planktivores (Johnson et al. 2010). Indeed, planktonic viruses are important resources for heterotrophic flagellates (González and Suttle 1993), and fungal zoospores that infect algae can be important resources for small grazers and filter feeders in coastal systems (Gleason et al. 2011). Trematodes in marine snails produce many free-swimming cercariae (Thieltges et al. 2008), with annual biomass production in temperate estuaries exceeding 20 kg per hectare (Kuris et al. 2008). These free-swimming cercariae are eaten by filter-feeding invertebrates and fishes (Hechinger et al. 2011b; Kaplan et al. 2009). These and other free-living infective stages could be an abundant energy source for low trophic-level consumers in marine systems.

More often, predators eat parasites by accident. In estuaries, tertiary consumers, like crabs (predating snails), fish, and birds (predating invertebrates and fish), eat parasites when they eat their hosts. When a host is eaten, 71 percent of parasite species suffer concurrent predation and are digested. Concurrent predation accounts for 31 percent of all links in estuary food webs. Therefore, including parasites as potential resources reveals one unexpected way that parasites affect food-web structure (Dunne et al. 2013). As an example, at Palmyra Atoll, blacktip reef sharks (*Carcharhinus melanopterus*) eat three mullet (Mugilidae) species. These mullet are, in turn, host to over a dozen parasites that cannot use blacktip sharks as hosts (Figure 2.2D) (McLaughlin 2018). Blacktip sharks consume and digest these parasites along with their mullet hosts. Although eating parasites does not provide nutrition to the shark, being digested by a shark is as much an impact on the mullet's parasites as it is on the mullet.

Being eaten is not always bad for a parasite. Some parasites use predation to facilitate transmission to new hosts (Figure 2.2E). These parasites can survive their host being eaten by infecting the predator in a process known as trophic transmission (Lafferty and Shaw 2013). Trophic transmission occurs in 48 percent of Palmyra Atoll trematodes, which require their intermediate fish host to be eaten by a Jack (Carangidae) to complete their life cycle (Figure 2.2D). Such parasites have complex life cycles adapted to moving through food webs, with larval stages infecting prey species and adult species infecting predator species.

Although traveling through food webs can be treacherous, some parasites bend food webs to their advantage. The estuarine trematode *Euhaplorchis californiensis* encysts on the killifish (*Fundulus parvipinnis*) brain (Shaw et al. 2010) and tilts the trophic transmission odds in its favor. To complete its life cycle, the trematode must navigate the estuarine food web to get from the fish's brain to a bird's gut. Encysted on the fish's brain, the parasite manipulates monoamine neurotransmitters (Shaw et al. 2009), causing its fish host to exhibit behaviors that increase bird predation (Lafferty and Morris 1996). This behavior manipulation (Kuris 2003; Lafferty and Shaw 2013) may be common in marine systems (Poulin 2010), and by changing predation rates, parasites could alter energy flow through marine systems.

Putting parasites in food webs shows how parasite diversity and host use adds to food-web complexity. Furthermore, substantial parasite biomass alters energy flows. More relevant to marine biologists might be the extent to which parasites affect free-living species. Parasites infect plants and compete with herbivores. Parasites infect herbivores, releasing plants from grazing. Parasites infect predators, leading to trophic cascades, and even manipulating prey susceptibility to predators. Parasites even contribute to food webs when they release edible infectious stages. These are all reasons why ecologists should consider parasites in food webs. But to what extent should marine disease ecologists consider food webs when trying to understand marine diseases? Next, we consider how food webs create challenges and opportunities for parasites, and how changes to food webs can increase or decrease infectious diseases.

2.5 How food webs affect parasites

Food webs present opportunities and challenges for parasites. First, free-living species represent an opportunity as potential hosts and dictate the parasite species that a food web can support. As a result, host diversity begets parasite diversity (Hechinger and Lafferty 2005). Second, free-living species consume one another and those trophic links present a challenge that parasites must navigate to ensure transmission while avoiding being eaten (Figure 2.2F). Counts of species and trophic links are often interpreted as measures of food-web complexity (May 1973). In this sense, parasite diversity should track food-web complexity (Lafferty 1997). Below, we examine how empirical changes in food-web complexity affect parasites. In particular, we discuss how food-web simplification through disturbance, the addition of novel links that accompany invasions, and the non-random trophic truncation associated with fishing, affect parasites. There are, of course, many other forces that can affect food-web structure. See Burge and Hershberger (Chapter 5, this volume) for a discussion about climate change and marine disease and Bojko et al. (Chapter 6, this volume) on the interactions between anthropogenic pollution and marine-disease dynamics. Comparing the role of parasites across food webs with and without these changes in complexity can help us to better understand how food webs affect parasites.

Frequent or strong disturbance tends to simplify food webs (Connell 1978) and parasites appear to be more sensitive to disturbance than their hosts. Sites protected from disturbance have more parasites (Lafferty 1997), and parasites are often the last thing to recover after disturbance. In a restored estuary, the trematode community infecting the California horn snail (*Cerithideopsis californica*) took 6 years to recover (Huspeni and Lafferty 2004). A similar pattern was observed in the Yucatan Peninsula. Snails returned to a coastal lagoon 6 months after a hurricane, but it took another 8 months for any trematodes to be found in the snails and it took 4 years for the trematode community to recover (Aguirre-Macedo et al. 2011). Similarly, fish parasites in the Gulf of Mexico took over 2 years to recover from Hurricane Katrina (Overstreet 2007). Parasites are often the first species lost and the last to return following perturbations to food webs.

Invasions have less-predictable effects on parasites than strong disturbances, like hurricanes, because species invasions introduce novel interactions (Pagenkopp Lohan et al. Chapter 7, this volume). For example, native oysters (*Crassostrea virginica*) in the Chesapeake Bay succumb to introduced pathogens

(*Haplosporidium nelsoni* and *Perkinsus marinus*) (Ruiz et al. 1999). These invasive protists limit oyster populations and this has cascading impacts, because oysters function as hard-substrate habitat and play important roles as resources and consumers (Ruiz et al. 1999). Sometimes invaders can bring their parasites with them. For example, Aral Sea sturgeon (*Acipenser nudiventris*) populations were reduced by a monogene (*Nitzschia sturionis*) that was introduced with its host, the Caspian Sea sturgeon (*Huso huso*) (Prenter et al. 2004). But, when invaders displace native hosts, native parasites can suffer. The invasive Japanese mud snail (*Batillari cumingi*) can exclude native snails (*C. californica*) from California estuaries (Torchin et al. 2005). *Cerithideopsis californica* is the only first-intermediate host for more than twenty native trematode species, so when the snail is excluded, all its parasites go locally extinct (Torchin et al. 2005). Many invaders leave their parasites behind, and in places like the San Francisco Bay, where most free-living species are exotic, parasites are rare (Foster 2012; Torchin et al. 2003). Thus, how parasites respond to species invasions is tied to how invaders affect free-living diversity (Pagenkopp Lohan et al. Chapter 7, this volume).

Not all parasites respond the same way to food-web changes. Generalist parasites are more robust to changes in food-web structure, whereas specialists, and parasites with complex life cycles that function as serial specialists, are sensitive to host extinctions (Lafferty and Kuris 2009; Rudolf and Lafferty 2011; Strona and Lafferty 2016; Wood and Lafferty 2015). Because natural disturbances can be common, parasites evolve to specialize on dependable hosts (Strona and Lafferty 2016). Although parasite diversity declines with free-living species loss and food-web simplification, the details are important. If some hosts benefit from food-web simplification, their parasites will have increased opportunities for transmission (Lafferty 2004). Furthermore, it remains difficult to predict how parasite prevalence and intensity respond to food-web complexity. Complex marine food webs with intact trophic structure (such as protected reefs) should promote diverse parasite faunas (Lafferty et al. 2008), whereas simplified or less-densely connected webs (such as disturbed or overfished environments) should be dominated by generalist parasites, or by parasites that infect weedy species (with lower parasite diversity overall) (Lafferty and Kuris 1999).

Changes in food-web structure can affect parasites, and fishing provides a natural experiment in which we can observe parasite responses to truncations in trophic structure (Behringer et al. Chapter 10, this volume). Fishing targets larger, older organisms and higher trophic-level species first, so the system becomes dominated by smaller, younger organisms and lower trophic-level species, with a trend toward food-web simplification (Pauly et al. 1998). So, we expect fishing to reduce some parasite populations. Within marine protected areas in Chile, there are more parasites per square meter than in fished areas. However, on a per-host basis, parasite densities are the same within and without protected areas (only one parasite species was more abundant per host) (Wood et al. 2013). A meta-analysis found parasite abundance in fished species was lower in fished areas than protected areas, suggesting that fishing reduced parasite populations by reducing host populations (habitat and resources for parasites) and/or by disrupting complex parasite life cycles (Wood and Lafferty 2015). Differences in top-predator abundance probably explain why parasite diversity was higher in reef fish at unfished Palmyra Atoll than at fished Kiritimati Atoll (Lafferty et al. 2008). Wood et al. (2014) investigated this idea further by contrasting parasite abundance at fished versus unfished sites spanning six Line Islands. As fishing pressure increased across islands, parasites with direct life cycles increased in abundance, while trophically transmitted parasites decreased in abundance (Wood et al. 2014). Thus, parasite responses to changes in food-web structure can vary with parasite life cycle, host specificity, and transmission strategy (Wood et al. 2014). Although fishing can decrease parasitism in direct response to host loss, other parasites might increase in abundance due to compensatory increases in host abundance.

Fishing may have cascading indirect effects on parasites, especially when fishing for predators increases the abundance of their prey. For example,

directly transmitted parasites may increase with prey abundance at fished sites, and this has been reported in urchin populations released from predation by the spiny lobster fishery in southern California (Behrens and Lafferty 2004). By the same mechanism, removing grazers through fishing might increase disease in plants. Sea turtle declines are a hypothesized factor contributing to seagrass wasting disease. Low water flow and accumulated detritus facilitate infection in dense seagrass beds, conditions that occur when seagrasses are not cropped short by grazing sea turtles (Jackson et al. 2001). Marine protected areas may have a role to play in the management and mitigation of marine disease (Raymundo et al. Chapter 9, this volume). Greater knowledge of food-web complexity and parasite life cycles (see Frasca et al. Chapter 11, this volume, for a discussion of future diagnostic techniques) is necessary to predict how particular parasites will respond to food-web changes. For instance, fishing removes sea urchin predators from Galapagos reefs, so it was predicted that the denser urchin populations in fished areas would be infected with more parasitic snails. However, sea urchin predators also eat mutualistic crabs, which eat the parasitic snails (Figure 2.3). Parasitic snails were therefore less abundant at fished sites because fishing indirectly increased crab predation pressure on parasitic snails (Sonnenholzner et al. 2011). Although indirect effects are easier to anticipate for simple food chains, in more complex food webs predictions are harder to make.

Ironically, their sensitivity to food-web perturbations can make parasites positive indicators of ecosystem "health." There are two reasons for this counterintuitive role. First, the complex life cycles that advantage parasites when navigating the trophic webs also make them more susceptible to perturbations. If a host link critical to the life cycle is lost, a parasite cannot persist. Second, in order to persist, parasites must find hosts. If host densities fall below the threshold where this can happen, then parasites can disappear before their hosts. For these reasons, parasite richness tends to follow food-web complexity. Next, we explore another factor that can affect parasites in food webs: intraspecific variation in host quality.

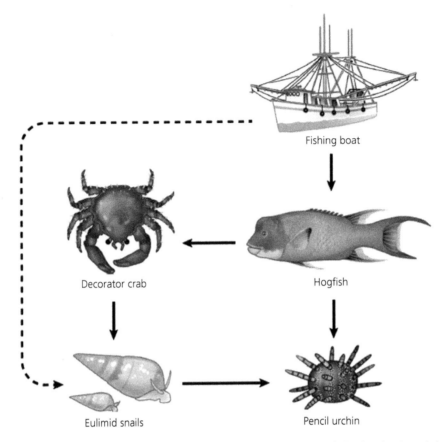

Fishing boat

Decorator crab

Hogfish

Eulimid snails

Pencil urchin

Figure 2.3 Fishing affects food webs and food webs affect parasites (figure from Lafferty (2017), with data based on Sonnenholzner et al. 2011). In the Galapagos, comparisons between fished areas and marine reserves show how fishing has a net negative effect on Eulimid snails that parasitize pencil urchins. Fishing reduces predators that eat urchins and spider crabs. There are more urchins and crabs at fished sites. Urchins at these fished sites had fewer snail parasites, probably due to crabs eating the snails. Removing the spider crabs from the food web would probably reverse how fishing affects the parasitic snail. This example indicates how food-web complexity can alter parasite success in ways that are sometimes difficult to predict.

2.6 How host quality affects parasites

In food-web diagrams, nodes often represent species comprised of abstract, identical individuals. However, there can be considerable variation in the ability of individuals to function as hosts. An individual parasite's growth and fecundity depends on host quality. Well-fed hosts may have surplus energetic stores that parasites can tap. Such reserves may be important to parasites with high energetic demands and/or facing strong within-host competition (Mideo 2009). Trematodes in starved snails produce fewer and poorer-quality transmission stages (Seppälä et al. 2008), acanthocephalans in

amphipods grow to a smaller size when hosts are food deprived (Labaude et al. 2015), and parasitic mussel larvae grow smaller in fish hosts in worse condition (Österling and Larsen 2013). Well-fed hosts grow more and can attain a larger size, which leads to more habitat for parasites (Lo et al. 1998). Resource quality can also influence host behavior and alter parasite transmission rates. Zooplankton fed low-quality food grow to smaller sizes, have lower size-corrected feeding rates, and thus encounter fewer parasite spores (Penczykowski et al. 2014). In contrast, well-fed hosts live longer, in turn increasing their parasites' life spans and lifetime reproduction (Penczykowski et al. 2014). For example, infected snails survive better when they are well fed, so their trematode parasites also live longer (Krist et al. 2004). When choosing among individual hosts that vary in quality, a parasite's success can depend on whether one considers it as a parasite species or an individual.

Although host quality should benefit individual parasite success, food-web dynamics might lead to trade-offs between host quality and abundance. Predators, for instance, might keep prey at low densities, thereby preventing crowding and improving prey as hosts for parasites. However, this also makes these prey hosts harder for parasites to contact and infect. Further, it is not clear if parasites should prefer well-fed hosts or malnourished hosts, if the latter have weakened immune systems (Cohen et al. 1993). However, crowded phytoplankton

(Tillmann et al. 1999) and snails (Krist et al. 2004) are not necessarily better hosts, perhaps because benefits gained from weakened immune defenses are outweighed by the consequently limited resources for parasites and decreased host life span (Tillmann et al. 1999). The relative importance of host density and host quality for parasite fitness remains understudied and may be context-dependent. We can leverage our understanding of how parasites use hosts also to fill in missing information in food webs.

2.7 Using parasites to improve food webs

Trophically transmitted parasites give clues about what eats what, particularly for host species whose habits or habitats make them difficult to study.

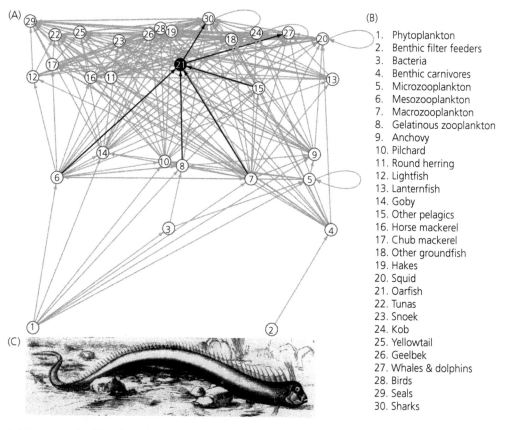

Figure 2.4 Benguela food web including oarfish. (A) Nodes represent free-living species or assemblages (white) and their trophic interactions (gray lines) (Yodzis 1998). Through parasitology we can integrate oarfish (black node) into the Benguela food web. Black lines indicate the feeding interactions involving oarfish that we can infer from parasitological analysis. Vertical height indicates consumer short-weighted trophic level. (B) Node list for the species represented in A. (C) An 1860 depiction of an oarfish that washed ashore and was originally described as a sea serpent (Ellis 2006).

For example, although oarfish (*Regalecus* spp.) are famous for being the largest bony fish (up to 10 m), most observations come from dead or dying individuals at the surface and a few midwater observations via remotely operated submersible, making it hard to tell what oarfish ate, or what ate them. For these reasons, although oarfish are present in the Benguela current ecosystem, their ecology was not well enough understood to include them in the food web. However, dissecting a few oarfish for parasitological analyses suggests how oarfish integrate into pelagic food webs (Kuris et al. 2015). Oarfish become infected with adult trematodes by eating planktonic crustaceans and gelatinous zooplankton (Kuris et al. In Preparation). Furthermore, large tetraphyllidean cestode larvae found in oarfish (Kuris et al. 2015) have a three- to four-host life cycle: crustacean to fish (sometimes squid) to shark. These tapeworms belong to *Clistobothrium*, which mature in large lamniform sharks, like the shortfin mako (*Isurus oxyrinchus*) (Ruhnke 1993), porbeagle (*Lamna nasus*), and great white (*Carcharodon carcharias*) (Klotz et al. 2018), suggesting sharks from this order are oarfish predators. Larval *Contracaecum* sp. nematodes from the oarfish gut mature in marine mammals, suggesting that toothed whales also eat oarfish (Kuris et al. 2015). By using these clues from parasite life cycles, we can integrate oarfish into the Benguela food web (Yodzis 1998) (Figure 2.4.) along with better-studied free-living species (Kuris et al. In Preparation).

2.8 Summary

Parasites have important and unique effects on marine food webs (McLaughlin 2018):

- In terms of richness, abundance, and biomass, parasite contributions to food-web structure are equivalent to comparable free-living groups.
- By infecting taxa across all trophic levels, parasites affect both bottom-up and top-down processes in marine systems.
- When host densities are high enough, parasites can regulate or even decimate their populations, causing regime shifts in marine systems from the tropics to temperate zones (Ben-Horin et al. Chapter 12, this volume).

- Free-living parasite stages can be important resources for free-living consumers, and filter feeders in particular.
- Through behavior manipulation, parasites may alter trophic interaction strength and the rate and energy flow direction in marine food webs.
- Their unique lifestyle renders parasites more susceptible to perturbations than their free-living hosts. As a result, parasites serve as useful ecosystem integrity indicators.
- By omitting parasites from marine ecology, we compromise our understanding of healthy marine systems, and our ability to design successful interventions to conserve diversity, preserve ecosystem services, or maintain commercial yields.

Future efforts to incorporate parasites into food webs can help us to better understand marine diseases (see Bateman et al. Chapter 1, this volume for direction in assembling the "pathobiome"):

- Combining parasitology with food webs can help marine-disease ecologists identify how parasites may threaten or contribute to marine biodiversity. Changes to food webs alter host diversity, abundance, and quality, and this has corresponding effects on parasite diversity, transmission success, and fitness. As food webs change, whether from fishing, climate disruption, or species invasions, we can expect parasite communities to change as well. Such changes might not be welcome. They could introduce new parasite species to which native hosts have little evolutionary history, or they could lead to parasite extinctions that add to biodiversity loss, and change host abundance.
- A theory for how food webs affect parasites will help us better understand why a particular infectious disease has become problematic, give insight into how restoration might reduce a costly marine disease, or let us use parasites as indicators to follow changes to food-web complexity.
- Although parasites can play important roles in marine food webs, and food webs affect marine diseases, studies on parasites in marine food webs are restricted to a few nearshore systems, which raises the question "What role do parasites play in other marine food webs?" This question can

only be addressed by assembling marine food webs that include parasites on the same empirical footing as their free-living counterparts.

Acknowledgments

We thank the Palmyra Atoll National Wildlife Refuge, U.S. Fish and Wildlife Service, Department of the Interior, The Nature Conservancy, and the U.S. Geological Survey for their support. This work benefited from grants from the Marisla Foundation and U.S. National Science Foundation (DEB-0224565). Any use of trade, product, or firm names in this publication is for descriptive purposes only and does not imply endorsement by the U.S. Government. This is PARC contribution 0151.

References

Aguirre-Macedo, M. L., Vidal-Martínez, V. M., and Lafferty, K. D. 2011. Trematode communities in snails can indicate impact and recovery from hurricanes in a tropical coastal lagoon. *International Journal for Parasitology* 41: 1403–8.

Allen, J. E., and Wynn, T. A. 2011. Evolution of th2 immunity: A rapid repair response to tissue destructive pathogens. *PLoS Pathogens* 7: e1002003.

Aronson, R. B., and Precht, W. F. 2001. White-band disease and the changing face of Caribbean coral reefs. In: J. W. Porter (ed.) *The Ecology and Etiology of Newly Emerging Marine Diseases*, pp. 25–38. Springer, Berlin.

Behrens, M. D., and Lafferty, K. D. 2004. Effects of marine reserves and urchin disease on southern Californian rocky reef communities. *Marine Ecology Progress Series* 279: 129–39.

Ben-Horin, T., Lenihan, H. S., and Lafferty, K. D. 2013. Variable intertidal temperature explains why disease endangers black abalone. *Ecology* 94: 161–8.

Bratbak, G., Egge, J. K., and Heldal, M. 1993. Viral mortality of the marine alga *Emiliania huxleyi* (Haptophyceae) and termination of algal blooms. *Marine Ecology Progress Series* 93: 39–48.

Brussaard, C. P. 2004. Viral control of phytoplankton populations—a review. *Journal of Eukaryotic Microbiology* 51: 125–38.

Buck, J. C., and Ripple, W. J. 2017. Infectious agents trigger trophic cascades. *Trends in Ecology & Evolution* 32: 681–94.

Cohen, J. E., Beaver, R. A., Cousins, S. H., DeAngelis, D. L., Goldwasser, L., Heong, K. L., Holt, R. D., Kohn, A. J.,

Lawton, J. H., Martinez, N. D., O'Malley, R., Page, L. M., Patten, B. C., Pimm, S. L., Polis, G. A., Rejmanek, M., Schoener, T. W., Schoenly, K., Sprules, W. G., Teal, J. M., Ulanowicz, R. E., Warren, P. H., Wilbur, H. M., and Yodzis, P. 1993. Improving food webs. *Ecology* 74: 252–8.

Connell, J. H. 1978. Diversity in tropical rain forests and coral reefs. *Science* 199: 1302–10.

Dobson, A., Lafferty, K. D., Kuris, A. M., Hechinger, R. F., and Jetz, W. 2008. Homage to Linnaeus: How many parasites? How many hosts? *Proceedings of the National Academy of Sciences* 105: 11482–9.

Dudgeon, S. R., Aronson, R. B., Bruno, J. F., and Precht, W. F. 2010. Phase shifts and stable states on coral reefs. *Marine Ecology Progress Series* 413: 201–16.

Dungan, M. L., Miller, T. E., and Thomson, D. A. 1982. Catastrophic decline of a top carnivore in the Gulf of California rocky intertidal zone. *Science* 216: 989–91.

Dunne, J. A., Williams, R. J., and Martinez, N. D. 2004. Network structure and robustness of marine food webs. *Marine Ecology Progress Series* 273: 291–302.

Dunne, J. A., Lafferty, K. D., Dobson, A. P., Hechinger, R. F., Kuris, A. M., Martinez, N. D., McLaughlin, J. P., Mouritsen, K. N., Poulin, R., Reise, K., Stouffer, D. B., Thieltges, D. W., Williams, R. J., and Zander, C. D. 2013. Parasites affect food web structure primarily through increased diversity and complexity. *PLoS Biology* 11: e1001579. doi:10.1371/journal.pbio.1001579.

Ellis, R. 2006. *Monsters of the sea*. Globe Pequot, Guilford, CT.

Estes, J. A., and Palmisano, J. F. 1974. Sea otters: Their role in structuring nearshore communities. *Science* 185: 1058–60.

Falkowski, P. G., Katz, M. E., Knoll, A. H., Quigg, A., Raven, J. A., Schofield, O., and Taylor, F. 2004. The evolution of modern eukaryotic phytoplankton. *Science* 305: 354–60.

Foster, N. L. 2012. *Reduced parasitism in a highly invaded estuary: San Francisco Bay*. PhD, University of California, Santa Barbara.

Fuhrman, J. A. 1999. Marine viruses and their biogeochemical and ecological effects. *Nature* 399: 541.

Gibson, D., Bray, R., and Harris, E. 2005. *Host–parasite database of the Natural History Museum, London*. http://www.nhm.ac.uk/research-curation/scientific-resources/taxonomy-systematics/host-parasites/database/index.jsp.

Gladfelter, W. B. 1982. White-band disease in *Acropora palmata*: Implications for the structure and growth of shallow reefs. *Bulletin of Marine Science* 32: 639–43.

Gleason, F. H., Küpper, F. C., Amon, J. P., Picard, K., Gachon, C. M., Marano, A. V., Sime-Ngando, T., and Lilje, O. 2011. Zoosporic true fungi in marine ecosystems: A review. *Marine and Freshwater Research* 62: 383–93.

González, J. M., and Suttle, C. A. 1993. Grazing by marine nanoflagellates on viruses and virus-sized particles: Ingestion and digestion. *Marine Ecology Progress Series* 94: 1–10.

Grutter, A. S. 1999. Cleaner fish really do clean. *Nature* 398: 672.

Guardo, G. D., Marruchella, G., Agrimi, U., and Kennedy, S. 2005. Morbillivirus infections in aquatic mammals: A brief overview. *Transboundary and Emerging Diseases* 52: 88–93.

Hammill, M., and Stenson, G. 2000. Estimated prey consumption by harp seals (*Phoca groenlandica*), hooded seals (*Cystophora cristata*), grey seals (*Halichoerus grypus*) and harbour seals (*Phoca vitulina*) in Atlantic Canada. *Journal of Northwest Atlantic Fishery Science* 26: 1–24.

Härkönen, L., Dietz, R., Reijnders, P., Teilmann, J., Harding, K., Hall, A., Brasseur, S., Siebert, U., Goodman, S. J., and Jepson, P. D. 2006. A review of the 1988 and 2002 phocine distemper virus epidemics in European harbour seals. *Diseases of Aquatic Organisms* 68: 115–30.

Harvell, C., Kim, K., Burkholder, J., Colwell, R., Epstein, P. R., Grimes, D., Hofmann, E., Lipp, E., Osterhaus, A., and Overstreet, R. M. 1999. Emerging marine diseases—climate links and anthropogenic factors. *Science* 285: 1505–10.

Hechinger, R. F., and Lafferty, K. D. 2005. Host diversity begets parasite diversity: Bird final hosts and trematodes in snail intermediate hosts. *Proceedings of the Royal Society of London B: Biological Sciences* 272: 1059–66.

Hechinger, R., Lafferty, K., Dobson, A. P., Brown, J., and Kuris, A. 2011a. A common scaling rule for abundance, energetics, and production of parasitic and free-living species. *Science* 333: 445–8.

Hechinger, R. F., Lafferty, K. D., McLaughlin, J. P., Fredensborg, B. L., Huspeni, T. C., Lorda, J., Sandhu, P. K., Shaw, J. C., Torchin, M. E., Whitney, K. L., and Kuris, A. M. 2011b. Food webs including parasites, biomass, body sizes, and life stages for three California/Baja California estuaries. *Ecology* 92: 791. doi:10.1890/10-1383.1.

Herrero-Pérezrul, M. 2008. *Diversity and abundance of reef macro invertebrates (Mollusca; Echinodermata) in the southern Gulf of California, México.* Paper presented at the Proceedings of the 11th International Coral Reef Symposium, Ft Lauderdale, Florida.

Huspeni, T. C., and Lafferty, K. D. 2004. Using larval trematodes that parasitize snails to evaluate a saltmarsh restoration project. *Ecological Applications* 14: 795–804.

Jackson, J. B., Kirby, M. X., Berger, W. H., Bjorndal, K. A., Botsford, L. W., Bourque, B. J., Bradbury, R. H., Cooke, R., Erlandson, J., and Estes, J. A. 2001. Historical overfishing and the recent collapse of coastal ecosystems. *Science* 293: 629–37.

Johnson, P. T., Dobson, A., Lafferty, K. D., Marcogliese, D. J., Memmott, J., Orlofske, S. A., Poulin, R., and Thieltges, D. W. 2010. When parasites become prey: Ecological and epidemiological significance of eating parasites. *Trends in Ecology & Evolution* 25: 362–71.

Kaplan, A. T., Rebhal, S., Lafferty, K., and Kuris, A. 2009. Small estuarine fishes feed on large trematode cercariae: Lab and field investigations. *Journal of Parasitology* 95: 477–80.

Klotz, D., Hirzmann, J., Bauer, C., Schöne, J., Iseringhausen, M., Wohlsein, P., Baumgärtner, W., and Herder, V. 2018. Subcutaneous merocercoids of Clistobothrium sp. in two Cape fur seals (*Arctocephalus pusillus pusillus*). *International Journal for Parasitology: Parasites and Wildlife* 7: 99–105.

Krist, A., Jokela, J., Wiehn, J., and Lively, C. 2004. Effects of host condition on susceptibility to infection, parasite developmental rate, and parasite transmission in a snail–trematode interaction. *Journal of Evolutionary Biology* 17: 33–40.

Kuris, A. 2003. Evolutionary ecology of trophically transmitted parasites. *Journal of Parasitology* 89: S96–100.

Kuris, A. M., Hechinger, R. F., Shaw, J. C., Whitney, K. L., Aguirre-Macedo, M., Boch, C. A., Dobson, A. P., Dunham, E. J., Fredensborg, B. L., Huspeni, T. C., Lorda, J., Mababa, L., Mancini, F. T., Mora, A., Pickering, M., Talhouk, N. L., Torchin, M. E., and Lafferty, K. 2008. Ecosystem energetic implications of parasite and free-living biomass in three estuaries. *Nature* 454: 515–18.

Kuris, A. M., Jaramillo, A. G., McLaughlin, J. P., Weinstein, S. B., Garcia-Vedrenne, A. E., Poinar Jr, G. O., Pickering, M., Steinauer, M. L., Espinoza, M., and Ashford, J. E. 2015. Monsters of the sea serpent: Parasites of an oarfish, *Regalecus russellii. Journal of Parasitology* 101: 41–4.

Labaude, S., Cézilly, F., Tercier, X., and Rigaud, T. 2015. Influence of host nutritional condition on post-infection traits in the association between the manipulative acanthocephalan *Pomphorhynchus laevis* and the amphipod *Gammarus pulex. Parasites & Vectors* 8: 403.

Lafferty, K. 1997. Environmental parasitology: What can parasites tell us about human impacts on the environment? *Parasitology Today* 13: 251–5.

Lafferty, K. D. 2004. Fishing for lobsters indirectly increases epidemics in sea urchins. *Ecological Applications* 14: 1566–73.

Lafferty, K. D. 2017. Marine infectious disease ecology. *Annual Review of Ecology, Evolution, and Systematics* 48: 473–96.

Lafferty, K. D., and Kuris, A. M. 1993. Mass mortality of abalone *Haliotis cracherodii* on the California Channel Islands: Tests of epidemiological hypotheses. *Marine Ecology Progress Series* 96: 239–48.

Lafferty, K. D., and Kuris, A. M. 1999. How environmental stress affects the impacts of parasites. *Limnology and Oceanography* 44: 925–31.

Lafferty, K. D., and Kuris, A. M. 2002. Trophic strategies, animal diversity and body size. *Trends in Ecology & Evolution* 17: 507–13.

Lafferty, K. D., and Kuris, A. M. 2009. Parasites reduce food web robustness because they are sensitive to secondary extinction as illustrated by an invasive estuarine snail. *Philosophical Transactions of the Royal Society B: Biological Sciences* 364: 1659–63.

Lafferty, K. D., and Morris, A. K. 1996. Altered behavior of parasitized killifish increases susceptibility to predation by bird final hosts. *Ecology* 77: 1390–7.

Lafferty, K. D., and Shaw, J. C. 2013. Comparing mechanisms of host manipulation across host and parasite taxa. *Journal of Experimental Biology* 216: 56–66.

Lafferty, K. D., and Suchanek, T. H. 2016. Revisiting Paine's 1966 sea star removal experiment, the most-cited empirical article in the *American Naturalist*. *American Naturalist* 188: 365–78.

Lafferty, K. D., Dobson, A. P., and Kuris, A. M. 2006. Parasites dominate food web links. *Proceedings of National Academy of Sciences* 103: 11211–16.

Lafferty, K., Shaw, J., and Kuris, A. 2008. Reef fishes have higher parasite richness at unfished Palmyra Atoll compared to fished Kiritimati Island. *EcoHealth* 5: 338–45.

Lessios, H. 1988. Mass mortality of *Diadema antillarum* in the Caribbean: What have we learned? *Annual Review of Ecology and Systematics* 19: 371–93.

Lo, C. M., Morand, S., and Galzin, R. 1998. Parasite diversity\host age and size relationship in three coral-reef fishes from French Polynesia. *International Journal for parasitology* 28: 1695–708.

May, R. M. 1973. *Stability and Complexity in Model Ecosystems*. Princeton University Press, NJ.

McLaughlin, J. P. 2018. The food web for the sand flats at Palmyra Atoll. Dissertation. University of Santa Barbara, CA.

Medawar, P. B., and Medawar, J. S. 1985. *Aristotle to Zoos: A Philosophical Dictionary of Biology*. Harvard University Press, MA.

Menge, B. A., Cerny-Chipman, E. B., Johnson, A., Sullivan, J., Gravem, S., and Chan, F. 2016. Sea star wasting disease in the keystone predator *Pisaster ochraceus* in Oregon: Insights into differential population impacts, recovery, predation rate, and temperature effects from long-term research. *PLoS One* 11: e0153994.

Mideo, N. 2009. Parasite adaptations to within-host competition. *Trends in Parasitology* 25: 261–8.

Miller, M., Gardner, I., Kreuder, C., Paradies, D., Worcester, K., Jessup, D., Dodd, E., Harris, M., Ames, J., and Packham, A. 2002. Coastal freshwater runoff is a risk factor for *Toxoplasma gondii* infection of southern sea otters (*Enhydra lutris nereis*). *International Journal for Parasitology* 32: 997–1006.

Miner, C. M., Altstatt, J. M., Raimondi, P. T., and Minchinton, T. E. 2006. Recruitment failure and shifts in community structure following mass mortality limit recovery prospects of black abalone. *Marine Ecology Progress Series* 327: 107–17.

Monier, A., Chambouvet, A., Milner, D. S., Attah, V., Terrado, R., Lovejoy, C., Moreau, H., Santoro, A. E., Derelle, É., and Richards, T. A. 2017. Host-derived viral transporter protein for nitrogen uptake in infected marine phytoplankton. *Proceedings of the National Academy of Sciences* 114: E7489–98.

Montecino-Latorre, D., Eisenlord, M. E., Turner, M., Yoshioka, R., Harvell, C. D., Pattengill-Semmens, C. V., Nichols, J. D., and Gaydos, J. K. 2016. Devastating transboundary impacts of sea star wasting disease on subtidal asteroids. *PLoS One* 11: e0163190.

Moret, Y., and Schmid-Hempel, P. 2000. Survival for immunity: The price of immune system activation for bumblebee workers. *Science* 290: 1166–8.

Muehlstein, L. 1989. *Perspectives on the wasting disease of eelgrass* Zostera marina. doi: 10.3354/dao007211.

Orth, R. J., Luckenbach, M. L., Marion, S. R., Moore, K. A., and Wilcox, D. J. 2006. Seagrass recovery in the Delmarva Coastal Bays, USA. *Aquatic Botany* 84: 26–36.

Österling, M. E., and Larsen, B. M. 2013. Impact of origin and condition of host fish (*Salmo trutta*) on parasitic larvae of *Margaritifera margaritifera*. *Aquatic Conservation: Marine and Freshwater Ecosystems* 23: 564–70.

Overstreet, R. M. 2007. Effects of a hurricane on fish parasites. *Parassitologia* 49: 161–8.

Paine, R. T. 1966. Food web complexity and species diversity. *American Naturalist* 100: 65–75.

Patterson, K. L., Porter, J. W., Ritchie, K. B., Polson, S. W., Mueller, E., Peters, E. C., Santavy, D. L., and Smith, G. W. 2002. The etiology of white pox, a lethal disease of the Caribbean elkhorn coral, *Acropora palmata*. *Proceedings of the National Academy of Sciences* 99: 8725–30.

Pauly, D., Christensen, V., Dalsgaard, J., Froese, R., and Torres, F. 1998. Fishing down marine food webs. *Science* 279: 860–3.

Penczykowski, R. M., Lemanski, B. C., Sieg, R. D., Hall, S. R., Housley Ochs, J., Kubanek, J., and Duffy, M. A. 2014. Poor resource quality lowers transmission potential by changing foraging behaviour. *Functional Ecology* 28: 1245–55.

Poorvin, L., Rinta-Kanto, J. M., Hutchins, D. A., and Wilhelm, S. W. 2004. Viral release of iron and its bioavailability to marine plankton. *Limnology and Oceanography* 49: 1734–41.

Poulin, R. 2010. Parasite manipulation of host behavior: An update and frequently asked questions *Advances in the Study of Behavior* 41: 151–86.

Prenter, J., MacNeil, C., Dick, J. T., and Dunn, A. M. 2004. Roles of parasites in animal invasions. *Trends in Ecology & Evolution* 19: 385–90.

Rudolf, V. H. W., and Lafferty, K. D. 2011. Stage structure alters how complexity affects stability of ecological networks. *Ecology Letters* 14: 75–9. doi:10.1111/j.1461-0248.2010.01558.x.

Ruhnke, T. 1993. A new species of Clistobothrium (Cestoda: Tetraphyllidea), with an evaluation of the systematic status of the genus. *Journal of Parasitology* 79: 37–43.

Ruiz, G. M., Hines, A. H., and Grosholz, E. D. 1999. Nonindigenous species as stressors in estuarine and marine communities: Assessing invasion impacts and interactions. *Limnology and Oceanography* 44: 950–72.

Sandin, S. A., Smith, J. E., DeMartini, E. E., Dinsdale, E. A., Donner, S. D., Friedlander, A. M., Konotchick, T., Malay, M., Maragos, J. E., and Obura, D. 2008. Baselines and degradation of coral reefs in the northern Line Islands. *PLoS One* 3: e1548.

Scheibling, R. 1986. Increased macroalgal abundance following mass mortalities of sea urchins (*Strongylocentrotus droebachiensis*) along the Atlantic coast of Nova Scotia. *Oecologia* 68: 186–98.

Seppälä, O., Liljeroos, K., Karvonen, A., and Jokela, J. 2008. Host condition as a constraint for parasite reproduction. *Oikos* 117: 749–53.

Shannon, L. J., Cury, P. M., and Jarre, A. 2000. Modelling effects of fishing in the southern Benguela ecosystem. ICES *Journal of Marine Science* 57: 720–2.

Shaw, J., Korzan, W., Carpenter, R., Kuris, A., Lafferty, K., Summers, C., and Øverli, Ø. 2009. Parasite manipulation of brain monoamines in California killifish (*Fundulus parvipinnis*) by the trematode *Euhaplorchis californiensis*. *Proceedings of the Royal Society of London B: Biological Sciences* 276: 1137–46.

Shaw, J., Hechinger, R., Lafferty, K. D., and Kuris, A. M. 2010. Ecology of the brain trematode *Euhaplorchis californiensis* and its host, the California killifish (*Fundulus parvipinnis*). *Journal of Parasitology* 96: 482–90.

Sonnenholzner, J. I., Lafferty, K. D., and Ladah, L. B. 2011. Food webs and fishing affect parasitism of the sea urchin *Eucidaris galapagensis* in the Galápagos. *Ecology* 92: 2276–84.

Steinbeck, J., and Ricketts, E. F. 2009. *Sea of Cortez: A Leisurely Journal of Travel and Research*. Penguin, London.

Strona, G., and Lafferty, K. D. 2016. Environmental change makes robust ecological networks fragile. *Nature Communications* 7: 12462.

Suttle, C. A. 2007. Marine viruses—major players in the global ecosystem. *Nature Reviews Microbiology* 5: 801–12.

Swartz, S. 1981. Cleaning symbiosis between topsmelt, *Atherinops affinis*, and gray whale, *Eschrichtius robustus*, in Laguna San Ignacio, Baja California Sur, Mexico. *Fishery Bulletin* 79: 316.

Tauber, A. I. 2003. Metchnikoff and the phagocytosis theory. *Nature Reviews Molecular Cell Biology* 4: 897.

Théodoridès, J. 1989. Parasitology of marine zooplankton. *Advances in Marine Biology* 25: 117–177.

Thieltges, D. W., de Montaudouin, X., Fredensborg, B., Jensen, K. T., Koprivnikar, J., and Poulin, R. 2008. Production of marine trematode cercariae: A potentially overlooked path of energy flow in benthic systems. *Marine Ecology Progress Series* 372: 147–55.

Tillmann, U., Hesse, K.-J., and Tillmann, A. 1999. Large-scale parasitic infection of diatoms in the Northfrisian Wadden Sea. *Journal of Sea Research* 42: 255–61.

Torchin, M. E., Lafferty, K. D., Dobson, A. P., McKenzie, V. J., and Kuris, A. M. 2003. Introduced species and their missing parasites. *Nature* 421: 628–30.

Torchin, M. E., Byers, J. E., and Huspeni, T. C. 2005. Differential parasitism of native and introduced snails: Replacement of a parasite fauna. *Biological Invasions* 7: 885–94.

Uthicke, S., Schaffelke, B., and Byrne, M. 2009. A boom–bust phylum? Ecological and evolutionary consequences of density variations in echinoderms. *Ecological Monographs* 79: 3–24.

Van Bressem, M.-F., Van Waerebeek, K., Jepson, P. D., Raga, J. A., Duignan, P. J., Nielsen, O., Di Beneditto, A. P., Siciliano, S., Ramos, R., and Kant, W. 2001. An insight into the epidemiology of dolphin morbillivirus worldwide. *Veterinary Microbiology* 81: 287–304.

Weinstein, S. B., Buck, J. C., and Young, H. S. 2018. A landscape of disgust. *Science* 359: 1213–14.

Windsor, D. A. 1998. Most of the species on Earth are parasites. *International Journal for Parasitology* 28: 1939–41.

Wood, C. L., and Lafferty, K. D. 2015. How have fisheries affected parasite communities? *Parasitology* 142: 134–44.

Wood, C. L., Sandin, S. A., Zgliczynski, B., Guerra, A. S., and Micheli, F. 2014. Fishing drives declines in fish parasite diversity and has variable effects on parasite abundance. *Ecology* 95: 1929–46.

Yodzis, P. 1996. *Food Webs and Perturbation Experiments: Theory and Practice*, pp. 192–200. Springer, Berlin.

Yodzis, P. 1998. Local trophodynamics and the interaction of marine mammals and fisheries in the Benguela ecosystem. *Journal of Animal Ecology* 67: 635–58.

CHAPTER 3

Disease can shape marine ecosystems

Joseph P. Morton, Brian R. Silliman, and Kevin D. Lafferty

3.1 Introduction

Although parasitism represents the most common consumer strategy on the planet (Dobson et al. 2008, Kuris et al. 2008) and infectious agents can control host population densities and modify host phenotypes (Lafferty 1993, Behrens and Lafferty 2004, Fredensborg et al. 2005, Hudson et al. 2007), there has been a comparatively greater focus on predators in the ecological literature. Inspired by examples where predators exert strong top-down control over community structure, marine disease ecologists have been asking whether disease agents could have similar effects (McLaughlin et al. Chapter 2, this volume, Harvell and Lamb Chapter 8, this volume). Parasites, like predators, can generate or modify trophic cascades, influence competitive outcomes by suppressing competitive dominants, and regulate important foundation species and ecosystem engineers via density-mediated or trait-mediated effects. In this chapter, we review the different pathways and mechanisms by which parasites can influence marine ecosystems, including their effects on species coexistence and ecosystem function.

3.2 Parasites generate trophic cascades

Trophic cascades occur when predators indirectly increase the productivity or biomass of their prey's prey (Estes and Palmisano 1974, Carpenter et al. 1985, Silliman and Bertness 2002, Myers et al. 2007,

Estes et al. 2011, Rosenblatt et al. 2013, Bertness et al. 2014). Like prey, hosts can drive top-down direct and indirect effects, suggesting parasites that infect and impact influential hosts cause trophic cascades (Buck and Ripple 2017). But almost all studies reporting trophic cascades caused by infectious agents are: (1) from terrestrial systems, (2) involve parasitoid wasps, or (3) increase plant biomass. The few studies demonstrating trophic cascades generated by infectious agents in marine ecosystems indicate that parasites, like predators, can alter diverse ecosystems by modifying host behavior or density. The similarity between predators and infectious agents is clear in California kelp forests where lobsters control sea urchins populations, preventing overgrazing (Lafferty 2004). When lobster fishing allows urchin populations to increase, bacterial outbreaks putatively caused by *Vibrio anguillarum* reduce urchin populations, albeit to a lesser extent than lobster predation does (Behrens and Lafferty 2004). It is not surprising that a lethal parasite like *Vibrio anguillarum* can have effects similar to a predator.

Most parasites differ from predators in that their consumption is partial and non-lethal. Non-lethal parasites might induce trophic cascades via *consumptive* trait-mediated indirect effects (Buck and Ripple 2017). For instance, in New England rocky intertidal communities, the digenean trematode *Cryptocotyle lingua* castrates the intertidal gastropod *Littorina littorea*. Infected snails feed less, releasing intertidal algae from grazing and altering ecosystem

Morton J.P., Silliman B.R., and Lafferty K.D., *Disease can shape marine ecosystems* In: *Marine Disease Ecology*. Edited by: Donald C. Behringer, Kevin D. Lafferty, and Brian. R. Silliman, Oxford University Press (2020). © Oxford University Press. DOI: 10.1093/oso/9780198821632.003.0003

structure and productivity (Wood et al. 2007). Such parasite trait-mediated consumptive effects are common in the literature (but see Philpott et al. 2004, Pardee and Philpott 2011).

Just as predators often cause trophic cascades through non-consumptive trait-mediated indirect effects like fear, parasites might cause trophic cascades through disgust. Many species have evolved behaviors to avoid disease transmission (Behringer et al. 2018), and the ecology of fear likely applies to both parasites and predators (Buck and Ripple 2017). In a well-documented marine example, uninfected Caribbean spiny lobsters, *Panulirus argus*, avoid juvenile conspecifics infected with *Panulirus argus* virus 1 (PaV1) (Behringer et al. 2006, Anderson and Behringer 2013). Lobster avoidance behavior intensifies competition for shelter between healthy and diseased lobsters, and increases predation on isolated, infected individuals (Behringer and Butler 2012), thereby creating potential for indirect effects. Because avoidance behaviors are common and can influence population dynamics (Anderson and Behringer 2013, Behringer et al. 2018), cascading indirect effects through disgust appear plausible.

When a disease affects a keystone species' abundance, size structure, or behavior, cascading effects can lead to long-term changes in ecosystem structure and function (Castello et al. 1995, Eviner and Likens 2008, Preston et al. 2016, Schultz et al 2016, Harvell et al 2019). For instance, from 1982 to 1984, a species-specific pathogen killed off the long-spined sea urchin (*Diadema antillarum*), an important grazer throughout the Caribbean. This sudden and prolonged near-extirpation resulted in a dramatic ecosystem phase shift from a coral-dominated system, characterized by minimal macroalgal cover, to an algal-dominated system (Figure 3.1). For example, where disease had reduced grazing urchin populations in Jamaica by 99 percent, coral cover decreased from 52 percent to 3 percent in the following decade, because fleshy macroalgal cover increased from 4 to 92 percent over the same period (Hughes et al. 2010). Similarly, in temperate ecosystems, sea star wasting disease extirpated sunflower sea stars (*Pycnopodia helianthoides*), allowing urchin populations to explode, after which kelp forests in British Columbia and California were turned into urchin barrens (Schultz et al. 2016, Harvell et al. 2019, discussed in Harvell and Lamb Chapter 8, this volume). These and other parasites that extirpate important species can have cascading effects that shift ecosystems to alternate stable states.

Figure 3.1 Mass mortality of *Diadema antillarum*-affected Caribbean coral reef community structure (figure by J. Morton). Urchin deaths between 1982 and 1984 due to a species-specific pathogen resulted in an ecosystem phase shift from a coral-dominated system to an algal-dominated system.

A common alternative stable state is lush vegetation. The Green World Hypothesis states that in three-level food chains, primary producers can flourish, because predators and parasites control herbivores (Hairston et al. 1960). For instance, herbivorous insects are often regulated by parasitoids and viruses in terrestrial environments (Zhu et al. 2014). Such parasites indirectly benefit plants in food chains with odd links, and negatively affect plants in food chains with even links. However, outcomes for the basal resource in trait-mediated cascades involving parasites can be negative or positive. A positive effect on consumption might occur if infected hosts need *more* food than uninfected conspecifics to compensate for the energetic losses caused by parasitic infestation, impacting the basal resource (Gérard and Théron 1997, Bernot and Lamberti 2008). Parasite-increased consumption has been shown in freshwater environments where larval trematodes (*Posthodiplostomum minimum*) have cascading impacts on periphyton (algae, cyanobacteria, heterotrophic microbes, and detritus attached to submerged surfaces) by increasing infected snail (*Physa acuta*) feeding rates (Bernot and Lamberti 2008). Currently, there are no comparable demonstrations in marine settings, but, given that parasites can have disparate effects on host feeding rates, this particular interaction likely occurs in marine settings as well (Wood et al. 2007, Bernot and Lamberti 2008). Parasites might not always promote a green world.

3.3 Infectious agents can suppress or facilitate competitive exclusion

Parasites can structure natural communities by altering the competitive interactions that affect coexistence. Coexistence among competitors can be increased or decreased by disease agents (Park 1948, Price et al. 1986, Mordecai 2011). Diseases can facilitate coexistence by impacting competitive dominants (Pennings and Callaway 1996, Calvo-Ugarteburu and McQuaid 1998), or cause exclusion when the dominant species is more disease tolerant than the subordinate(s). Either way, parasites that alter competitive interactions can influence marine communities, reducing or promoting biodiversity, with corresponding changes in ecosystem function.

Whether disease increases or decreases coexistence depends on how competing host species differ in susceptibility or tolerance. The *rickettsia*-like organism responsible for withering syndrome that collapsed black abalone populations in the California rocky intertidal zone also infects more-tolerant abalone species (discussed in Harvell and Lamb Chapter 8, this volume). These more-tolerant abalone act as disease reservoirs and increase infection in less-tolerant black abalone when they co-occur (Ben-Horin et al. 2013). In communities where black abalone have been eliminated, reduced grazing has shifted the habitat from rock covered with coralline algae—a condition conducive to abalone recruitment—to rock covered in macroalgae, invertebrates, and sea urchins (Miner et al. 2006). Thus, shifts in dominance hierarchies generated by disease may result in inhibitory feedbacks and state changes in ecosystem structure. When parasites and tolerant hosts team up, intolerant hosts suffer and diversity can decline.

Another way parasites can shift dominance hierarchies and alter ecosystem properties is by influencing species invasions. Non-native species might have an easier time invading where disease depresses otherwise dominant native species. For example, on Australian rocky shores, the dominant Sydney rock oyster (*Saccostrea glomerata*) suffers annual mass mortalities from the protozoan *Marteilia sydneyi* (QX disease) (Nell 2001). Annual disease outbreaks facilitate invasion by the non-native, QX-resistant Pacific oyster (*Crassostrea gigas*), which preempts space from native rock oysters. The result is a rocky intertidal zone dominated by a disease-tolerant non-native species. Similarly, the Mediterranean mussel (*Mytilus gallopovincialis*) was introduced to South Africa where it competes for space with the native mussel (*Perna perna*). Two native trematode species infect the native mussel but not the invader. In contrast to the previous example, these parasites have sublethal effects on their native hosts. One trematode reduces host growth, whereas the other castrates its host, reducing adductor muscle strength, and increasing water loss at low tide. Here, parasites increase invasion success and alter the intertidal community through trait-mediated indirect effects (Calvo-Ugarteburu and McQuaid 1998). Parasite-mediated invasions are further enhanced when

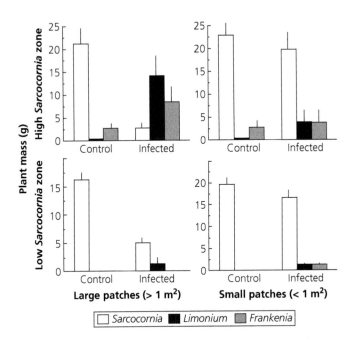

Figure 3.2 Impact of the parasitic plant *Cuscuta salina* (salt-marsh dodder) on the biomass of marsh plants (*Sarcocornia pacifica, Limonium californicum,* and *Frankenia salina*) in control areas and in large, older dodder patches and small areas recently infected by dodder, in the high-*Sarcocornia* and low-*Sarcocornia* zones. Uninfected control areas are dominated by *Sarcocornia*, whereas infected areas contain relatively high biomass of *Limonium* and *Frankenia*. The impact that dodder has on marsh vegetation is strongest in large patches in the high-*Sarcocornia* zone. Data are means with standard error. (Adapted from Pennings and Callaway (1996).)

marine invaders leave their parasites behind (Torchin et al. 2002).

Infectious diseases can also promote species coexistence and drive ecosystem dynamics by suppressing competitive dominants (Mordecai 2011, Freehan and Scheibling 2014). For instance, salt-marsh plants live in elevation zones with lower boundaries set by different tolerances for submergence and salinity (Janousek et al. 2018), and upper limits set by competition (Bertness and Ellison 1987). Competitive hierarchies are further affected by salt-marsh dodder (*Cuscuta salina*), a widespread parasitic plant in western US coastal marshes that prefers marsh pickleweed (*Sarcocornia pacifica*) as its host (Pennings and Callaway 1996). Sustained by nutritional uptake from its host via the haustoria (parasitic plant roots), dodder can overgrow and smother host plants. By reducing pickleweed biomass throughout the lower intertidal zone, dodder creates refugia for subordinate plants like *Limonium californicum* and *Frankenia salina* (Figure 3.2, Pennings and Callaway 1996). This parasite, therefore, increases plant diversity in the lower intertidal and drives cyclical vegetation dynamics. Density-dependent effects on ecosystem structure and dynamics also occur in the Sargasso Sea where phage infection can lead to long-term cycles in the bacterioplankton community (Parsons et al. 2012). Notably, phage are more likely to kill abundant bacterioplankton such as *Prochlorococcus*, indicating a "kill the winner" scenario in which phage infection bias prevents the fastest growing or most abundant bacterioplankton (the winner) from taking over the community (Thingstad and Lignell 1997, Thingstad 2000). In a sense, a parasite's tendency to reduce dominant species is akin to the Australian "tall poppy syndrome" in which people who are perceived as being superior are attacked, criticized, and "cut down."

3.4 Disease impacts on foundational species

Although density-dependent transmission can promote host diversity, it can also impact on abundant foundational species. Foundational organisms have high biomass density and play an outsized role in provisioning habitat and structuring communities. For these reasons, disease agents that affect foundational organisms can influence ecosystem structure and function. In particular, mass mortalities driven by disease can have a "flattening" effect, reducing the complex three-dimensional structure created by foundational organisms and impacting biodiversity—a familiar example being how the fungus *Cryphonectria parasitica* devastated the American chestnut tree *Castanea dentata* that once dominated North American forests. Similarly, in the 1930s, the pathogen *Labyrinthula zosterae* reduced eelgrass (*Zostera marina*) beds across the North Atlantic coast by 90 percent, decreasing fish and waterfowl richness and abundance (Hughes et al. 2000, Harvell and Lamb Chapter 8, this volume), and causing the eelgrass limpet (*Lottia alveus*) to go extinct (Carlton et al. 1991). In the Caribbean, successive disease outbreaks affecting dominant acroporid corals have shifted reef communities to the less complex, "weedy" species such as *Agaricia* spp. corals (Aronson and Precht 2001). This reduction in rugosity has diminished habitat quality. Such parasites that affect foundational species have substantial indirect bottom-up effects on marine communities.

Disease might also slow recovery of foundation species that have declined due to other causes such as overfishing or anthropogenic climate change. For instance, overfishing has reduced reefs formed by the eastern oyster *Crassostrea virginica* (Powell et al. 2012). Oyster reefs provide shelter for many species, modify estuarine physiochemistry through their water filtration, and play a role in denitrification, nutrient assimilation, and carbon sequestration. These functions increase with oyster density, which then increases their ability to reproduce and generate new reef structure (zu Ermgassen et al. 2012). However, two protozoan parasites, *Halosporidium nelsoni* (MSX) and *Perkinsus marinus* (dermo), can shrink reefs when parasite-induced death rates

exceed reef accretion rates (e.g., Powell et al. 2012, see Behringer et al. Chapter 10, this volume). Erosion increases further when harvest removes surficial shell on which larval oysters recruit. Thus, these diseases have stymied oyster reef recovery, along with its contributions to estuarine functioning and biodiversity. Given that foundational species are often key to habitat restoration, their parasites should figure in restoration strategies.

Increases in disease vectors can also affect foundational organisms. In southeastern US salt marshes, the salt-marsh periwinkle *Littoraria irrorata* grazes on cordgrass (*Spartina alterniflora*) leaves, making them a suitable substrate for opportunistic ascomycete fungal pathogens that the snails then feed on (Silliman and Zieman 2001, Silliman and Newell 2003). *Littoraria* further promote fungal growth on live cordgrass through depositing fecal pellets that are rich in nitrogen and fungal hyphae. Fungal infection can decrease grass growth by more than 60 percent (Silliman and Newell 2003), limiting salt marsh productivity (Silliman and Zieman 2001, Silliman and Bertness 2002, Silliman et al. 2005), with many knock-on effects on ecosystem structure and function, such as reducing marsh macroinvertebrate density, decomposition rate, and sediment infiltration rate (Hensel and Silliman 2013). Additionally, drought-induced soil stress can weaken plant defenses and stimulate overgrazing by snails (Silliman et al. 2005), and under such conditions, snails may concentrate in dense grazing fronts that convert salt-marsh habitat to mudflat (Silliman et al. 2005). In this way, interactions between pathogens and other consumers can control ecosystem functioning and successional dynamics at the ecosystem scale.

3.5 Parasites and ecosystem engineering

Parasites can affect hosts that are habitat for other species (Jones et al. 1994, 1997). Such ecosystem engineers modulate resource availability for other organisms, or cause state changes in the ecosystem. For example, in New Zealand mudflat communities, hard substrate is limited, but epibionts can live on cockle shells (*Austrovenus stutchburyi*) (Jones et al. 1997, Thomas et al. 1998). Normal cockles live

Figure 3.3 Effect of the manipulative trematode *Curtuteria australis* on community structure on a New Zealand mudflat (adapted from Lafferty and Kuris 2012 by J. Morton). Trematode metacercariae encyst in the cockle's foot, impeding burrowing and facilitating a novel epibiont community (Thomas et al. 1999).

just beneath the sediment surface, but those infected with the trematode *Curtuteria australis* have difficulty burrowing and leave their shells exposed. Exposure increases trophic transmission to predatory pied oystercatchers (*Haematopus longirostris*) (Thomas et al. 1998), and generates a hard substrate (Figure 3.3). Limpets prefer to settle on the increased shell area in manipulated cockles, whereas anemones show the opposite settlement preference due to desiccation stress associated with increased exposure during low tide. Thus, manipulation by trematodes generates distinct habitat types that can be occupied by different organisms, facilitating coexistence among competing epibionts and increasing species diversity on mudflats (Mouritsen and Poulin 2005). Furthermore, diminished burrowing activity by infected cockles reduces disturbance associated with bioturbation and stabilizes sediments (Mouritsen and Poulin 2005, Mouritsen and Poulin 2006, Mouritsen and Poulin 2010). Decreased bioturbation due to manipulative parasites was associated with increases in some benthic macroinvertebrates (Mouritsen and Poulin 2005,

Mouritsen and Poulin 2006). Because many benthic macroinvertebrates feed on algae, infected cockles also indirectly decreased primary productivity by enhancing top-down control (Mouritsen and Poulin 2006, Mouritsen and Poulin 2010). In another example, the parasitic rhizocephalan barnacle *Sacculina carcini* changes green crab (*Carcinus maenas*) behavior and physiology, and impairs growth (O'Brien and van Wyk 1985). Because an infected crab does not molt, its carapace can accumulate more epibiotic invertebrates over time than uninfected crabs (Mouritsen and Jensen 2006). These examples show that parasites can alter habitat condition and availability, in turn affecting non-host species, though it is not clear that these fascinating examples often apply in nature.

Every ecosystem engineer has parasites. Because ecosystem engineering relates directly to ecosystem engineer abundance, parasites that decrease engineer densities may indirectly affect habitat for other species. For instance, the amphipod *Corophium voluntator* is an ecosystem engineer on Danish intertidal mudflats that stabilizes sediments through burrowing activity. However, the digenean trematode *Microphallus claviformis* can induce mass mortalities in amphipods, thereby destabilizing the substrate, decreasing sediment silt content, and altering benthic community structure (Jensen and Mouritsen 1992, Mouritsen and Poulin 2005). Therefore, rather than create ecosystem engineers, it seems more plausible that parasites could decrease their abundance.

3.6 Conclusions and Future Directions

Ecologists have long focused on how species interactions like predation, competition, and facilitation can influence ecosystem structure and function. Similarly, parasites affect host abundance, traits, and behavior. We now recognize that parasites themselves can have strong impacts on marine populations, community biodiversity, and ecosystem function. For example, parasites that alter host phenotypes can change their host's role in the ecosystem and increase functional trait diversity (Frainer et al. 2018). Likewise, parasites can increase or decrease ecosystem functionality by regulating biodiversity or affecting influential hosts such as ecosystem engineers. Furthermore, when pathogens suppress foundational species such as corals and seagrass, their indirect impacts are even more likely to influence biodiversity and ecosystem function, including habitat provisioning, nutrient cycling, and carbon sequestration (Hughes et al. 2000, Aronson and Precht 2001). Many marine ecologists are now acknowledging how parasites affect the systems they study.

When considering potential roles for parasites as structuring agents in ecosystems, it has often been a useful heuristic to compare and contrast them with predators (Lafferty and Kuris 2002, Raffel et al. 2008). Like predators, parasites can influence ecosystems by affecting competition, and exerting top-down control through trait- and density-mediated means. Predators can also increase ecosystem resistance and resilience to disturbances by controlling their prey (Sala 2006, Wilmers et al. 2006). If parasites regulate top predators that buffer stress by keeping grazer populations in check, they might decrease ecosystem resistance and resilience (McLaughlin et al. Chapter 2, this volume). However, parasites might buffer disturbance when top predators are missing by regulating the predator's prey. For instance, many organisms form "fronts" in response to anthropogenic stressors and generate widespread, cascading loss to ecosystems (Silliman et al. 2013). Such front-forming species are known to harbor parasites that can alter their behavior or survivorship and might mitigate their impacts (Wood et al. 2007, Clausen et al. 2008, Hoover et al. 2011, Shi et al. 2014). We suggest that parasites can sometimes have effects on a par with predators in marine systems.

Although disease can affect marine ecosystems by killing off important organisms and initiating trophic cascades, few studies have quantified indirect effects. For instance, most studies describing trophic cascades involving parasites do not look beyond the main interaction chain. Determining how parasites affect ecosystems would benefit from experiments that manipulate parasite prevalence or intensity. Parasites might be excluded through treatment with pharmaceuticals (e.g., anthelmintic drugs) or by manipulating vectors or dispersive agents (Marvier 1998, Hudson et al. 2007, Pedersen and Antonovics 2013). Alternatively, one might add infected organisms to a system (Wood et al. 2007). Such experiments are helpful in determining cause and effect, so long as they are not too contrived. Also, although most studies to date have examined the role played by single parasites in ecosystems, future studies should consider that parasites are diverse, and different parasites, like different predators, may combine in complex ways to affect prey (host) populations and lead to different outcomes (McLaughlin et al. Chapter 2, this volume). Here, we hope to have given background and inspiration for such future studies.

3.7 Summary

- Parasites can modify ecosystem functionality by regulating biodiversity or influential host species.
- Parasites can generate or modify trophic cascades, thereby changing ecosystem structure and function.
- When parasites regulate foundational species or ecosystem engineers, they may have powerful indirect impacts on biodiversity and ecosystem function.

- Parasites that suppress or enhance dominant species can affect coexistence and biodiversity.
- Future studies should draw inspiration from studies of predator top-down effects and might try to use experimental manipulations where possible.

Acknowledgments

We would like to thank our colleagues who gave valuable edits and commentary on earlier drafts that improved the chapter: in particular, C. Chen, and A. Paxton. Any use of trade, product, or firm names in this publication is for descriptive purposes only and does not imply endorsement by the U.S. Government.

References

Anderson, J.R., Behringer, D.C. 2013. Spatial dynamics in the social lobster *Panulirus argus* in response to diseased conspecifics. *Marine Ecology Progress Series* 474: 191–200

Aronson, R.B. and Precht, W.F. 2001. White-band disease and the changing face of Caribbean coral reefs. *Hydrobiologia* 460: 25–38

Behrens, M.D. and Lafferty, K.D. 2004. Effects of marine reserves and urchin disease on southern Californian rocky reef communities. *Marine Ecology Progress Series* 279: 11

Behringer, D.C., Butler, M.J. 2012. Disease avoidance influences shelter use and predation in Caribbean spiny lobster. *Behavioral Ecology and Sociobiology* 64: 747–55

Behringer, D.C., Butler, M.J., Shields, J.D. 2006. Ecology: avoidance of disease by social lobsters. Nature 441: 421

Behringer, D.C., Karvonen, A., Bojko, J. 2018. Parasite avoidance behaviors in aquatic environments. *Philosophical Transactions of the Royal Society B* 373: 20170202

Ben-Horin, T., Lenihan, H.S., Lafferty, K.D. 2013. Variable intertidal temperature explains why disease endangers black abalone. *Ecology* 94: 161–8

Bernot, R.J., Lamberti, G.A. 2008. Indirect effects of a parasite on a benthic community: an experiment with trematodes, snails and periphyton. *Freshwater Biology* 53: 322–9

Bertness, M.D., Eillison, A.M. 1987. Determinants of pattern in a New England salt marsh plant community. *Ecological Monographs* 57: 129–47

Bertness, M.D., Brisson, C.P., Coverdale, T.C., Bevil, M.C., Crotty, S.M., Suglia, E.R. 2014. Experimental predator removal causes rapid salt marsh die-off. *Ecology Letters* 17: 830–5

Buck, J.C., Ripple, W.J. 2017. Infectious agents trigger trophic cascades. *Trends in Ecology and Evolution* 32: 681–94

Calvo-Ugarteburu, G., McQuaid, C.D. 1998. Parasitism and invasive species: effects of digenetic trematodes on mussels. *Marine Ecology Progress Series* 169: 149–63

Carlton, J.T., Vermeij, G.J., Lindberg, D.R., Carlton, D.A., Dubley, E.C. 1991. The first historical extinction of a marine invertebrate in an ocean basin: the demise of the eelgrass limpet *Lottia alveus*. *Biological Bulletin* 180: 72–80

Carpenter, S.R., Kitchell, J.F., Hodgson, J.R. 1985. Fish predation and herbivory can regulate lake ecosystems. *BioScience* 35: 634–9

Castello, J.D., Leopole, D.J., Smallidge, P.J. 1995. Pathogens, patterns, and processes in forest ecosystems. *BioScience* 45: 16–24

Clausen, K.T., Larsen, M.H., Iversen, N.K., Mouritsen, K.N. 2008. The influence of trematodes on the macroalgae consumption by the common periwinkle *Littorina littorea*. *Journal of the Marine Biological Association of the United Kingdom* 88: 1481–5

Dobson, A., Lafferty, K.D., Kuris, A.M., Hechinger, R.F., Jetz W. 2008. Homage to Linnaeus: How many parasites? How many hosts? *Proceedings of the National Academy of Sciences* 105: 11482–9

Estes, J.A., Palmisano, J.F. 1974. Sea otters: their role in structuring nearshore communities. *Science* 185: 1058–60

Estes, J.A., Terborgh, J., Brashares, J.S., Power, M.E., Berger, J., Bond, W.J., Carpenter, S.R., Essington T.E., Holt, R.D., Jackson, J.B.C., Marquis, R.J., Oksanen, L., Oksanen, T., Paine, R.T., Pikitch, E.K., Ripple, W.J., Sandin, S.A., Scheffer, M., Schoener, T.W., Shurin, J.B., Sinclair, A.R.E,. Soulé, M.E, Virtanen, R., Wardle, D.A. 2011. Trophic downgrading of planet Earth. *Science* 333: 301–6

Eviner, V.T., Hawkes C.V. 2008. Embracing variability in the application of plant–soil interactions to the restoration of communities and ecosystems. *Restoration Ecology* 16: 713–29

Frainer, A., Brendan, G.M., Amundsen, P., Knudsen, R., Lafferty, K.D. 2018. Parasitism and the biodiversity–functioning relationship. *Trends in Ecology and Evolution* 33: 260–8

Fredensborg, B.L., Mouritsen, K.N., Poulin, R. 2005. Impact of trematodes on host survival and population density in the intertidal gastropod *Zeacumantus subcarcinatus*. *Marine Ecology Progress Series* 290: 109–17

Freehan, C.J., Scheibling, R.E. 2014. Disease as a control of sea urchin populations in Nova Scotian kelp beds. *Marine Ecology Progress Series* 500: 149–58

Gérard, C., Théron, A. 1997. Age/size- and time-specific effects of *Schistosoma mansoni* on energy allocation patterns of its snail host *Biomphalaria glabrata*. *Oecologia* 112: 447–52

Hairston, N.G., Smith, F., Slobodkin, L.B. 1960. Community structure, population control, and competition. *American Naturalist* 94: 421–5

Harvell, C.D., Montecino-Latorre, D., Caldwell, J.M., Burt, J.M., Bosley, K., Keller, A., Heron, S.F., Salomon, A.K., Lee, L., Pontier, O., Pattengill-Semmens, C., Gaydos, J.K. 2019. Disease epidemic and a marine heat wave are associated with the continental-scale collapse of a pivotal predator (*Pycnopodia helianthoides*). *Science Advances* 5: eaau7042

Hensel, M.J.S., Silliman, B.R. 2013. Consumer diversity across kingdoms supports multiple functions in a coastal ecosystem. *Proceedings of the National Academy of Sciences* 110: 20621–6

Hoover, K., Grove, M., Gardner, M., Hughes, D.P., McNeil, J., Slavicek, J. 2011. A gene for an extended phenotype. *Science* 333: 1401

Hudson, P.J., Dobson, A.P., Newborn, D. 2007. Prevention of population cycles by parasite removal. *Science* 282: 2256–8

Hughes, J.E., Deegan, L.A., Wyda, J.C., Weaver, M.J., Wright, A. 2000. The effects of eelgrass habitat loss on estuarine fish communities of southern New England. *Estuaries* 25: 235–49

Hughes, T.P., Graham, N.A., Jackson, J.B., Mumby, P.J., Steneck, R.S. 2010. Rising to the challenge of sustaining coral reef resilience. *Trends in Ecology and Evolution* 25: 633–42

Janousek, C.N., Thorne, K.M., Takekawa, J.Y. 2018. Vertical zonation and niche breadth of tidal marsh plants along the northeast Pacific Coast. *Estuaries and Coasts* July 23, 1–14

Jensen, K., Mouritsen, K. 1992. Mass mortality in two common soft-bottom invertebrates, *Hydrobia ulvae* and *Corophium volutator*—the possible role of trematodes. *Helgoländ Meeresunters* 46: 329–39

Jones, C. G., Lawton, J. H., Shachak, M. 1994. Organisms as ecosystem engineers. *Oikos* 69: 373–86

Jones, C.G., Lawton, J.H., Shachak, M. 1997. Positive and negative effects of organisms as physical ecosystem engineers. *Ecology* 78: 1946–57

Kuris, A.M., Hechinger, R.F. Shaw, J.C., Whitney, K.L., Aguirre-Macedo, L., Boch, C.A., Dobson, A.P., Dunham, E.J., Fredensborg, B.L., Huspeni, T.C., Lorda, J., Mababa, L., Mancini, F.T., Mora, A.B., Pickering, M., Talhouk, N.L., Torchin, M.E., Lafferty, K.D. 2008. Ecosystem energetic implications of parasite and free-living biomass in three esturaries. *Nature* 454: 515–18

Lafferty, K.D. 1993. Effects of parasitic castration on growth, reproduction and population dynamics of the marine snail *Cerithidea californica*. *Marine Ecology Progress Series* 96: 229–37

Lafferty, K.D. 2004. Fishing for lobsters indirectly increases epidemics in sea urchins. *Ecological Applications* 14: 1566–73

Lafferty, K.D., Kuris, A.M. 2002 Trophic strategies, animal diversity and body size. *Trends in Ecology & Evolution* 17: 507–13

Lafferty, K.D., Kuris, A.M. 2012. Ecological consequences of manipulative parasites. In: D.B. Hughes, J. Brodeur, and F. Thomas (eds) *Host Manipulation by Parasites*, pp. 158–71. Oxford: Oxford University Press

Marvier, M.A. 1998. Parasite impacts on host communities: plant parasites in a California coastal prairie. *Ecology* 79: 2616–23

Miner, C.M., Altstatt, J.M, Raimondi, P.T., Minchinton, T.E. 2006. Recruitment failure and shifts in community structure following mass mortality limit recovery prospects of black abalone. *Marine Ecology Progress Series* 327: 107–17

Mordecai, E.A. 2011. Pathogen impacts on plant communities: unifying theory, concepts, and empirical work. *Ecological Monographs*. 81: 429–41

Mouritsen, K.N., Tomas Jensen, T. 2006. The effect of *Sacculina carcini* infections on the fouling, burying behaviour and condition of the shore crab, *Carcinus maenas*. *Marine Biology Research* 2: 270–75

Mouritsen, K.N., Poulin, R. 2005. Parasites boost biodiversity and changes animal community structure by trait-mediated indirect effects. *Oikos* 108: 344–50

Mouritsen, K.N., Poulin, R. 2006. A parasite indirectly impacts both abundance of primary producers and biomass of secondary producers in an intertidal benthic community. *Journal of Marine Biological Association of the UK* 86: 221–6

Mouritsen, K.N., Poulin, R. 2010. Parasitism as a determinant of community structure on intertidal flats. *Marine Biology* 157: 201–7

Myers, R.A., Baum, J.K., Shepherd, T.D., Powers, S.P., Peterson, C.H. 2007. Cascading effects of the loss of apex predatory sharks from a coastal ocean. *Science* 315: 1846–50

Nell, J.A. 2001. The history of oyster farming in Australia. *Marine Fisheries Review* 63: 14–25

O'Brien, J., van Wyk, P. 1985. Effects of crustacean parasitic castrators (epicaridean isopods and rhizocephalan barnacles) on growth of their crustacean hosts. In: A.M. Wenner (ed.), *Crustacean Issues, Vol. 3, Factors in Adult Growth*, pp. 191–218. Rotterdam: A.A. Balkema Press.

Pardee, G.L., Philpott, S.M. 2011. Cascading indirect effects in a coffee agroecosystem: effects of parasitic phorid flies on ants and the coffee berry borer in a high-shade and low-shade habitat. *Environmental Entomology* 40: 581–8

Park, T. 1948. Interspecies competition in populations of *Trilobium confusum* Duval and *Triobium castaneum* Herbst. *Ecological Monographs* 18: 265–307

Parsons, R.J., Breitbart, M., Lomas, M.W., Carlson, C.A. 2012. Ocean time-series reveals recurring seasonal patterns of virioplankton dynamics in the northwestern Sargasso Sea. *ISME Journal* 6: 273–8

Pedersen A.B., Antonovics J. 2013. Anthelmintic treatment alters the parasite community in a wild mouse host. *Biology Letters* 9: 20130205

Pennings, S.C., Callaway, R.M. 1996. Impact of a parasitic plant on the structure and dynamics of salt marsh vegetation. *Ecology* 77: 1410–19

Philpott, S.M., Maldonado, J., Vandermeer, J., Perfecto, I. 2004. Taking trophic cascades up a level: behaviorally-modified effects of phorid flies on ants and ant prey in coffee agroecosystems. *Oikos* 105: 141–7

Powell, E.N., Klinck, J.M., Ashton-Alcox, K., Hofmann, E.E., Morson, J. 2012. The rise and fall of *Crassostrea virginica* oyster reefs: the role of disease and fishing in their demise and a vignette on their management. *Journal of Marine Research* 70: 505–58

Preston, D.L., Mischler, J.A., Townsend, A.R., Johnson, P.T.J. 2016. Disease ecology meets ecosystem science. *Ecosystems* 19: 737–48

Price, P.W., Westoby, M., Rice B., Atsatt, P.R., Fritz, R.S., Thompson, J.N., Mobley, K. 1986. Parasite mediation in ecological interactions. *Annual Review of Ecology and Systematic* 17: 487–505

Raffel, T.R., Martin, L.B., Rohr, J.R. 2008. Parasites as predators: unifying natural enemy ecology. *Trends in Ecology and Evolution* 23: 610–18

Rosenblatt, A., Heithaus, M., Mather, M., Matich, P., Nifong, J., Ripple, W., Silliman B.R. 2013. The roles of large top predators in coastal ecosystems: new insights from long term ecological research. *Oceanography* 26: 156–67

Sala, E. 2006. Top predators provide insurance against climate change. *Trends in Ecology and Evolution* 21: 479–80

Schultz, J.A., Cloutier, R.N., Côté, I.M. 2016. Evidence for a trophic cascade on rocky reefs following sea star mass mortality in British Columbia. *Peer Journal* 4: e1980

Shi, W., Guo, Y., Xu, C., Tan, S., Miao, J., Feng, Y., Zhao, H., St Leger, R.J., Fang, W. 2014. Unveiling the mechanism by which microsporidian parasites prevent locust swarm behavior. *Proceedings of the National Academy of Sciences* 111: 1343–8

Silliman, B.R., Bertness, M.D. 2002. A trophic cascade regulates salt marsh primary production. *Proceedings of the National Academy of Sciences* 99: 10500–5

Silliman, B.R., Newell S.Y. 2003. Fungal-farming in a snail. *Proceedings of the National Academy of Science* 100: 15643–8

Silliman, B.R., Zieman, J.C. 2001. Top-down control of *Spartina alterniflora* production by periwinkle grazing in a Virginia salt marsh. *Ecology* 82: 2830–45

Silliman, B.R., Van De Koppel, J., Bertness, M.D., Stanton, L.E., Mendelssohn I.A. 2005. Drought, snails, and large-scale die-off of southern U.S. salt marshes. *Science* 5755: 1803–6

Silliman, B.R., McCoy, M.W., Angelini, C., Holt, R.D., Griffin, J.N., van de Koppel, J. 2013. Consumer fronts, global change, and runaway collapse in ecosystems. *Annual Review of Ecology, Evolution, and Systematics* 44: 503–38

Thingstad, T.F. 2000. Elements of a theory for the mechanisms controlling abundance, diversity, and biogeochemical role of lytic bacterial viruses in aquatic systems. *Limnology and Oceanography* 45: 1320–8

Thingstad, T.F., Lignell, R. 1997. Theoretical models for the control of bacterial growth rate, abundance, diversity and carbon demand. *Aquatic Microbial Ecology* 13: 19–27

Thomas, F., Renaud, F., de Meêus, T., Poulin, R. 1998. Manipulation of host behaviour by parasites: ecosystem engineering in the intertidal zone? *Procceedings of the Royal Society B* 265: 1091–6

Thomas, F., Poulin, R., De Meeüs, T., Guégan, J., Renaud, F. 1999. Parasites and ecosystem engineering: what roles could they play? *Oikos* 84: 167–71

Torchin, M.E., Lafferty, K.D., Kuris, A.M. 2002 Parasites and marine invasions. *Parasitology* 124:137–51

Wilmers, C.C., Post, E., Peterson, R.O., Vucetich, J.A. 2006. Predator disease out-break modulates top-down, bottom-up and climatic effects on herbivore population dynamics. *Ecology Letters* 9: 383–9

Wood, C.L., Byers, J.E., Cottingham, K.L., Altman, I., Donahue, M.J., Blakeslee, A.M.H. 2007. Parasites alter community structure. *Proceedings of the National Academy of Sciences* 104: 9335–9

Zhu, F., Poelman, E.H., Dicke, M. 2014. Insect herbivore-associated organisms affect plant responses to herbivory. *New Phytologist* 204: 315–21

zu Ermgassen, P.S.E., Spalding, M.D., Blake, B., Coen, L.D., Dumbauld, B., Geiger, S., Grabowski, J.H., Grizzle, R., Luckenbach, M., McGraw, K.A., Rodney, B., Ruesink, J.L., Powers, S.P., Brumbaugh, R.D. 2012. Historical ecology with real numbers: past and present extent and biomass of an imperiled estuarine ecosystem. *Proceedings of the Royal Society B* 279: 3393–400

SECTION 2

Drivers of Marine Disease

CHAPTER 4

Bacteriophage can drive virulence in marine pathogens

Mark Little*, Maria Isabel Rojas*, and Forest Rohwer

4.1 Introduction

Marine ecosystems around the globe are in dramatic decline due to anthropogenic impacts such as pollution, overfishing, climate change, and increasingly prevalent diseases in ecologically important macro-organisms (Bateman et al. Chapter 1, this volume, Burge and Hershberger Chapter 5, this volume, Bojko et al. Chapter 6, this volume). Many marine disease ecologists are reconsidering the strategies used to understand and investigate the etiology of these diseases. Recent studies incorporate the notion that an array of stressors can disrupt natural holobiont communities, leading to a variety of detrimental ecological outcomes often potentiated by microbial pathogenesis (Egan and Gardiner 2016; Morrow et al. 2018; Bateman et al. Chapter 1, this volume; Morton et al. Chapter 3, this volume). While microbial diseases can be caused by a variety of organisms, most bacteria involved in pathogenicity in marine environments contain horizontally acquired elements that are largely overlooked and play essential ecological and evolutionary roles. These elements are often carried by bacteriophage, or simply "phage", genomes integrated in the genome of the bacterial host. The expression of these prophage-encoded genes can confer pathogenicity and dysbiosis, the latter defined by an unbalanced composition of the host-associated microbial community.

Phages are viruses that infect bacteria and have the unique ability to undergo one of two lifestyles, lytic or lysogenic. In the lytic cycle, upon infection, the phage uses the cellular machinery of the bacterial host to replicate, synthesize new viral particles, and release its progeny, often killing the bacterial host by cell lysis (Echols 1972). During the lysogenic cycle, the phage genome is integrated into the bacterial genome and its replication occurs only as part of the normal cell cycle (reviewed in Young 1992). Interestingly, different environmental and cellular cues can trigger the switch from one cycle to the other (Wommack and Colwell 2000).

Through infection, phages have the capability to move host genes between bacteria. When fragments of the host chromosome are packaged within the viral particles, bacterial DNA is shared through infection to a recipient bacterial cell. This horizontal transfer is termed transduction and it can be the generalized transfer of a DNA sequence from a random position in the bacterial genome, or the specialized transfer of DNA from a specific location in the bacterial chromosome. In either case, the amount of packaged DNA is limited by the size of

* These authors contributed equally to this work

Little, M., Rojas, M.I., and Rohwer, F., *Bacteriophage can drive virulence in marine pathogens* In: *Marine Disease Ecology*. Edited by: Donald C. Behringer, Kevin D. Lafferty, and Brian. R. Silliman, Oxford University Press (2020). © Oxford University Press. DOI: 10.1093/oso/9780198821632.003.0004

the viral capsid and hence of the original viral genome (Cui et al. 2014). In addition to phages, other mechanisms of horizontal gene transfer between bacteria occur via the transfer of transposons and plasmids. Transposons are DNA sequences that can jump from one location in the bacterial genome to another, generating gene duplications or truncating genes when their insertion interferes with the coding sequence of the gene. Plasmids, on the other hand, are small DNA molecules, often circular and double-stranded, that can replicate independently of the bacterial genome and can be transferred between bacterial hosts through conjugation.

Phages, generalized transducing agents, transposable elements, and plasmids are some of the major drivers of microbial evolutionary processes and therefore likely play a key role in microbial pathogenicity and dysbiosis. This chapter focuses on the role of phage-encoded elements in the context of the etiology of economically and ecologically relevant marine pathogenesis and dysbiosis. This section also provides a meta-analysis of all known, fully sequenced, marine bacterial host-associated pathogenic and non-pathogenic genomes that serve as a baseline for understanding how the ecology of horizontal gene-transfer carried out by phage contributes to the evolution of marine pathogens.

Transduction in the marine environment occurs at high rates and has been suggested to have the minimum capacity to move 10^{24} genes from viruses to host per year globally (Rohwer and Thurber 2009). Within the past 5 years, the influence of horizontally acquired genetic elements from viruses has gained traction in the field of marine disease ecology. Recent analyses of the genomes of multiple pathogenic *Vibrio* strains revealed prophage-encoded elements that contribute to the pathogenicity of the bacteria (Weynberg et al. 2015; Box 4.1). These tripartite eukaryote–microbe–phage interactions likely determine many marine disease mechanisms. When a bacterium incorporates in its genome a viral genome, or acquires a prophage through infection, it is called a lysogen. Viral replication and survival occur through bacterial cell division, thus producing more lysogens as progeny. Phages that can initiate their incorporation into the chromosome of the host are known as temperate phages. It is well known that temperate phages have considerable gene repertoires that may enhance bacterial host fitness, and since phages are the most abundant biological entities on the planet, with an estimated 4.80×10^{31} phages on Earth, it is reasonable to predict that these viruses strongly influence the unfolding of marine pathogenesis across a variety

---

Box 4.1 Case study: The role of prophage in a coral pathogen

Vibrio coralliilyticus is a bacterial pathogen implicated in coral and oyster disease (see Behringer et al. Chapter 10, this volume). All five documented strains of *V. coralliilyticus* harbor at least one or multiple prophages and are implicated in diseases of the corals *Pocillopora damicornis*, *Montipora aequituberculata*, and *Acropora cytherea*, and of the Pacific oyster *Crassotrea gigas*. These bacterial strains are associated with different pathogenic phenotypes across coral species, such as bleaching and white syndrome (WS) (Kimes et al. 2012; Ushijima et al. 2014). Traditionally, 16S rDNA is used as taxonomic marker gene for bacterial species-level identification; however, the field of metagenomics and whole-genome sequencing is expanding the ability to analyze these genomes and understand the functional capacity of horizontal gene transfer between prophages and bacteria. Bioinformatic analyses show that *V. coralliilyticus* can carry integrated in its genome a complete prophage that codes for a zonal occludens toxin (ZOT) (Rohwer and Thurber 2009; Weynberg et al. 2015). This exotoxin gene has striking homology to that of *Vibrio cholerae* prophage-encoded cholera toxin (CTX) gene. *Vibrio cholerae* infection in humans causes cholera, an acute diarrheal illness that results from the incorporation and expression of the prophage CTXϕ in the bacterial genome (Waldor and Makalanos 1996). This homology suggests that ZOT could play a role in coral and oyster disease through a mechanism similar to that of CTX (Figure 4.1).

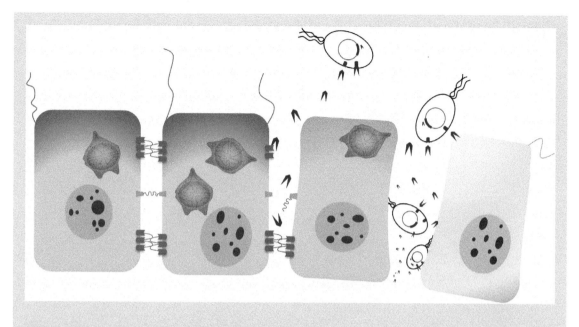

Figure 4.1 Prophage-encoded ZOT toxin of *Vibrio coralliilyticus* disrupting intercellular occluding junctions (tight junctions) that maintain integrity of epithelium during coral infection and disease.

of organisms and ecosystems (Cobián Güemes et al. 2016).

While prophages encompass around 25 percent of phages in the global phage gene pool, only forty-one prophage-mediated phenotypes have been observed or experimentally demonstrated (Bondy-Denomy and Davidson 2014; Casjens 2005). Prophages or temperate phages can enhance the fitness of their bacterial hosts in a variety of ways (Figure 4.2A–E), such as (1) conferring metabolic capacities through the acquisition of photosynthetic genes in *Cyanobacteria* (Rohwer and Thurber 2009), (2) encoding functional proteins such as anti-CRISPR systems in *Pseudomonas aeruginosa* which allow the bacteria to outcompete other bacteria (Bondy-Denomy et al. 2014), and (3) exclusion factors like the Imm protein of the famous phage T4 in *Escherichia coli* that prevents other phages from infecting the lysogen (Lu and Henning 1994; Obeng et al. 2016). Horizontally acquired mutualistic viruses therefore allow lysogens to broaden their ecological niche space

(Figure 4.2A–E). In many instances, the prophage can encode exotoxins that directly affect the host (Figure 4.2B). In addition, some prophage-encoded proteins have been shown to inhibit predation from bacterivorous protists (Figure 4.2D). We are only starting to shed light on the functional roles of integrated phage, but the fact that most bacterial genomes harbor about one to two prophages (Casjens 2003) indicates that these are significant players in a plethora of ecological dynamics. An assessment of temperateness, or the ability to initiate lysogeny or a lyosgenic conversion of viruses in seawater, revealed that within phage communities, 80 percent of the members contain the potential for a temperate lifestyle (Breitbart et al. 2004). Clearly, there are more functions to be discovered considering the high prevalence of lysogeny, where most bacterial genomes harbor multiple prophage and at a maximum have been observed to comprise 20 percent of bacterial genomic sequence space (Canchaya et al. 2003; Casjens 2005).

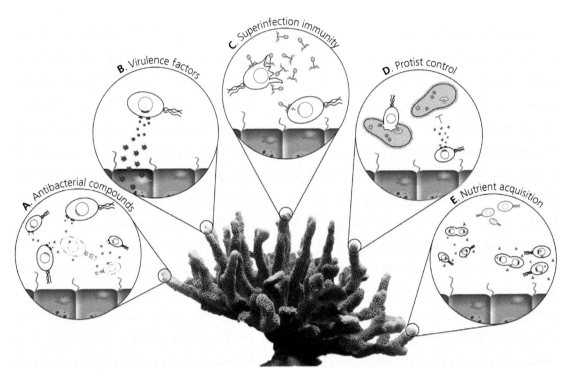

Figure 4.2 Conceptual overview of genes encoded by temperate phages that enhance lysogen fitness in ecologically relevant marine holobionts. **A.** Prophage-encoded proteins that inhibit competing bacteria and enable expansion of the lysogen's niche space. **B.** Prophage-encoded virulence factors that directly affect the host. **C.** Superinfection immunity prevents lytic control of lysogens by lytic phages. On mucosal metazoan epithelium the progeny of spontaneous prophage inductions can cause lytic infection in competing commensal or pathogenic non-lysogens (Barr et al 2013; Silveira and Rohwer 2016). **D.** Prophage-encoded proteins inhibit predation by unicellular protists that could have a negative secondary effect on the multicellular host. **E.** Prophage-encoded genes that allow bacteria to expand their metabolic repertoire and niche.

4.2 The role of prophages in disease

In 1951, the first report of phage-mediated virulence was described in the bacterium *Corynebacterium diphtheriae*, the disease-causing agent of diphtheria. When non-virulent strains of *C. diphtheriae* were challenged with phages, the next generation of bacterial progeny presented a virulent phenotype (Freeman 1951). Since this first discovery of prophage-mediated bacterial fitness enhancements, it has been revealed that a large portion of strain-to-strain differences are due to phage-mediated horizontal gene transfer (Lawrence 2002).

Bacterial strains exhibiting pathogenicity have been shown to contain a higher proportion of phage genes compared to non-pathogenic strains, and currently twelve prophages encoding virulence genes have been discovered among seven relevant bacterial

pathogens including *C. diphtheriae*, *E. coli*, *S. enterica*, *P. aeruginosa*, *S. mitis*, *C. jejuni*, and *V. cholerae* (Busby et al. 2013; Davies et al. 2016). Temperate phages can produce a variety of exotoxins such as cholera, Shiga toxin, and botulism, and these types of prophage-mediated functions are extremely relevant in the case of marine disease pathogenesis. Since the first phage-mediated phenotype was observed in diphtheria, the *E. coli* prophage system has been studied extensively. Certain *E. coli* prophages encode Shiga toxins (Stx) whose production is independent of phage lytic activity but require phage-mediated bacteriolysis for secretion. Conversely, in the case of *C. diphtheriae*, the production and secretion of the toxin does not require lysis of the lysogen (Holmes 2000).

Some bacterial toxins, many of which contribute to pathogenicity, likely evolved to evade predation from other micro-organisms, such as protists (Figure 4.2D).

An example of this survival strategy is the afore-mentioned Shiga toxin, which confers *E. coli* anti-predatorial defense against the bacteriovore *Tetrahymena thermophila* (Lainhart et al. 2009). In the marine environment, a study on *Serratia marcescens* challenged a population of this bacterium against two bacterivorous predators with different feeding mechanisms—*Acanthamoeba castellanii*, a surface feeder, and *Tetrahymena thermophila*, a particle feeder—and observed that the *S. marcescens* population became more resistant to the infection by lytic phages, presumably due to the acquisition of a prophage in their genome (Örmälä-Odegrip et al. 2015). These findings may suggest that predation pressure by bacterivores selects for bacteria carrying prophage in their genomes. These prophages potentially encode for proteins that either are directly toxic to bacteriovores or indirectly deter them from predating on the bacterial host. These phenomena are relevant when considering the ability of pathogens to evade protist predation as well as infection by lytic phages in the marine environment.

4.3 Prophages in marine diseases

Understanding the distribution and role of temperate phages in marine bacterial pathogens is of high relevance considering their ability to exhibit superinfection exclusion to other phages targeting the same host range. This is important to consider because it allows for pathogenic lysogens to prevent lytic control (i.e., bacterial death via cell lysis) by other phages naturally present or introduced to the community as phage therapeutics. These dynamics are likely at large in instances of bacterially mediated marine disease and pathogenesis. To test this hypothesis, we searched for predicted temperate phages in publicly available marine bacterial pathogen and non-pathogen genomes from host-associated marine environments on a global scale. This meta-analysis utilized bioinformatic tools to demonstrate significantly higher proportion of prophages in pathogenic than in non-pathogenic marine host-associated bacteria (Figure 4.3).

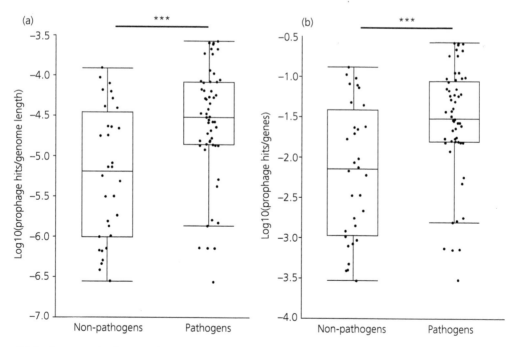

Figure 4.3 Prophage gene abundance in publicly available host-associated bacterial genomes in marine ecosystems. **A** Comparison of number of hits to predicted prophage-encoded genes in marine bacterial genomes normalized by the respective genome length for host-associated non-pathogens and pathogens. **B** Comparison of number of hits to predicted prophage-encoded genes in marine bacterial genomes normalized by total number of bacterial genes in the respective non-pathogen or pathogen classification. Data were accessed from public databases in March 2018. *** p ≤ 0.001; two-sample Wilcoxon test.

The results mentioned above were not unexpected since virus-like particles have been identified as playing a role in ecologically relevant marine diseases. For example, putative phage hyperparasites have been described to be associated with the bacterium *Candidatus Xenohaliotis californiensis* (WS-RLO), which causes withering syndrome in abalone (Friedman and Crosson 2012). These virus-like particles have similar morphology to that of the *Siphoviridae* family of phages, although their genomes have not been sequenced (Cruz-Flores et al. 2016). While the function of these intracellular viruses in pathogenesis is not yet understood, due to their 50-nm size and the pleomorphic traits conferred to bacteria, it is possible that they constitute generalized transducing agents. Furthermore, this system may be an example of how temperate phages play a role in bacterial pathogenesis.

4.4 Evolutionary implications of prophages in marine diseases

Temperate phages are present in forty to fifty percent of all known microbial genomes and in twenty-one of thirty known bacterial phyla (Canchaya et al. 2003; Touchon et al. 2016). In addition, lysogen abundance is more prevalent in pathogens and negatively correlated with spacer acquisition across CRISPR-Cas systems in respective host microbial genomes (Touchon et al. 2016). The contribution of phages in structuring microbial communities through predation and lysogeny is apparent, but the need for understanding functional roles in microbial disease ecology is often overlooked. Researchers have only begun to understand the roles of lysogens in nature in either pathogenic or ecological contexts. In this chapter we seek to shed light on what is known about phage-encoded function in bacterial pathogenesis and dysbiosis in marine disease.

While bacteriovore anti-predation conferred by horizontal gene transfer or phage infection has yet to be assessed in marine host-associated pathogenic bacteria, it is likely to explain the selection of systems that can lead to microbial disease in macro-organismal hosts. This process would require receptor-mediated endocytosis (RME), via cell-surface receptors of the eukaryotic host, where the exo-toxins against bacteriovore predation have a secondary effect on the macro-organismal host. These dynamics should be considered as marine disease ecologists investigate and attempt to address problems and develop or apply therapies in aquaculture and the environment. Here, prophages, the microbiological Trojan horse, are clearly an important overlooked component of disease etiologies. Opportunely, genomics and bioinformatics advance rapidly and allow for the evaluation of the contribution of prophage-encoded traits to the global pool of marine diseases (Frasca et al. Chapter 11, this volume).

The events associated with the acquisition of a potentially mutualistic phage require lytic-to-temperate switching during an active infection event and this is defined as lysogeny. Lysogeny can introduce novel phenotypes to the bacterial host via gene expression (Brüssow et al. 2004). Recent metagenomic-based studies on lytic-to-temperate switching dynamics in marine ecosystems demonstrate that at high microbial abundances and within certain environments, there is a higher presence of integrase and excisionase genes, suggesting an ecological lytic-to-temperate switch in these viral communities (Knowles et al. 2016). Integrases are phage-encoded proteins that enable viruses to incorporate their genome into bacterial chromosomes, while excisionases allow prophages to exit the chromosome during induction to the lytic cycle. Therefore, the relative level of integrase and excisionase genes in viral communities suggests the functional potential for lysogeny (Knowles et al. 2016). Mathematical modelling further demonstrates that temperate phage lifestyles can be more prevalent under environmental conditions that favor bacterial growth (Maslov and Sneppen 2015). Although marine bacterial pathogeneses are currently increasing at unprecedented rates, these are relatively isolated phenomena when considering the spatial scale at which organisms exist in any given ecosystem.

The life history of temperate phages can shed light on the understanding of how prophages are distributed across bacterial phyla. The abundance of bacterial mechanisms that protect against phage infection, such as CRISPR-Cas systems, has been negatively correlated with the presence of prophages in a global dataset of bacterial genomes (Touchon et al. 2016). In this study, the bacteria with minimal doubling time, meaning those with genomes with the fastest growing capability, were strongly correlated with the occurrence of lysogeny (Touchon et al. 2016). This finding adds further support to the observed increases of lysogeny in certain environments, as described in the Piggyback-the-Winner (PtW) model, where lysogen occurrence is correlated with higher microbial abundance (Knowles et al. 2016). In contrast, an alternative scenario can take place when virion numbers increase with bacterial abundance via lytic activity, and these dynamics are explained by the "Kill-the-Winner" model (Thingstad 2000). In either case, bet-hedging strategies of allocating bacterial resources to viral lytic production and others to lysogeny greatly influence ecological dynamics where phage impact bacterial growth and function (Morton et al. Chapter 3, this volume).

4.5 Meta-analysis of prophage-encoded functions

Prophage-encoded functions have yet to be assessed in pathogenic and dysbiotic host-associated system dynamics. This meta-analysis covers a total of eighty-nine publicly available complete marine genomes of host-associated bacteria, with sequences coming from thirty-two non-pathogenic and fifty-seven pathogenic isolates found in algae, invertebrates, and vertebrates (Klemetsen et al. 2018). The reference database used is the most extensive marine bacterial genome resource to date and includes comprehensive metadata on the source of the isolates, organismal pathology at the species level, bacterial genome length, data on encoded proteins per genome, and many other useful metadata (Klemetsen et al. 2018). The results of our analysis revealed a significantly higher abundance of prophage in marine pathogens when compared to non-pathogens (Figure 4.3).

A comparison of the number of hits to prophage-encoded genes between pathogenic and non-pathogenic host-associated marine bacteria revealed a significantly higher level of prophages per genome across two different normalization methods (Figure 4.3). The log abundance of prophage hits was divided by either the respective bacterial genome length (Figure 4.3A) or the predicted number of bacterial-encoded proteins in the respective genomes (Figure 4.3B). Both normalization methods revealed a significantly higher number of prophage genetic elements in the pathogenic host-associated bacteria (Figure 4.3).

To understand the functional relevance of these predicted prophage-encoded elements, we analyzed the results with a computer program (PhiSpy) designed to find prophages in bacterial genomes. The predicted prophage-encoded gene hits were annotated against the SEED project subsystems database, which integrates the data generated by the Rapid Annotation of microbial genomes using Subsystems Technology (RAST), to gain insight into what known functions may be prophage-encoded in these ecologically relevant marine pathogens. The SEED database uses a system that organizes families of functional genes into categories. To determine which functions were positively enriched in prophage-encoded genes carried by pathogenic bacteria, the relative abundance of subsystems was averaged, and the averaged abundance of non-pathogenic bacteria was subtracted from that of pathogenic bacteria (Figure 4.4). The three highest positively enriched subsystems in pathogens were carbohydrate utilization, membrane transport, and virulence. The observation of carbohydrate utilization and membrane transport suggests that prophages within pathogenic marine bacteria encode other functions, beyond virulence factors, that are ecologically interesting (e.g., fitness enhancement), and that have a direct or indirect effect on host disease. Interestingly, ~ 50 percent of the prophage hits did not match to any known function in the SEED subsystems database, which confirms that there is still much more to learn about these ecologically relevant prophages and prophage-encoded elements. The results suggest PtW dynamics, where the functional genes acquired via lysogeny would enable these pathogenic strains to outcompete the natural holobiont bacterial community members and ultimately lead to diseased states.

The tripartite eukaryote–microbe–phage dynamics discussed in this chapter are essential for marine biologists to consider when trying to understand the

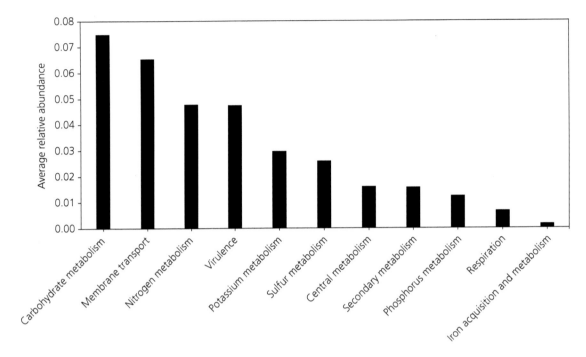

Figure 4.4 Subsystems level 1 functional profiles of prophage-encoded functions in publicly available host-associated bacterial genomes between non-pathogens and pathogens. Average abundance of subsystem level 1 genes of pathogens relative to the corresponding genes in non-pathogenic bacteria. Data were accessed from public databases in March 2018.

etiology and ecology of marine infectious diseases. While this chapter considers all fully sequenced host-associated marine bacterial pathogens at a more general level across global scales, it offers a baseline for future targeted studies on marine bacterial pathogenesis and prophage-encoded phenotypes. Future approaches include determining the specific protein or function that these genes encode to predict their role in fitness, pathogenicity, metabolism, competition, and microbial as well as ecological dynamics. This knowledge could inform researchers and clinicians when designing strategies to enrich desired bacterial traits, for instance to reverse dysbiosis in different ecosystems or directly defeat pathogenesis.

4.6 Summary

- Phages, the viruses that infect bacteria, can confer functions to the bacterial host that contribute to pathogenicity and dysbiosis through the lysogenic lifestyle.

- Bacterial pathogens in the marine environment contain higher abundances of prophages in their genome than non-pathogenic bacteria.
- A comparison of prophage genetic content between marine pathogens and non-pathogens revealed that pathogen-associated prophages are enriched in genes encoding for carbohydrate metabolism, membrane transport, nitrogen metabolism, virulence, and others.
- Horizontally acquired prophage-encoded DNA regions may play a large role in the ecology and evolution of marine diseases, due to the functions they confer.
- Future studies on the bacteria that are associated with diseases should examine the regions of prophages in the genome for insights into the etiology and ecology of pathogenicity.
- These approaches are relevant to non-genome-associated datasets (e.g., metagenomic data), and much is to be discovered about the ecology and evolution of prophages in diseases as these types of analyses become better explored and implemented.

Acknowledgments

We would like to thank Taylor O'Connell for developing the bioinformatic pipeline used to assess the genomes of interest for this chapter. This research was sponsored by the GBMF Investigator Award 3781 (to Forest Rohwer).

References

Barr, J. J., Auro, R., Furlan, M., Whiteson, K. L., Erb, M. L., Pogliano, J., Stotland, A., Wolkowicz, R., Cutting, A. S., Doran, K. S., Salamon, P., Youle, M., and Rohwer, F. L. 2013. Bacteriophage adhering to mucus provide a non-host-derived immunity. Proc Natl Acad Sci U S A, 110(26), 10771–6.

Bondy-Denomy, J. and Davidson, A. R. 2014. When a virus is not a parasite: the beneficial effects of prophages on bacterial fitness. J Microbiol 52, 235–42, doi:10.1007/s12275-014-4083-3.

Breitbart, M., Felts, B., Kelley, S., Mahaffy, J. M., Nulton, J., Salamon, P., and Rohwer, F. L. 2004. Diversity and population structure of a near-shore marine-sediment viral community. Proc Biol Sci 271, 565–74, doi:10.1098/rspb.2003.2628.

Brüssow, H., Canchaya, C., and Hardt, W. D. 2004. Phages and the evolution of bacterial pathogens: from genomic rearrangements to lysogenic conversion. Microbiol Mol Biol Rev 68, 560–602, table of contents, doi:10.1128/MMBR.68.3.560–602.2004.

Busby, B., Kristensen, D. M., and Koonin, E. V. 2013. Contribution of phage-derived genomic islands to the virulence of facultative bacterial pathogens. Environ Microbiol 15, 307–12, doi:10.1111/j.1462-2920.2012.02886.x.

Canchaya, C., Proux, C., Fournous, G., Bruttin, A., and Brüssow, H. 2003. Prophage genomics. Microbiol Molec Biol Rev 67(2): 238–76.

Casjens, S. R. 2003. Prophages and bacterial genomics: what have we learned so far? Molec Microbiol 49(2), 277–300.

Casjens, S. R. 2005. Comparative genomics and evolution of the tailed-bacteriophages. Curr Opin Microbiol 8(4): 451–8.

Cobián Güemes, A. G., Youle, M., Cantu, V. A., Felts, B., Nulton, J., and Rohwer, F. L. 2016. Viruses as winners in the game of life. Annu Rev Virol 3(1): 197–214.

Cruz-Flores, R., Caceres-Martinez, J., Munoz-Flores, M., Vasquez-Yeomans, R., Hernandez Rodriguez, M., Angel Del Rio-Portilla, M., Rocha-Olivares, A., and Castro-Longoria, E. 2016. Hyperparasitism by the bacteriophage (Caudovirales) infecting Candidatus Xenohaliotis californiensis (Rickettsiales-like prokaryote) parasite of wild abalone Haliotis fulgens and Haliotis corrugata from the peninsula of Baja California, Mexico. J Invertebr Pathol 140: 58–67.

Cui, J., Schlub, T. E., and Holmes, E. C. 2014. An allometric relationship between the genome length and virion volume of viruses. J Virol 88(11): 6403–10.

Davies, E. V., James, C. E., Williams, D., O'Brien, S., Fothergill, J. L., Haldenby, S., Paterson, S., Winstanley, C., and Brockhurst, M. A. 2016. Temperate phages both mediate and drive adaptive evolution in pathogen biofilms. Proc Natl Acad Sci U S A 113(29): 8266–71.

Echols, H., 1972. Developmental pathways for the temperate phage: lysis vs lysogeny. Annu Rev Genet 6(1): 157–90.

Egan, S. and Gardiner, M. 2016. Microbial dysbiosis: rethinking disease in marine ecosystems. Front Microbiol 7: 991.

Freeman, V.J. 1951. Studies on the virulence of bacteriophage-infected strains of Corynebacterium diphtheriae. J Bacteriol 61(6): 675.

Friedman, C. S. and Crosson, L. M. 2012. Putative phage hyperparasite in the rickettsial pathogen of abalone, "Candidatus Xenohaliotis californiensis". Microb Ecol 64(4): 1064–72.

Holmes, R.K. 2000. Biology and molecular epidemiology of diphtheria toxin and the tox gene. J Infect Dis 181(Suppl 1): S156–67.

Kimes, N. E., Grim, C. J., Johnson, W. R., Hasan, N. A., Tall, B. D., Kothary, M. H., Kiss, H., Munk, A. C., Tapia, R., Green, L., Detter, C., Bruce, D. C., Brettin, T. S., Colwell, R. R., and Morris, P. J. 2012. Temperature regulation of virulence factors in the pathogen Vibrio coralliilyticus. ISME J 6(4): 835–46. http://dx.doi.org/10.1038/ismej.2011.154.

Klemetsen, T., Raknes, I. A., Fu, J. Agafonov, A., Balasundaram, S. V., Tartari, G., Robertsen, E., and Willassen, N. P. 2018. The Mar databases: development and implementation of databases specific for marine metagenomics. Nucleic Acids Res 46(D1): D692–99.

Knowles, B., Silveira, C. B., Bailey, B. A., Barott, K., Cantu, V. A., Cobian-Guemes, A. G., Coutinho, F. H., Dinsdale, E. A., Felts, B., Furby, K. A., George, E. E., Green, K. T., Gregoracci, G. B., Haas, A. F., Haggerty, J. M., Hester, E. R., Hisakawa, N., Kelly, L. W., Lim, Y. W., Little, M., Luque, A., McDole-Somera, T., McNair, K., de Oliveira, L. S., Quistad, S. D., Robinett, N. L., Sala, E., Salamon, P., Sanchez, S. E., Sandin, S., Silva, G. G., Smith, J., Sullivan, C., Thompson, C., Vermeij, M. J., Youle, M., Young, C., Zgliczynski, B., Brainard, R., Edwards, R. A., Nulton, J., Thompson, F., and Rohwer, F. L. 2016. Lytic to temperate switching of viral communities. Nature 531, 466–70, doi:10.1038/nature17193.

Lainhart, W., Stolfa, G., and Koudelka, G. B. 2009. Shiga toxin as a bacterial defense against a eukaryotic predator, Tetrahymena thermophila. J Bacteriol 191(16): 5116–22.

Lawrence, J. G. 2002. Gene transfer in bacteria: speciation without species? Theor Popul Biol 61(4): 449–60.

Lu, M.J., and Henning, U. 1994. Superinfection exclusion by T-even-type coliphages. Trends Microbiol 2(4): 137–9.

Maslov, S. and Sneppen, K. 2015. Well-temperate phage: optimal bet-hedging against local environmental collapses. Sci Rep 5: 10523.

Morrow, K.M., Muller, E., and Lesser, M.P. 2018. How does the coral microbiome cause, respond to, or modulate the bleaching process? In: van Oppen, M.J.H. and Lough, J.M. (eds), Coral Bleaching: Patterns, Processes, Causes and Consequences, pp. 153–188. Cham, Switzerland: Springer International Publishing.

Obeng, N., Pratama, A. A., and van Elsas, J. D. 2016. The significance of mutualistic phages for bacterial ecology and evolution. Trend Microbiol 24(6): 440–9.

Örmälä-Odegrip, A. M., Ojala, V., Hiltunen, T., Zhang, J., Bamford, J. K., and Laakso, J. 2015. Protist predation can select for bacteria with lowered susceptibility to infection by lytic phages. BMC Evol Biol 15: 81.

Rohwer, F. L. and Thurber, R. V. 2009. Viruses manipulate the marine environment. Nature 459(7244): 207–12.

Silveira, C. B. and Rohwer, F. L. 2016. Piggyback-the-Winner in host-associated microbial communities. NPJ Biofilms Microbiomes 2: 16010.

Thingstad, T.F. 2000. Elements of a theory for the mechanisms controlling abundance, diversity, and biogeochemical role of lytic bacterial viruses in aquatic systems. Limnol Oceanogr 45(6): 1320–8.

Touchon, M., Bernheim, A., and Rocha, E. P. 2016. Genetic and life-history traits associated with the distribution of prophages in bacteria. ISME J 10(11): 2744–54.

Ushijima, B., Videau, P., Burger, A. H., Shore-Maggio, A., Runyon, C. M., Sudek, M., Aeby, G. S., and Callahan, S. M. 2014. *Vibrio coralliilyticus* strain Ocn008 is an etiological agent of acute montipora white syndrome. Appl Environ Microbiol 80(7): 2102–9.

Weynberg, K. D., Voolstra, C. R., Neave, M. J., Buerger, P., and van Oppen, M. J. 2015. From cholera to corals: viruses as drivers of virulence in a major coral bacterial pathogen. Sci Rep 5: 17889.

Wommack, K. E. and Colwell, R. R. 2000. Virioplankton: viruses in aquatic ecosystems. Microbiol Molec Biol Rev 64(1): 69–114.

Young, R. Y. 1992. Bacteriophage lysis: mechanism and regulation. Microbiol Molec Biol Rev 56(3): 430–81.

CHAPTER 5

Climate change can drive marine diseases

Colleen A. Burge and Paul K. Hershberger

5.1 Introduction

Disease is an episodic component of marine and terrestrial ecosystems where hosts and pathogens typically exist in homeostasis (Lafferty and Harvell 2014). Stable host–pathogen–environment relationships underlie the principles of disease ecology, but periodic disease outbreaks can result from host, pathogen, and environmental disruptions (Burge et al. 2018). Climate change operates in the "environment" component of the host–pathogen–environment triad, with changes in the environment affecting the host and/or pathogen (Figure 5.1). Additional anthropogenic factors act in the "environment" component (and can occur synergistically), such as habitat degradation (Raymundo et al. Chapter 9, this volume), pollution (Bojko et al. Chapter 6, this volume), and invasions (Pagenkopp Lohan et al. Chapter 7, this volume). Resulting disease outbreaks can affect natural and managed systems (Lafferty et al. 2015).

Changes in global and regional climate patterns became apparent in the mid-1900s, evidenced by independent observations of decreasing sea ice coverage, receding glaciers, reduced snow cover, and increases in sea level, ocean heat content, sea surface temperature, temperatures over the oceans, water vapor, and troposphere temperatures (Kennedy et al. 2010, Meillo et al. 2014). Many of these changes are driven by human activities associated with the Industrial Revolution that resulted in accelerated releases of heat-trapping greenhouse gases into the atmosphere, including carbon dioxide, methane, and nitrous oxide (Feely et al. 2009, Huber and Knutti 2012). Climate models forecast a 2.5–10 $^{\circ}$C increase in atmospheric temperatures over the next century. Ongoing landscape-level effects are observed as changes in sea ice cover; frequency of severe weather events; ocean acidity, temperature, and circulation patterns; wildfire frequency; and terrestrial growing seasons.

These ongoing and forecasted climate-driven changes will certainly influence the delicate balance that typically occurs between the host, pathogen, and environment triad. For many marine species, a shift in environment can drive disease, as seen for diseases impacting ectothermic non-mobile marine invertebrates such as corals, abalone, and oysters. Climate change may affect water quality and disrupt host–pathogen interactions (Burge et al. 2014), sometimes creating beneficial conditions for pathogen amplification and spread (Harvell et al. 2002, Altizer et al. 2013, Burge et al. 2014) or for microbial and ecological dysbiosis (Egan and Gardiner 2016). Water quality disruptions stemming from climate change that may have consequences on marine disease outbreaks include changes in temperature (increased or decreased), hypoxia, CO_2 accumulation

Burge, C.A., and Hershberger, P.K., *Climate change can drive marine diseases* In: *Marine Disease Ecology*. Edited by: Donald C. Behringer, Kevin D. Lafferty, and Brian. R. Silliman, Oxford University Press (2020). © Oxford University Press. DOI: 10.1093/oso/9780198821632.003.0005

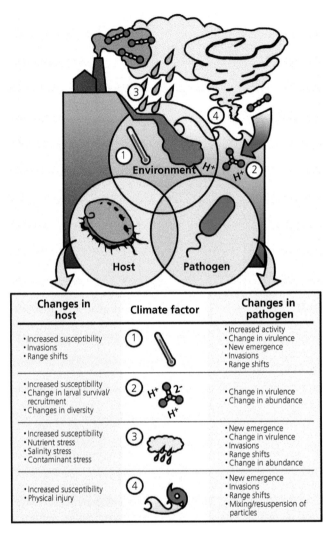

Changes in host	Climate factor	Changes in pathogen
• Increased susceptibility • Invasions • Range shifts	①	• Increased activity • Change in virulence • New emergence • Invasions • Range shifts
• Increased susceptibility • Change in larval survival/ recruitment • Changes in diversity	② H^+ $2-$ H^+	• Change in virulence • Change in abundance
• Increased susceptibility • Nutrient stress • Salinity stress • Contaminant stress	③	• New emergence • Change in virulence • Invasions • Range shifts • Change in abundance
• Increased susceptibility • Physical injury	④	• New emergence • Invasions • Range shifts • Mixing/resuspension of particles

Figure 5.1 Climate change impacts on marine host–pathogen–environment relationships. Shifts in the global environment are leading to physical ocean changes, including changes in temperature, increases in CO_2 concentrations/decreases in pH, changes in precipitation (leading to changes in salinity), and exposure to storms and cyclones. All of these factors are shifting the host–pathogen–environment equilibrium. (Adapted from Burge et al. (2014) with permission from the *Annual Review of Marine Science*, Volume 6, © 2014 by Annual Reviews, http://www.annualreviews.org; drawn by Reyn Yoshioka.)

(reduced pH), precipitation (leading to increased or decreased salinity), and storm and cyclone frequencies and intensities (Burge et al. 2014) (Figure 5.1). Changes in water quality may result in immediate impacts to host–pathogen equilibriums or long-term impacts such as range expansion of hosts (primary and secondary) and pathogens into new regions. For example, the range expansion of

Perkinsus marinus (the causative agent of dermo disease in *Crassostrea virginica*) coincided with a pronounced warming period in the mid-1980s (Burge et al. 2014). This chapter uses examples such as these to illustrate how climate change affects marine disease. Specifically, we identify several marine disease drivers associated with climate change that have the potential to alter marine host–pathogen interactions.

5.2 Marine host–pathogen ecophysiology

Host–pathogen interactions occur as part of the natural seascape, and as such the environment, including natural variability, influences marine disease processes. Here, we are most concerned with situations where anthropogenic change (climate change, coastal pollution, fishing, habitat fragmentation, etc.) might increase marine diseases. Host–pathogen interactions are commonly described as a coevolutionary arms race in which each partner is adapting to maximize its own reproductive output and fitness (Roy et al. 2009). However, currently little is known about the role host and pathogen evolution will play in changing environmental conditions (Altizer et al. 2013). Evolutionary outcomes might depend on host factors—including genetics, diversity of beneficial microbiome/mutualists, immunocompetency, nutrition, social behavior, life history, and density—or pathogen factors—including virulence, exposure level, strain, specificity, and ability to evolve. For both hosts and pathogens, changing abiotic factors and range shifts can translocate species, thereby leading to novel host–parasite interactions. Because climate change alters the conditions under which current host–pathogen interactions have evolved, we can expect the unexpected.

All organisms have evolved to fit a physiological niche. As such, understanding the ecophysiology, or the physiological optima within the host–pathogen relationships, is critical for identifying and forecasting how climate drives disease patterns (Lafferty 2009b). Changing conditions can benefit or impair hosts or pathogens, or both. Where physiological data for hosts and pathogens are available, climate models (Ben-Horin et al. Chapter 12, this volume) may be useful in understanding potential long-term impacts on marine host–pathogen interactions and may be used to estimate impacts on host susceptibilities, pathogen virulence, and host or pathogen range expansions. For example, Cohen et al. (2018) used climate projections provided by the Intergovernmental Panel on Climate Change (IPCC) and monthly sea surface temperature projections generated by the ECHAM5/MPI-OM model (Max-Planck Institute for Meteorology, Hamburg, Germany) to construct likely future disease scenarios

for several marine host–pathogen interactions in invertebrates and mammals in tropical, temperate, and polar regions (SRES A2; Meehl et al. 2007; IPCC 2007; broadly comparable to the IPCC 2014 RCP 8.5). Emission strategies such as the SRES (Special Report on Emissions Scenarios) A2 used by Cohen et al. (2018) are widely used in assessments of future climate change (IPCC 2007); recent IPCC reports provide updated scenarios presented as Representative Concentration Pathways (IPCC 2014).

Physiological optima are often defined by performance. In considering a host–pathogen relationship, pathogen growth and host immune or stress responses may be measured within a defined range of an abiotic parameter, often temperature, but factors such as salinity, pH, and hypoxia (and combinations of) have also been studied. For example, in sea fan corals, both maximal pathogen growth (Alker et al. 2004, Ward et al. 2007, Burge et al. 2012) and host immune response (Ward et al. 2007, Mydlarz et al. 2008) were detected at higher temperatures. However, increased temperatures in the Caribbean Sea likely favored the pathogens more than the sea fans (Harvell et al. 2009, Altizer et al. 2013). Defining physiological optima of both hosts and pathogens may be useful in understanding disease.

Trade-offs and constraints in oxygen- and capacity-limited thermal tolerance may limit species distribution and define fitness levels (reviewed by Pörtner et al. 2006). Though temperature is often considered a driving factor in both host and pathogen physiology, functional oxygen capacity is also an important aspect underlying thermal tolerance. For example, the concentrations of dissolved oxygen in seawater are inversely related to temperature, so high temperatures can effectively limit the aerobic capacity of respiring organisms. Water quality factors such as hypoxia, increased CO_2, and pollutants may alter performance optima (Pörtner 2012) and should be considered in addition to temperature.

The interaction between environmental factors and disease can be illustrated by comparing temperature optima for pathogens and potential host species (Figure 5.2; see Figure 2, Shields 2019). For a cold- or warm-water pathogen, a window of opportunity exists for the progression from low-level or covert

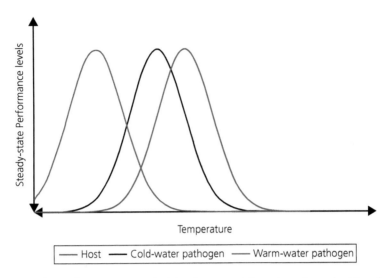

Figure 5.2 Theoretical marine host–pathogen temperature physiology. Temperature is increasing on the x-axis, while performance ranges from low to high on the y-axis. In blue, the "cold-water" pathogen has high performance at low temperature where both the host and warm-water pathogen have low performance; in red, the warm-water pathogen has the best performance at higher temperatures. Curve fit may vary based on host or pathogen type (Angilleta 2006) and a Gaussian curve (i.e., "bell-curve") has been used here for simplicity.

infection to overt disease and/or partial or mass mortality. Though disease is typically associated with temperature changes, more extreme temperature shifts (i.e., above optima for both the host and pathogen) may limit infection (see: Granja et al. 2003, Delisle et al. 2018). In the event of a mass mortality, other opportunistic pathogens may also be present and contribute to pathology and/or mortality (i.e., ostreid herpesvirus 1 (OsHV-1) infections/Pacific oyster mortality syndrome: De Lorgeril et al. 2018). Defining optima might be especially helpful in understanding diseases with strong seasonal fluctuations (i.e., occurring when mean or maximum temperatures are the greatest), which are common among marine ectotherms and plants. Acute disease expression may be linked with a seasonal change in temperatures (i.e., rapid virus- or bacteria-induced mortalities), but sustained disease and/or mortalities may also be linked with prolonged periods of enhanced winter temperatures (Burge et al. 2014, Thomas et al. 2018). For some pathogens, simple kinetics may describe an increase of disease with each proximate stressor (i.e., bacteria or fungi), while for others, passage through a specific stressor threshold may trigger disease progression (i.e., viral replication genes triggered enzyme activity). If the performance

profiles of the host and pathogen(s) mis-match, then the effects of a perturbation on a disease system are relatively straightforward to understand. However, if there is considerable diversity in tolerance (either phenotypic or genotypic) within and among host species, and/or among pathogen strains, the effect of a specific physiological stressor can be complex and difficult to forecast.

5.3 Proximate disease drivers influenced by climate change

5.3.1 Temperature

Temperature is the best-studied climate-related driver of marine disease. Although environmentally relevant changes in temperature rarely cause direct mortality to finfishes, subtle temperature changes affect physiological processes and various metrics of health and performance, including cellular membrane fluidity; membrane phospholipid and cholesterol compositions; water and solute balances; membrane passive permeability; Na^+/K^+–ATPase and gill morphologies; capacities for oxidative metabolism (enzymes, cell ultrastructure, myoglobin, etc.); oxygen solubility and transport; energy processes (microtubule dynamics and kerocyte movements); alphastat regulation; and swimming performance (Crockett and Londraville 2006). These and other physiological adaptations, along with their consequential impacts on the host transcriptome, influence host susceptibility to disease. Fishes and other ectotherms are most susceptible within a narrow temperature range that is disease- and host-specific (Figure 1 in Fryer and Pilcher 1974). These temperature-dependent susceptibilities are mediated by functional changes in the specific and non-specific host immune responses (Le Morvan et al. 1998, Abram et al. 2017) and pathogen regeneration times. As with other co-variables, these temperature-dependent changes can include a sudden shift towards overt disease.

Changing ocean temperatures influence disease processes by altering metabolic processes for pathogens and hosts. For example, warmer temperatures can decrease generation times and increase net reproductive rates for sea lice, thereby enhancing the potential for epizootics on marine fishes (Groner et al. 2014). From a host perspective, temperature directly influences ectothermic metabolic and bioenergetic processes. For example, a lack of sunlight and photosynthesis in polar and subpolar regions during overwinter periods creates a natural dearth of available food for planktivorous marine fishes. As a result, Pacific herring typically recede to cool deep waters during winter periods, lowering their metabolic rates to basal levels and subsisting on lipid reserves that were accumulated during active feeding periods in the previous summer and fall. Overwinter starvation can occur if the basal energetic demands during this fasting period outpace the available lipid reserves (Schultz and Conover 1999, Beamish and Mahnken 2001). The potential for overwinter starvation increases with warmer temperatures, which increase basal metabolic rates without providing a corresponding increase in food availability. Starvation potential is further exacerbated by parasitic infections; for example, infections with the endemic parasite *Ichthyophonus* sp. interrupt fat storage mechanisms and cause infected fish to begin their overwinter fasting period with 30 percent lower lipid reserves than uninfected fish (Vollenweider et al. 2011). As a result, the combined energetic demands of climate change and parasitic infections are likely to tip the bioenergetic balance in polar and subpolar marine systems in favor of host starvation.

There are many examples where temperature changes affect marine diseases (see Table 1 in Burge et al. 2014), with elevated temperatures resulting in disease outbreaks, including vibriosis in a variety of organisms including fish (e.g., *Vibrio ordalii* in marine salmonids; Actis et al. 2011), shellfish (i.e., *Vibrio harveyi* infections in European abalone; Travers et al. 2009), and corals (i.e., *Vibrio shiloi*-induced coral bleaching; Kushmaro et al. 1998). Cooler temperatures or so-called cold-water diseases predispose populations to other diseases, including viral hemorrhagic septicemia (VHS) of marine fishes including Pacific herring (Hershberger et al. 2016) and Denman Island disease of Pacific oysters (Hervio et al. 1996). For multiple disease syndromes of corals and temperature-induced (non-infectious) bleaching, disease outbreaks increase with warming (Lafferty et al. 2004, Harvell et al. 2009), though both warm and cold temperatures may be linked with coral disease outbreaks (Ruiz-Moreno et al. 2012). Marine heat waves (Hobday et al. 2016) can also be related to disease outbreaks in oysters; Green et al. (2019) recently demonstrated how a simulated marine heat wave could shift the microbiome of Pacific oysters and cause mass mortalities associated with *Vibrio*. Taken together, changes in temperature—both warming and cooling temperature—may impact marine disease outbreaks in the future.

Climate change is expected to affect the annual and decadal variation in coastal ocean temperatures driven by natural climatic phenomena including El Niño/La Niña Southern Oscillations (ENSO) and Pacific Decadal Oscillation (Royer 1998). These regime changes influence marine disease processes (Vikas and Dwarakish 2015) and result in ecological consequences that cascade to all organizational levels, ranging from individuals through ecosystems. Among animals, these impacts are realized first in ectotherms, whose physiological processes are closely tied to ambient water temperatures. For example, withering syndrome (WS) in California abalone was first observed in black abalone populations, shortly after the 1982–1983 ENSO event (Tissot 1995). WS was originally thought to be linked to starvation, but patterns of spread suggested an infectious agent (Lafferty and Kuris 1993). Die-offs proceeded during warm and cold periods, though mortality rates were higher at high temperatures. Subsequent warm-water or ENSO events were associated with severe losses in farmed red abalone and elevated disease signs in wild species (reviewed in Burge et al. 2014). Further, laboratory studies demonstrated a link between temperature variability and transmission of WS and the causative agent, *"Candidatus Xenohaliotis californiensis"* (Ben-Horin et al. 2013), and between mean temperature and mortality rate in infected individuals (Moore et al. 2000, Braid et al. 2005, Vilchis et al. 2005). An increased frequency in ENSO events under climate change could lead to increased mortality rates due to disease in both abalone affected by WS and other marine organisms in areas affected by large-scale regime changes.

5.3.2 Other proximate drivers of disease

In addition to temperature, other drivers of marine disease in the host, pathogen, and environment triad are affected by climate change, including salinity, ocean acidification, and storm frequency/intensity (Box 5.1).

Recent analyses of temporal patterns in ocean salinities indicate that the global water cycle is intensifying rapidly (Durack et al. 2012), largely because increased atmospheric temperatures create elevated air moisture contents. Changes in seawater salinities are expected to impact the susceptibilities of marine fishes and shellfishes to diseases by affecting bioenergetic demands, osmoregulatory functions, and ion transport mechanisms. Some impacts have already been realized; for example, recruitment failures of Pacific herring in Prince William Sound, AK (USA) consistently occur during years of high freshwater discharge into the nearby Gulf of Alaska (Ward et al. 2017). Non-mobile marine invertebrates are particularly influenced by changing salinity regimes, with oyster diseases such as MSX and dermo typically becoming exacerbated during periods of drought that are accompanied by high salinities (Ford 1985, Bushek et al. 2012). Similarly, in Pacific oysters exposed to OsHV-1, higher salinity conditions are associated with increased mortalities

Box 5.1 Diseases affect organisms in transition

Diseases can occur when organisms undergo rapid transitions between environmental temperature extremes, such as when anadromous fishes transition from their seawater rearing environment into the freshwater spawning realm. This environmental transition is accompanied by osmoregulatory changes, bioenergetic shifts from immunological defenses towards reproductive efforts, and other metabolic stressors. In the case of Pacific salmon and other semelparous fishes, this terminal transition can start a physiological race between successful spawning and pre-programmed adult mortality. Elevated freshwater temperatures and/or pathogens can tip the conditions of this race in favor of pre-spawning mortality. For example, late-run sockeye salmon entered the Fraser River, BC (Canada) 2–6 weeks earlier than expected during 1999 and 2000; despite this early river entry, spawn timing remained unchanged. Pre-spawn mortality from *Parvicapsula minibicornis* disease occurred, presumably because the extended freshwater transience period provided sufficient incubation time for the infections to progress to lethal levels (St-Hilaire et al. 2002). Extended energy depletion for these non-feeding adults could have also played a role in pre-spawning mortality. Similarly, returning adult Chinook salmon in the Yukon River, AK (USA) experienced pre-spawning mortality during the late 1990s and early 2000s in association with an epizootic of *Ichthyophonus* sp. disease. The rapid progression of this marine disease during the terminal freshwater phase was hypothesized to be associated with elevated riverine temperatures (Kocan et al. 2004).

(Fuhrmann et al. 2016). Non-infectious diseases of fishes and other marine organisms, including those caused by harmful algal blooms (HABS), are highly dependent on marine salinity profiles (Fu et al. 2002); interestingly, phytoplankton, including those that cause HABS, are the taxa showing the largest latitudinal range shift concurrent with climate change (> 400 km per decade; reviewed in Howes et al. 2015). Changes in salinity, though less well understood than temperature, will both directly and indirectly affect marine disease.

Climate-driven increases in storm frequency and intensity may lead to increases in freshwater input from terrestrial runoff or overcapacity issues at wastewater treatment facilities. Increasing terrestrial runoff can lead to reduced salinity, resuspension of sediments, increased sewage pollution, and increased nutrients. *Vibrio* pathogens of marine organisms may be more pervasive after storm events as favorable conditions are created for the bacteria (reviewed in Burge et al. 2014). For marine hosts, lower salinity waters may affect host susceptibility and create more hospitable conditions for the pathogen. For organisms in highly variable environments impacted by freshwater releases, such as estuaries, an influx of low salinity and pathogens may impact host physiology and survival (Fuhrmann et al. 2016, 2018). Epizootic releases of excess freshwater and untreated/undertreated sewage can represent strong risk factors for some coral diseases (Wear and Thurber 2015, Shore-Maggio et al. 2018). Severe storms can also impact the structural integrity of ecosystem engineers such as corals and oysters. For example, coral skeletal structures may be weakened through direct breakage, abrasion, and surface injuries which may provide entry points for pathogens (Burge et al. 2014).

Ocean acidification (OA), a direct effect of climate change, can impact various aspects of organism health. OA is primarily caused by increased CO_2 in the atmosphere becoming dissociated in seawater, driving a decrease in ocean pH and shifts in carbonate speciation (Doney et al. 2009). The ocean is absorbing 25 percent of the atmospheric CO_2 (Le Quéré et al. 2018), thereby impacting the seawater carbonate system, which is also influenced by temperature, salinity, and pressure (Howes et al. 2015). In marine organisms, OA can affect membrane transport, calcification, photosynthesis, neuronal processes, growth, reproductive success, survival, and behavior (Kroeker et al. 2013, Howes et al. 2015). The effects of OA on disease or host–pathogen interactions are less well documented and not included in recent meta-analyses of OA effects on organisms or microbial processes in the ocean (Kroeker et al. 2013, Nagelkerken and Connell 2015, Nagelkerken and Munday 2016); however, exposure to low pH is known to affect immune and physiological responses in various organisms including marine invertebrates such as species of urchins (Brothers et al. 2016), mussels (Ellis et al.

2015), oysters (Cao et al. 2018, Fuhrmann et al. 2019), finfishes (Bresolin de Souza et al. 2014, Kreiss et al. 2015, Mota et al. 2019), and seagrasses (Arnold et al. 2014, Groner et al. 2018). Calcifying organisms including corals, bivalves, and pteropods are particularly vulnerable to OA (reviewed in Doney et al. 2009, Kroeker et al. 2013), where effects are mediated by a reduced saturation state (Ω) of calcium carbonate ($CaCO_3$) leading to abnormal deposition of calcium carbonate or increasing dissolution (Gazeau et al. 2007).

The re-emergence of the larval bacterial pathogen *Vibrio tubiashii* (syn. *V. corallyticus*) in 2006 and 2007 in the Northeast Pacific coincided with mixing of unusually warm surface seawater and cooled upwelled waters enriched with CO_2, nutrients, and *Vibrio* (Elston et al. 2008); a follow-up study indicated that growth of *V. tubiashii* is greater at low pH (Dorfmeier 2012). Interestingly, in another study focused on *V. tubiashii* infections of mussels (*Mytilus edulis*), *V. tubiashii* was more successful in infecting both mussels and hemocytes exposed to low pH conditions (Asplund et al. 2014), though no impact was detected on growth of *V. tubiashii* and/or measured immune parameters. In marine finfish, elevated partial pressures of CO_2 (hypercarbia) can impact innate immune responses by activating the complement system and stimulating the inflammatory response. When in combination with low ambient temperatures, hypercarbia can also influence adaptive immune responses by downregulating the IgM heavy chain constant region (Bresolin de Souza et al. 2014). The concomitant exposure of marine fish to hyperbarbia and low temperature can also result in an increased gill tissue mass (Kreiss et al. 2015) and decreased epidermal thickness (Mota et al. 2019). By affecting these first and second lines of defense in finfish, these changes likely influence host susceptibility to infectious and parasitic disease.

Unfortunately, assigning causation between climate change, these proximate factors, and the resulting disease processes is virtually impossible using a traditional, reductionist scientific approach (Box 5.2); therefore, creative investigative approaches are required before effective adaptive mitigation strategies can be offered.

Box 5.2 Limitations in linking climate change and marine disease patterns

Limitations and constraints in the scientific method make it hard to confirm causation between climate change and disease patterns (Lafferty 2009a). First, it is difficult to demonstrate how climate change influences proximate disease factors such as temperature, salinity, and host assemblages (e.g., Thomas et al. 2018). Furthermore, cause-and-effect relationships linking these factors to disease are difficult to demonstrate, and scientists must often rely on a weight-of-evidence approach involving epidemiological associations. Nevertheless, some current and forecasted effects of climate change will likely affect the appearance, frequency, and magnitude of marine diseases. Therefore, long-term ecological studies examining the consequences of climate–disease interactions at community and ecosystem scales are crucial for documenting disease impacts and developing management strategies, including forecasting models, to mitigate disease impacts (Altizer et al. 2013).

5.4 Summary

Drivers are logical but evidence is limited:

- Large-scale forces, including climate change, can interrupt the host–pathogen–environment homeostasis, leading to disease.
- Linking a proximate "climate driver" to a specific disease is difficult to demonstrate, often relying on a weight-of-evidence approach.
- Range shifts or translocation of hosts or pathogens to new locations may lead to increased disease.
- Disease affects organisms when changes move the host–pathogen–environment triad out of balance.
- Multi-stressor syndromes and/or microbial dysbiosis can make assigning disease causation difficult, if not impossible.

Specific needs to improve our understanding of climate-related drivers of marine disease include:

- Improved understanding of host–pathogen optima as they relate to climate change-related water quality factors such as changes in temperature, CO_2, and salinity.

- Conducting long-term ecological studies examining the consequences of climate–disease interactions on the scale of communities and ecosystems.
- Improved monitoring and detection of disease to enhance management and forecasting models.
- Employing adaptive management strategies to mitigate disease impacts, including climate–disease forecasting models.

References

Abram, Q.H., B. Dixon, and B.A. Katzenback. 2017. Impacts of low temperature on the teleost immune system. *Biology* 6(4): pii: E39.

Actis, L.A., M.E. Tolmasky, and J.H. Crosa. 2011. Vibriosis. In: P.T.K. Woo and D.W. Bruno (eds) *Fish Diseases and Disorders Vol 3: Viral, Bacterial and Fungal Infections*, 2nd edition, pp. 570–605. CAB International, Cambridge, MA.

Alker, A.P., K. Kim, D.H. Dube, and C.D. Harvell. 2004. Localized induction of a generalized response against multiple biotic agents in Caribbean Sea fans. *Coral Reefs* 23(3): 397–405. doi:10.1007/s00338-004-0405-y.

Altizer, S., R.S. Ostfeld, P.T. Johnson, S. Kutz, and C.D. Harvell. 2013. Climate change and infectious diseases: from evidence to a predictive framework. *Science* 341(6145): 514–19.

Angilletta Jr, M.J. 2006. Estimating and comparing thermal performance curves. *Journal of Thermal Biology* 31(7): 541–5.

Arnold, T., G. Freundlich, T. Weilnau, A. Verdi, and I.R. Tibbetts. 2014. Impacts of groundwater discharge at Myora Springs (North Stradbroke Island, Australia) on the phenolic metabolism of eelgrass, *Zostera muelleri*, and grazing by the juvenile rabbitfish, *Siganus fuscescens*. *PLoS One* 9(8): e104738.

Asplund, M.E., S.P. Baden, S. Russ, R.P. Ellis, N. Gong, and B.E. Hernroth. 2014. Ocean acidification and host–pathogen interactions: blue mussels, *Mytilus edulis*, encountering *Vibrio tubiashii*. *Environmental Microbiology* 16:1029–39.

Beamish, R.J. and C. Mahnken. 2001. A critical size and period hypothesis to explain natural regulation of salmon abundance and the linkage to climate and climate change. *Progress in Oceanography* 49: 423–43.

Ben-Horin, T., H.S. Lenihan, and K.D. Lafferty. 2013. Variable intertidal temperature explains why disease endangers black abalone. *Ecology* 94: 161–8.

Braid, B.A., J.D. Moore, T.T. Robbins, R.P. Hedrick, R.S. Tjeerdema, and C.S. Friedman. 2005. Health and survival of red abalone, *Haliotis rufescens*, under varying temperature, food supply, and exposure to the agent of withering syndrome. *Journal of Invertebrate Pathology* 89(3): 219–31.

Bresolin de Souza, K., F. Jutfelt, P. Kling, L. Forlin, and J. Sturve. 2014. Effects of increased CO_2 on fish gill and plasma proteome. *PLoS One* 9: e102901.

Brothers, C., J. Harianto, J. McClintock, and M. Byrne. 2016. Sea urchins in a high-CO_2 world: the influence of acclimation on the immune response to ocean warming and acidification. *Proceedings of the Royal Society B: Biological Sciences* 283: 20161501.

Burge, C.A., N.L. Douglas, I. Conti-Jerpe, E. Weil, S. Roberts, C.S. Friedman, and C.D. Harvell. 2012. Friend or foe: the association of Labyrinthulomycetes with the Caribbean sea fan *Gorgonia ventalina*. *Diseases of Aquatic Organisms* 101: 1–12.

Burge, C.A., C.M. Eakin, C.S. Friedman, B. Froelich, P.K. Hershberger, E.E. Hoffman, L.E. Petes, K.C. Prager, E. Weil, B.L. Willis, S.E. Ford, and C.D. Harvell. 2014. Climate change influences on marine infectious disease: implications for management and society. *Annual Review in Marine Science* 6: 249–77.

Burge, C.A., A. Shore-Maggio, and N.D. Rivlin. 2018. Ecology of emerging infectious diseases of invertebrates. In: A. Hajek and D. Shapiro (eds) *Ecology of Invertebrate Diseases*. Wiley, Chichester.

Bushek, D., S.E. Ford, and I. Burt. 2012. Long-term patterns of an estuarine pathogen along a salinity gradient. *Journal of Marine Research* 70: 225–51.

Cao, R., Y. Liu, Q. Wang, D. Yang, H. Liu, W. Ran, Y. Qu, and J. Zhao. 2018. Seawater acidification reduced the resistance of *Crassostrea gigas* to *Vibrio splendidus* challenge: an energy metabolism perspective. *Frontiers in Physiology* 9: 880.

Cohen, R.E., C.C. James, A. Lee, M.M. Martinelli, W.T. Muraoka, M. Ortega, R. Sadowski, L. Starkey, A.R. Szesciorka, S.E. Timko, E.L. Weiss, and P.J.S. Franks. 2018. Marine host–pathogen dynamics: Influences of global climate change. *Oceanography* 31(2): 182–93.

Crockett, E.L. and R.L. Londraville. 2006. Temperature. In: D.H. Evans and J.B. Claiborne (eds) *The Physiology of Fishes*, 3rd edition, pp. 231–70. Taylor & Francis, Boca Raton, FL.

Delisle, L., B. Petton, J.F. Burguin, B. Morga, C. Corporeau, and F. Pernet. 2018. Temperature modulate disease susceptibility of the Pacific oyster *Crassostrea gigas* and virulence of the ostreid herpesvirus type 1. *Fish & Shellfish Immunology* 80: 71–9.

De Lorgeril, J., A. Lucasson, B. Petton, E. Toulza, C. Montagnani, C. Clerissi, J. Vidal-Dupiol, C. Chaparro, R. Galinier, J.M. Escoubas, P. Haffner, L. Degremont, G.M. Charriere, M. Lafont, A. Delort, A. Vergnes, M. Chiarello, N. Faury, T. Rubio, M.A. Leroy, A. Perignon, D. Regler, B. Morga, M. Alunno-Bruscia, P. Boudry, F. Le Roux, D. Gestoumieux-Garzon, Y. Gueguen, and G. Mitta. 2018. Immune-suppression by OsHV-1 viral infection causes fatal bacteraemia in Pacific oysters. *Nature Communications* 9(1): 4215. doi:10.1038/s41467-018-06659-3.

Doney, D.C., V.J. Fabry, R.A. Feely, and J.A. Kleypas. 2009. Ocean acidification: the other CO_2 problem. *Annual Review of Marine Science* 1: 169–92.

Dorfmeier, E.M. 2012. Ocean acidification and disease: how will a changing climate impact *Vibrio tubiashii* growth and pathogenicity to Pacific oyster larvae? Master thesis. University of Washington, Seattle.

Durack, P.J., S.E. Wijffels, and R.J. Matear. 2012. Ocean salinities reveal strong global water cycle intensification during 1950 to 2000. *Science* 336: 455–8.

Egan, S., and M. Gardiner. 2016. Microbial dysbiosis: rethinking disease in marine ecosystems. *Frontiers in Microbiology* 7: 991.

Ellis, R.P, S. Widdicombe, H. Parry, T.H. Hutchinson, and J.I. Spicer. 2015. Pathogenic challenge reveals immune trade-off in mussels exposed to reduced seawater pH and increased temperature. *Journal of Experimental Marine Biology and Ecology* 462: 83–9.

Elston, R.A., H. Hasegawa, K.L. Humphrey, I.K. Polyak, and C.C. Häse. 2008. Re-emergence of *Vibrio tubiashii* in bivalve shellfish aquaculture: severity, environmental drivers, geographic extent and management. *Diseases of Aquatic Organisms* 82(2): 119–34.

Feely, R.A., S.C. Doney, and S.R. Cooley. 2009. Ocean acidification: present conditions and future changes in a high CO_2 world. *Oceanography* 22: 36–47.

Ford, S.E. 1985. Effects of salinity on survival of the MSX parasite *Haplosporidium nelson* (Haskin, Stauber, and Mackin) in oysters. *Journal of Shellfish Research* 5: 85–90.

Fryer, J.S. and K.S. Pilcher. 1974. Effects of temperature on diseases of salmonid fishes. US Environmental Protection Agency, Ecological Research Series EPA-660/73–020, Washington, DC.

Fu, F.X., A.O. Tatters, and D.A. Hutchins. 2002. Global change and the future of harmful algal blooms in the ocean. *Marine Ecology Progress Series* 470: 207–33.

Fuhrmann, M., B. Petton, V. Quillien, N. Faury, B. Morga, and F. Pernet. 2016. Salinity influences disease-induced mortality of the oyster *Crassostrea gigas* and infectivity of the ostreid herpesvirus 1 (OsHV-1). *Aquaculture Environment Interactions* 8: 543–52.

Fuhrmann, M., L. Delisle, B. Petton, C. Corporeau, and F. Pernet. 2018. Metabolism of the Pacific oyster, *Crassostrea gigas*, is influenced by salinity and modulates survival to the ostreid herpesvirus OsHV-1. *Biology Open* 7(2): bio028134.

Fuhrmann, M., G. Richard, C. Quere, B. Petton, and F. Pernet. 2019. Low pH reduced survival of the oyster

Crassostrea gigas exposed to the ostreid herpesvirus 1 by altering the metabolic response of the host. *Aquaculture* 503: 167–74.

Gazeau, F., C. Quiblier, J.M. Jansen, J.P. Gattuso, J.J. Middelburg, and C.H. Heip. 2007. Impact of elevated CO_2 on shellfish calcification. *Geophysical Research Letters* 34: 1–5.

Granja, C.B., L.F. Aranguren, O.M. Vidal, L. Aragon, and M. Salazar. 2003. Does hyperthermia increase apoptosis in white spot syndrome virus (WSSV)-infected *Litopenaeus vannamei*? *Diseases of Aquatic Organisms* 54: 73–8.

Green, T.J., N. Siboni, W.L. King, M. Labbate, J.R. Seymour, and D. Raftos. 2019. Simulated marine heat wave alters abundance and structure of *Vibrio* populations associated with the Pacific oyster resulting in a mass mortality event. *Microbial Ecology* 77(3): 736–47.

Groner, M.L., G. Gettinby, M. Stormoen, C.W. Revie, and R. Cox. 2014. Modelling the impact of temperature-induced life history plasticity and mate limitation on the epidemic potential of a marine ectoparasite. *PLoS One* 9: e88465.

Groner, M.L., C.A. Burge, R. Cox, N.D. Rivlin, M. Turner, K.L. Van Alstyne, S. Wyllie-Echeverria, J. Bucci, P. Staudigel, and C.S. Friedman. 2018. Oysters and eelgrass: potential partners in a high pCO_2 ocean. *Ecology* 99(8): 1802–14.

Harvell, C.D., C.E. Mitchell, J.R. Ward, S. Altizer, A.P. Dobson, R.S. Ostfeld, and M.D. Samuel. 2002. Climate warming and disease risk for terrestrial and marine biota. *Science* 296: 2158–62.

Harvell, C.D., S. Altizer, I.M. Cattadori, L. Harrington, and E. Weil. 2009. Climate change and wildlife diseases: when does the host matter the most? *Ecology* 90: 912–20.

Hershberger, P.K., K.A. Garver, and J.R. Winton. 2016. Principles underlying the epizootiology of viral hemorrhagic septicemia in Pacific herring and other fishes throughout the North Pacific Ocean. *Canadian Journal of Fisheries and Aquatic Sciences* 73: 853–9.

Hervio, D., S.M. Bower, and G.R. Meyer. 1996. Detection, isolation, and experimental transmission of *Mikrocytos mackini*, a microcell parasite of Pacific oysters *Crassostrea gigas* (Thunberg). *Journal of Invertebrate Pathology* 67(1): 72–9.

Hobday, A.J., L.V. Alexander, S.E. Perkins, D.A. Smale, S.C. Straub, E.C. Oliver, J.A. Benthuysen, M.T. Burrows, M.G. Donat, M. Feng, and N.J. Holbrook. 2016. A hierarchical approach to defining marine heatwaves. *Progress in Oceanography* 141: 227–38.

Howes, E.L., F. Joos, M. Eakin, and J.P. Gattuso. 2015. An updated synthesis of the observed and projected impacts of climate change on the chemical, physical and biological processes in the oceans. *Frontiers in Marine Science* 2: 36.

Huber, M. and R. Knutti. 2012. Anthropogenic and natural warming inferred from changes in Earth's energy balance. *Nature Geoscience* 5: 31–6.

IPCC. 2007. Climate Change 2007: Synthesis Report. Contribution of Working Groups I, II and III to the Fourth Assessment Report of the Intergovernmental Panel on Climate Change. Core Writing Team, R.K. Pachauri, and A. Reisinger, eds. IPCC, Geneva.

IPCC. 2014. Climate Change 2014: Synthesis Report. Contribution of Working Groups I, II and III to the Fifth Assessment Report of the Intergovernmental Panel on Climate Change. Core Writing Team, R.K. Pachauri, and L.A. Meyer, eds. IPCC, Geneva.

Kennedy, J.J., P.W. Thorne, T.C. Peterson, R.A. Reudy, P.A. Stott, D.E. Parker, S.A. Good, H.A. Titchner, and K.M. Willett. 2010. How do we know the world has warmed? State of the Climate in 2009. *Bulletin of the American Meteorological Society* 91: S26–7

Kocan, R.M., P.K. Hershberger, and J. Winton. 2004. Ichthyophoniasis: an emerging disease of Chinook salmon (*Oncorhynchus tshawytscha*) in the Yukon River. *Journal of Aquatic Animal Health* 16: 58–72.

Kreiss, C.M., K. Michael, M. Lucassen, F. Julfelt, R. Motyka, S. Dupont, and H.O. Pörtner. 2015. Ocean warming and acidification modulate energy budget and gill ion regulatory mechanisms in Atlantic cod (*Gadus morhua*). *Journal of Comparative Physiology B* 185: 767–81.

Kroeker, K.J., R.L. Kordas, R. Crim, I.E. Hendriks, L. Ramajo, G.S. Singh, C.M. Duarte, and J.P. Gattuso. 2013. Impacts of ocean acidification on marine organisms: quantifying sensitivities and interaction with warming. *Global Change Biology* 19(6); 1884–96.

Kushmaro, A., E. Rosenberg, M. Fine, Y.B. Haim, and Y. Loya, 1998. Effect of temperature on bleaching of the coral *Oculina patagonica* by Vibrio AK-1. *Marine Ecology Progress Series* 171: 131–7.

Lafferty, K.D. 2009a. Calling for an ecological approach to studying climate change and infectious diseases. *Ecology* 90: 932–3.

Lafferty, K.D. 2009b. The ecology of climate change and infectious diseases. *Ecology* 90: 888–900.

Lafferty, K.D. and C.D. Harvell. 2014. The role of infectious disease in marine communities." In: M.D. Bertness, J.F. Bruno, B.R. Silliman, and J.J. Stachowicz (eds) *Marine Community Ecology and Conservation*, pp. 85–108. Sinauer Associates, Sunderland, MA.

Lafferty, K.D. and A.M. Kuris. 1993. Mass mortality of abalone *Haliotis cracherodii* on the California Channel Islands: tests of epidemiological hypotheses. *Marine Ecology—Progress Series* 96: 239.

Lafferty, K.D., J.W. Porter, and S.E. Ford. 2004. Are diseases increasing in the ocean? *Annual Review of Ecology, Evolution and Systematics* 35: 31–54.

Lafferty, K.D., C.D. Harvell, J.M. Conrad, C.S. Friedman, M.L. Kent, A.M. Kuris, E.N. Powell, D. Rondeau, and S.M. Saksida. 2015. Infectious diseases affect marine fisheries and aquaculture economics. *Annual Reviews in Marine Science* 7: 471–96.

Le Morvan, C., D. Troutaud, and P. Deschaux. 1998. Differential effects of temperature on specific and non-specific immune defenses in fish. *Journal of Experimental Biology* 201: 165–8.

Le Quéré, C., R.M. Andrew, P. Friedlingstein, S. Sitch, J. Pongratz, A.C. Manning, et al. 2018. Global carbon budget 2017. *Earth System Science Data* 10: 405–48.

Meehl, G.A., C. Covey, T. Delworth, M. Latif, B. McAvaney, J.F. Mitchell, R.J. Stouffer, and K.E. Taylor. 2007. The WCRP CMIP3 multimodel dataset: a new era in climate change research. *Bulletin of the American Meteorological Society* 88(9): 1383–94.

Melillo, J.M., T.C. Richmond, and G.W. Yohe (eds). 2014. *Climate Change Impacts in the United States: The Third National Climate Assessment*. US Global Change Research Program, Washington, DC.

Moore, J.D., T.T. Robbins, and C.S. Friedman. 2000. Withering syndrome in farmed red abalone *Haliotis rufescens*: thermal induction and association with a gastrointestinal rickettsiales-like prokaryote. *Journal of Aquatic Animal Health* 12(1): 26–34.

Mota, V.C., T.O. Nilsen, J. Gerwins, M. Gallo, E. Ytteborg, G. Baeverfjord, J. Kolarevic, S.T. Summerfelt, and B.F. Terjesen. 2019. The effects of carbon dioxide on growth performance, welfare, and health of Atlantic salmon post-smolt (*Salmo salar*) in recirculating aquaculture systems. *Aquaculture* 498: 578–86.

Mydlarz, L.D., S.F. Holthouse, E.C. Peters, and C.D. Harvell. 2008. Cellular responses in sea fan corals: granular amoebocytes react to pathogen and climate stressors. *PLoS One* 3(3).

Nagelkerken, I. and S.D. Connell. 2015. Global alteration of ocean ecosystem functioning due to increasing human CO_2 emissions. *Proceedings of the National Academy of Sciences* 112(43): 13272–7.

Nagelkerken, I. and P.L. Munday. 2016. Animal behavior shapes the ecological effects of ocean acidification and warming: moving from individual to community-level responses. *Global Change Biology* 22: 974–89.

Pörtner, H.O. 2012. Integrating climate-related stressor effects on marine organisms: unifying principles linking molecule to ecosystem-level changes. *Marine Ecology Progress Series* 470: 273–90.

Pörtner, H.O., A.F. Bennett, F. Bozinovic, A. Clarke, M.A. Lardies, M. Lucassen, B. Pelster, F. Schiemer, and J.H. Stillman. 2006. Trade-offs in thermal adaptation: the need for a molecular to ecological integration. *Physiological and Biochemical Zoology* 79(2): 295–313.

Roy, H.E., R.S. Hails, H. Hesketh, D.B. Roy, and J.K. Pell. 2009. Beyond biological control: non-pest insects and their pathogens in a changing world. *Insect Conservation and Diversity* 2(2): 65–72.

Royer, T. 1998. Coastal ocean processes in the northern North Pacific. In: K.H. Brink and A.R. Robinson (eds) *The Sea*, Vol 11, pp. 395–414. Wiley, New York.

Ruiz-Moreno, D., B.L. Willis, A.C. Page, E. Weil, A. Cróquer, B. Vargas-Angel, A.G. Jordan-Garza, E. Jordán-Dahlgren, L. Raymundo, and C.D. Harvell. 2012. Global coral disease prevalence associated with sea temperature anomalies and local factors. *Diseases of Aquatic Organisms* 100(3): 249–61.

Schultz, E.T. and D.O. Conover. 1999. The allometry of energy reserve depletion: test of a mechanism for size-dependent winter mortality. *Oecologia* 119(4): 474–83.

Shields J.D. 2019. Climate change enhances disease processes in crustaceans: case studies in lobsters, crabs, and shrimps, *Journal of Crustacean Biology*, 1-11 https://doi.org/10.1093/jcbiol/ruz072

Shore-Maggio, A., G.S. Aeby, and S.M. Callahan. 2018. Influence of salinity and sedimentation on *Vibrio* infection of the Hawaiian coral *Montipora capitata*. *Diseases of Aquatic Organisms* 128: 63–71.

St-Hilaire, S., M. Boichuk, D. Barnes, M. Higgins, R. Devlin, R. Withier, J. Khattra, S. Jones, and D. Kieser. 2002. Epizootiology of *Parvicapsula minibicornis* in Frasier River sockeye salmon, *Oncorhynchus nerka* (Walbaum). *Journal of Fish Diseases* 25: 107–20.

Thomas, Y., C. Cassou, P. Gernez, and S. Pouvreau. 2018. Oysters as sentinels of climate variability and climate change in coastal ecosystems. *Environmental Research Letters* 13: 104009.

Tissot, B.N. 1995. Recruitment, growth, and survivorship of black abalone on Santa Cruz Island following mass mortality. *Bulletin of the Southern California Academy of Sciences* 94(3): 179–89.

Travers, M.A., O. Basuyaux, N. Le Goïc, S. Huchette, J.L. Nicolas, M. Koken, and C. Paillard. 2009. Influence of temperature and spawning effort on *Haliotis tuberculata* mortalities caused by *Vibrio harveyi*: an example of emerging vibriosis linked to global warming. *Global Change Biology* 15(6): 1365–76.

Vikas, M. and G.S. Dwarakish. 2015. El Niño: a review. *International Journal of Earth Sciences and Engineering* 8: 130–7.

Vilchis, L.I., M.J. Tegner, J.D. Moore, C.S. Friedman, K.L. Riser, T.T. Robbins, and P.K. Dayton. 2005. Ocean warming effects on growth, reproduction, and survivorship of southern California abalone. *Ecological Applications* 15(2): 469–80.

Vollenweider, J.J., J. Gregg, R.A. Heintz, and P.K. Hershberger. 2011. Energetic cost of *Ichthyophonus* infection in juvenile Pacific herring (*Clupea pallasii*). *Journal of Parasitology Research* 2011: 926812.

Ward, E.J., M. Adkison, J. Couture, S.C. Dressel, M.A. Litzow, S. Moffitt, T. Hoem Neher, J. Trochta, and R. Brenner. 2017. Evaluating signals of oil spill impacts, climate, and species interactions in Pacific herring and Pacific salmon populations in Prince William Sound and Copper River, Alaska. *PLoS One* 12(3): e0172898.

Ward, J.R., K. Kim, and C.D. Harvell. 2007. Temperature affects coral disease resistance and pathogen growth. *Marine Ecology Progress Series* 329:115–21.

Wear, S.L. and R. Vega Thurber. 2015. Sewage pollution: mitigation is key for coral reef stewardship. *Annals of the New York Academy of Sciences* 1355: 15–30.

CHAPTER 6

Pollution can drive marine diseases

Jamie Bojko, Erin K. Lipp, Alex T. Ford, and Donald C. Behringer

6.1 Introduction

The marine environment covers over two-thirds of the planet, is a substantial food source for humans, and includes some of the most iconic ecosystems on Earth (Powles et al. 2000; Stentiford et al. 2012). Ecological research taking place in marine environments began as early as the 1850s, with Charles Darwin's interest in barnacle evolution and distribution (Bertness et al. 2014). Our early understanding of fisheries stocks and viability also began early in the field of marine ecology, with published research available around the 1910s (Bertness et al. 2014). Continued marine research by Smith (1941), who researched sponge disease, and by ZoBell and Feltham (1942), who identified marine bacteria as important ecological components, identified that disease influenced ecosystem health and function in the ocean. This formed the beginnings of modern-day marine disease ecology. Further development of the field (reviewed by Ketchum (1970)) linked human pollution with chemical and biological contamination of ocean fisheries, increasing our initial understanding of human impacts and health threats related to marine disease ecology.

Disease is influential in marine ecosystems and can alter ecosystem services, cause fluctuations in biomass, and change the dynamics of succession and ecological stability (Preston et al. 2016). Disease can in turn be influenced by environmental variation, such as changes in temperature, dissolved oxygen, and organic material (Ward and Lafferty 2004). Pollution from human activities has resulted in dramatic changes to environmental variables in the ocean and thus has the potential to heavily impact marine disease (Islam and Tanaka 2004).

In this chapter, we review how human-derived pollution affects the epidemiology and ecology of marine diseases. Our scope includes: 1) the zoonotic and pathological effects of pathogens and parasites influenced by pollution; 2) changes to disease susceptibility and prevalence in marine species; and 3) how human impacts have changed natural host–parasite/pathogen dynamics and disease ecology more broadly—all in reference to biological, chemical, and physical pollution (Figure 6.1). This chapter excludes information about human impacts on disease from fisheries and aquaculture (Behringer et al. Chapter 10, this volume) and climate change (Burge and Hershberger Chapter 5, this volume). One of the major highlights of our review is that while it is clear that environmental variation, disease, and pollution interact in many ways, there is little strong empirical evidence—with the proper use of controls and an adequate combination of data sources—to support causal linkages between pollution and disease outbreaks (Wear and Vega-Thurber 2015). To remedy this, there has been a call for collaboration between ecologists, chemists, and pathologists to "battle a common enemy" and explore the interactions

Bojko, J., Lipp, E.K., Ford, A.T., and Behringer, D.C., *Pollution can drive marine diseases* In: *Marine Disease Ecology*. Edited by: Donald C. Behringer, Kevin D. Lafferty, and Brian. R. Silliman, Oxford University Press (2020). © Oxford University Press.
DOI: 10.1093/oso/9780198821632.003.0006

Figure 6.1 Conceptual diagram of the three types of pollution (biological, chemical, and physical) presented in this chapter, and the main constituents of each. Biological pollution consists of toxins (including toxin-producing pathogens and contaminants), biohazard (feces/sewage-originating pathogens), and excessive nutrient input (sewage and fertilizer runoff). Chemical pollution is discussed using examples of industrial waste (e.g., plastics), pharmaceutical waste, and heavy/radioactive metal waste. Physical pollution is discussed using examples of garbage, excessive light and sound, and excessive freshwater waste due to river damning, sewage release, and storm drain release.

between pollution and disease from multiple angles using rigorous experimental, environmental, and pathological approaches (Wear 2019).

6.2 Biological pollution

Biological pollution is the excessive release of untreated water and waste, which can introduce disease agents or spur their proliferation in marine environments. A by-product of human civilization is a "cocktail" of waste that includes numerous biological components (Wear and Vega-Thurber 2015). Health concerns for humans and ecosystems include the production, disposal, and treatment of fecal waste derived from humans and domesticated animals. As high as two-thirds of the human population does not have access to clean water, highlighting the high risk of contaminated water to human health (Eliasson 2015). Although fecal waste is a resource for biofuel and agriculture (i.e., soil nutrient amendments), it also introduces excessive nutrients, heavy metals, toxic contaminants, and infectious agents to receiving waters (Section 6.3). Fecal waste in marine waters can introduce disease-causing agents (Section 6.2.1); drive food web transfer of infectious agents (Section 6.2.2); and cause a destabilization of coastal ecosystems (i.e., through eutrophication), resulting in the emergence of disease (Section 6.2.3).

6.2.1 Direct effects of waste-associated pathogens

Pathogens can enter the marine environment through sewage and fecal waste. Globally, 2,212 km^3

(2.12×10^{15} L) of fecal waste is released into the environment each year (Connor et al. 2017). In the US, 85 percent of effluent is discharged into coastal waters (primarily bays and estuaries) (National Research Council 1993). Wastewater directly discharged to US surface water requires treatment to at least secondary standards (i.e., reducing biological oxygen demand, suspended solids, and fecal indicator bacteria), but septic systems, informal cess pits, and direct runoff from land can introduce significant levels of untreated or minimally treated waste directly to streams, rivers, and coastal zones (National Research Council 1993).

Sewage and fecal waste contain on average 10^7–10^9 cells of bacteria ml^{-1} (Ewert et al. 1980), $> 10^8$ ml^{-1} virus-like particles (Bettarel et al. 2000), and $> 10^4$ parasite cysts or oocysts L^{-1} (Sharafi et al. 2012). These include potentially pathogenic microbes that can have significant impacts on both human and animal health and marine disease transmission dynamics. Humans are exposed to enteric pathogens, generally causing gastrointestinal illness, through recreation (i.e., swimming), work (i.e., fishing), and contaminated seafood. Fecal indicator bacteria are used to establish risk thresholds for gastrointestinal illness associated with swimming and the safety of harvesting shellfish from marine waters (according to standards set by the EPA, FDA, WHO, or EU). Most human illnesses associated with exposure to marine waters are due to viral infections. Norovirus, the most common cause of gastroenteritis, has been found at $> 10^5$ L^{-1} of untreated wastewater (Kazama et al. 2016) and presents the greatest risk of infection associated with swimming (McBride et al. 2013). Viruses, such as norovirus and hepatitis A, can also accumulate in shellfish and have resulted in 83.7 percent and 12.3 percent of human outbreaks of these viruses worldwide, respectively (Bellou et al. 2013). These examples show that protists, bacteria, and viruses in sewage and fecal matter can contaminate the marine environment and make it back into the human food chain.

Given the volume of waste introduced into marine waters, even a small number of pathogens or their toxins (biochemicals) suggests a non-trivial risk for exposure and impact. While most microbes in sewage are non-pathogenic, given difficulties in assigning clear etiologies to many marine diseases

(Sutherland et al. 2016) it is likely that we do not yet fully understand the direct potential for disease associated with human waste in marine systems. For example, at least one variant of white pox disease in the threatened Caribbean elkhorn coral (*Acropora palmata*) is caused by the opportunistic bacterium *Serratia marcescens* (Sutherland et al. 2010; Sutherland et al. 2010) and molecular typing has shown that disease-associated strains in the Florida Keys matched those in human wastewater (Sutherland et al. 2010). Similarly, the bacterium *Edwardsiella tarda* introduced from wastewater causes "hemorrhagic and necrotic disease" or "piscine septicemia" in fish (Mohanty and Sahoo 2007). Such examples reveal that marine species can sustain diseases introduced through sewage and fecal waste.

6.2.2 Indirect effects of waste-associated pathogens and toxins on marine food webs

Human sewage and animal waste constitute a sustained source of pathogens and toxins to marine systems, where they can further be transferred through the food web. This can include the accumulation of human pathogens in shellfish. In some cases, pathogens are preferentially retained, as observed among filter feeders, thereby increasing contamination risk. For example, various species of marine bacteria from the genus *Vibrio* have been shown to easily bioaccumulate and proliferate in oysters and other shellfish (Murphree and Tamplin 1995; Warner and Oliver 2008; Froelich et al. 2013). Furthermore, norovirus receptors in oysters increase specific attachment of viral particles to the digestive tract tissue, benefitting the bioaccumulation of this pathogen (Le Guyader et al. 2013). Enteric bacteria (*Escherichia coli* and *Salmonella*) can undergo temporary accumulation and release via oyster pseudofeces (Timoney and Abston 1984). There is also some evidence that enteric bacteria and viruses may preferentially attach to algal cells (Gentry et al. 2009), and increase the risk of bioaccumulation.

Bioaccumulation of pathogens in shellfish does not only affect humans. One notable example of pathogen transmission among marine fauna from consumption of bivalve shellfish is toxoplasmosis, which causes brain infections in threatened southern

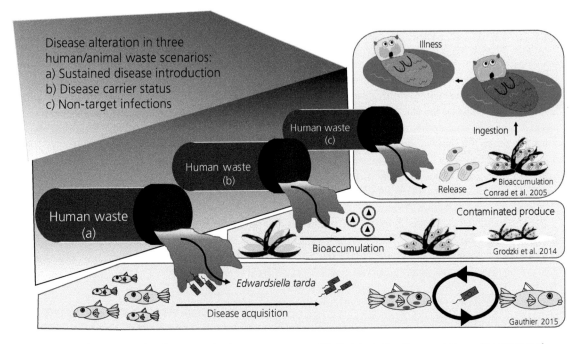

Figure 6.2 A graphic representation of some ways that human sewage can result in the introduction, bioaccumulation. and transmission of disease in the marine environment. Scenario (a) uses an example of *Edwardsiella tarda* (bacterial agent), which can be released from untreated human sewage into the marine environment and result in a self-sustaining piscine disease known as "hemorrhagic and necrotic disease" or "piscine septicemia" (Gauthier 2015). Scenario (b) explains the bioaccumulation of hepatitis E virions from human sewage by marine shellfish, which may then pass the disease back to humans via ingestion of undercooked shellfish (Grodzki et al. 2014). Scenario (c) explores how *Toxoplasma* sp. infections can end up in non-target hosts (sea otters) after release of runoff containing infected cat feces (Conrad et al. 2005).

sea otters in California (*Enhydra lutris nereis*) (Miller et al. 2002). The apicomplexan parasite *Toxoplasma gondii* is a broad zoonotic agent, which uses the domestic cat as a definitive host. Evidence suggests that urban freshwater runoff (including feces from feral and outdoor cats) has introduced *T. gondii* to coastal waters, where the oocysts bioaccumulate in bivalves, which are consumed by sea otters (Conrad et al. 2005) (Figure 6.2). The sea otters thus act as an alternative definitive host in the marine environment. Additionally, the oocysts can be aggregated with marine particles, as has been seen for bacteria and viruses, which enhances their offshore transport dynamics and uptake by filter feeders (Shapiro et al. 2012).

6.2.3 Nutrient pollution and disease

Nutrient enrichment, or eutrophication, results from excessive nutrients entering aquatic ecosystems and leading to increased primary production by plants or algae. Nutrients from human and animal waste are important but are often dwarfed in magnitude by the introduction of nutrients from agricultural production. The annual, global application of the three main fertilizer nutrients (nitrogen, phosphorus, and potassium) was nearly 190 million metric tons in 2015 (FAO 2017). Excessive nutrient inputs have drastically altered biogeochemical cycles, especially for nitrogen (Galloway et al. 2004; Galloway et al. 2008). In the future, this problem is likely to be most pronounced in tropical and subtropical waters, where nutrients are normally low, enrichment is pervasive, and pathogen abundance is considered to be relatively high (Johnson et al. 2010). While we must acknowledge that fertilizers have increased agricultural production and decreased human malnutrition, thereby reducing human disease, the excessive use of fertilizers imposes consequences for coastal marine environments by altering pathogen or disease dynamics.

Eutrophication has direct and indirect effects on disease dynamics, as summarized by Johnson et al. (2010) and more recently by Civitello et al. (2018). Effects include: 1) altered host/parasite traits (e.g., increased pathogen survival, increased host susceptibility to disease); 2) altered host/pathogen population densities; 3) altered parasite productivity or pathogenicity; 4) increased tolerance or duration of the parasite; and 5) increased pathogenicity due to increased parasite resources or altered host tolerance. These effects are interdependent and not mutually exclusive (Table 6.1). For example, a herbivorous host benefitting from an increase in nutritional condition resulting from greater algal abundance may be more tolerant of pathogens and parasites. However, an increase in host nutritional condition may also be of benefit to the parasite, increasing parasite density

(Johnson and Carpenter 2008). These factors may vary in response to the type, size, and immune system of the host. Most pronounced is the effect of nutritional condition on parasite abundance in vertebrate versus invertebrate hosts. Parasites are more likely to respond to the increased nutritional condition of invertebrate hosts rather than vertebrate hosts, likely due to the vertebrate adaptive immune system; however, the smaller relative size of invertebrates, and their greater likelihood of resource limitation, may factor into this pattern (Cressler et al. 2013).

Our understanding of how the process of nutrient enrichment drives patterns in disease outbreaks suffers from limited experimental evidence. In early 2000, it was reported that *Aspergillus sydowii*, a fungus of terrestrial origin, was the causative agent

Table 6.1 Direct and indirect effects of nutrient supplementation on disease dynamics of marine organisms across levels of organization (adapted from Civitello et al. 2018). "Level of effect" refers to an individual, population-level, or entire ecological community-level impact of the ecological change. "Host traits" refers to the susceptibility or niche of the host(s) related to the system described. The final three columns explore the effects of increased nutrient input, the ecological outcomes of nutrient input, and effects on disease dynamics.

Level of effect	Host traits	Effect of increased nutrient input	Ecological outcomes	Disease-related outcomes
Individual	Susceptible host	Increased survival, susceptibility, or fecundity	Increased resources Increased condition	Increased resistance to infection Increased production of susceptible hosts
Individual	Infected host	Increased survival, infectiousness, or fecundity	Increased resources Increased condition	Bears heavier parasite burden Survives longer Infects more susceptible hosts Increases production of susceptible hosts
Population	Susceptible host	Increased aggregation and density	Abundance increases Aggregate where food is abundant	Density-dependent infection increases Prevalence increases
Population	Infected host	Increased aggregation and density	Abundance increases Aggregate where food is abundant	Density-dependent transmission increases
Community	Susceptible host	Increase in competitors or predators	Competition for resources increases Predation increases	Population decreases Density-dependent infection decreases
Community	Infected host	Increase in competitors or predators	Competition for resources increases Predation increases	Population decreases Density-dependent transmission decreases
Community	Competitor	Increased density	Competition increases	Population increases or decreases depending on outcome of competition
Community	Predator	Increased density	Population increases Condition increases Reproduction increases	Population increases, furthering impact on host populations

behind epizootics of aspergillosis among sea fans *Gorgonia ventilina* throughout the Caribbean, with mortality at some locations > 50 percent (Kim and Harvell 2004). Nutrient enrichment experiments using time-release fertilizer demonstrated increased aspergillosis severity in sea fans, but also accelerated tissue loss from yellow band disease in the scleractinian corals *Orbicella franksi* and *Orbicella annularis* (Bruno et al. 2003). Similar links between nutrient enrichment and disease in *Siderastrea siderea* coral have been noted (Vega Thurber et al. 2013). These links suggest a complicated relationship between excessive nutrient presence and the progression of disease.

When nutrient enrichment is combined with another stressor, such as overfishing, the combination can be devastating to coral reefs. This can occur via: increasing algae cover; destabilizing normal microbiomes (the bacterial community naturally present on or in an organism); enhancing the load of putative coral pathogens; and increasing the likelihood of disease-induced mortality from injuries, such as parrotfish grazing bites (Zaneveld et al. 2016). Eutrophication in combination with further stressors constitutes an additional worry to marine animal pathology.

Nutrient enrichment has been shown to drive disease dynamics in other marine ecosystems, such as seagrass meadows. Wasting disease caused by the heterokont parasite *Labyrinthula zosterae* has been reported since the 1930s and is linked to pollution (including nutrient enrichment) and to a loss of biodiversity (Orth et al. 2006). Nitrate enrichment has been experimentally shown to increase infection susceptibility of eelgrass *Zostera marina* to infection by *L. zosterae*, but not by another heterokont parasite, *Aplanochytrium yorkensis*; however, reasons why are still being explored (Hughes et al. 2018). These examples provide important evidence for the alternation of disease epidemiology dynamics in marine communities associated with eutrophication.

The effects of nutrient enrichment can also be exacerbated or mitigated by biotic (e.g., predation, competition, density dependence) or abiotic (e.g., temperature, rainfall, CO_2) factors (Table 6.1). For example, from a bottom-up community control perspective, nutrient enrichment might increase the abundance of parasites if their herbivorous host

increases in abundance, but this could be tempered with a concomitant increase in predators of that host. Abiotic factors, some driven by climate change, could directly or indirectly affect nutrient inputs and tip the balance in favor of host or parasite. Drought or increased rainfall in a region could indirectly affect disease by altering the volume of nutrient-laden runoff or riverine input, or the physiochemical conditions of the water (Section 6.4.2).

Finally, runoff from farming efforts also results in the introduction of herbicides (e.g., triazine), fungicides (e.g., captan), and pesticides (e.g., malathion), which have all been shown to have ecotoxicological effects through suppression of immune function (Galloway and Depledge 2001). Experimental studies demonstrating links between exposure and altered disease susceptibility among marine organisms are increasing, but field studies that demonstrate cause and effect in a natural setting are uncommon and complicated by the many interactive factors that drive disease susceptibility. Simultaneous exposure to the herbicide "Diuron" increased infection by both seagrass disease species (*L. zosterae* and *A. yorkensis*), presumably because the herbicide weakened the plants and increased their susceptibility to infection (Hughes et al. 2018). Runoff therefore contains multiple components (such as herbicides) that can affect marine disease via altered host susceptibility.

Nutrient enrichment and pesticide runoff are problems for coastal ecosystems and are projected to get worse, especially in low-latitude regions where marine communities are least often exposed to high-nutrient concentrations. The direct effects of eutrophication on ecosystem health and community composition are clearly present (Vitousek et al. 1997; Pandolfi et al. 2005), but we are now beginning to understand how these factors influence disease dynamics. These effects are often indirect, which highlights the importance of experimental studies using adequate control systems to disentangle the complex interactions behind empirical observations.

6.3 Chemical pollution

Chemical pollution is the release of industrial or pharmaceutical compounds at higher than natural levels into the marine environment, which can influence the development and transmission of disease. For decades there has been an understanding that pollution can alter host–parasite interactions and epidemiology (Khan and Thulin 1991; Vidal-Martinez et al. 2010; Marcogliese and Pietrock 2011). These interactions either increase the effects of parasites and disease agents through direct tissue damage and immunosuppression of hosts, or conversely reduce parasitism by way of adversely affecting sensitive parasite life stages (Khan and Thulin 1991; MacKenzie 1999; Sures 2004; Blanar et al. 2009). Synopses of available data suggest that endoparasites generally decrease under pollution pressures and ectoparasites increase, likely due to the susceptibility of their various transmission stages, whereby ectoparasites are in contact with the environment and endoparasites only spend a short time in the environment in which they transmit and may be less tolerant due to this (Williams and MacKenzie 2003). Studies highlight that there are several exceptions to these rules (Sueiro et al. 2017) that vary with taxa and are thought to be associated with parasite tolerance and physiology of the transmissive stages (Blanar et al. 2009).

In this section, we explore chemical pollution and its link to disease emergence, disease prevalence, and overall ecological consequences from the introduction of chemical waste.

6.3.1 Heavy metals and radioactive materials

Heavy metals are one of the more ubiquitous pollutants in the marine environment. During the early stages of environmental parasitology, it emerged that parasites have a capacity to accumulate metals at concentrations far greater than their environments, and sometimes many times higher than their hosts (Nachev and Sures 2016). Lafferty (1997) highlighted that the relationships between heavy metals and parasites (cestodes, Digenea, and Acanthocephala) were generally negative, directly affecting the life histories and transmissibility of the parasites studied. Several assessments exploring metal pollution, host health, and disease epidemiology have since identified several interactions.

Some key studies in the field include a meta-analysis by Blanar et al. (2009), which revealed that the "effect" sizes of combined parasite and pollution factors often result in significantly negative effects upon the Digenea and Monogenea parasite groups, specifically in response to metal pollution. On the host side, experimental assessments by Boyce and Yamada (1977) revealed that infected fish can be more susceptible to zinc (Zn), indicating that disease susceptibility may be linked to pollution. Gheorgiu et al. (2006) observed guppies infested with *Gyrodactylus turnbulli* in an environment spiked with low (15 μg L^{-1}) and intermediate (120 μg L^{-1}) concentrations of Zn and found an increased parasite burden, whereas higher concentrations (240 μg L^{-1}) of the metal pollutant resulted in lower burdens of the parasite and was attributed to parasite sensitivity to the pollutant. The combined effects of parasites and metal pollution resulted in an overall greater mortality rate in the laboratory-reared guppy population.

A second common marine pollutant is tributyltin (TBT), a constituent of many anti-fouling marine paints. Anderson et al. (1996) highlighted that oysters (*Crassostrea virginica*) exposed to environmental concentrations of TBT had augmented infections of the protozoan *Perkinsus marinus* and *Perkinsus*-related mortality. The same pollutant (TBT) was banned following the observed incidences of imposex (females developing male sex organs) and subsequent collapse of multiple gastropod populations (Champ 2000), which have additionally been linked with immunosuppression in various marine species. Ford et al. (2006) reported an increased prevalence of intersexuality in amphipod crustaceans from clean and industrially polluted coastal locations around Scotland. These incidences of intersexuality were linked with feminizing microsporidian parasites, which led the authors to speculate whether industrial pollution was increasing the prevalence of the parasite. Further studies have identified a variety of industrial pollutants, including TBT and perfluorooctanesulfonic acid (PFOS), that can increase microsporidian infections, as observed in mesocosm experiments using Baltic amphipods (Jacobson et al. 2010; Jacobson et al. 2011). Thus, the presence of chemical pollutants can impact both host susceptibility and parasite viability.

Metal pollution can occur in combination with low to high levels of radioactive pollution. Accidents resulting in spills of radioactive material are rare but have occurred in the marine environment, including the Fukushima Daiichi Nuclear Power Plant (Japan) spill following an earthquake and tsunami in 2011 (Won et al. 2015). To date, no data exist on the specific effects of radioactive pollution on marine organisms and disease epidemiology; however, some studies demonstrate links between radiation and immunosuppression in wildlife, including fish and aquatic invertebrates (Morley 2012). It is possible that radiation induces changes in disease epidemiology via an increased susceptibility to disease. Low-dose, chronic exposure studies with freshwater fish have revealed a weakened immune response to bacterial infection and a greater rate of mortality induced by parasites (Sazykina and Kryshev 2003). Morley (2012) further highlights that acute radiation exposure is not principally associated with a risk of parasitic infection in invertebrates but tends to be associated with radiation damage to tissue structure and physiology rather than immunosuppression. The dearth of information on the effects of radioactive pollution on disease dynamics underscores the need for further study.

6.3.2 Medical waste

Medical waste, such as pharmaceuticals, can enter aquatic ecosystems from wastewater treatment plants (WWTPs), septic tanks, or diffusely through unlined landfills (Santos et al. 2010). Most pharmaceuticals entering WWTPs are not fully degraded or leave as toxic secondary metabolites. In addition to human pharmaceutical waste, an increasing number of drugs entering aquatic ecosystems come from veterinary waste, which often take the form of antiparasitic drugs used to combat diseases in farmed animals (Boxall et al. 2004). Aquatic organisms are typically not subjected to pharmaceutical waste in isolation but are instead exposed to contaminants including industrial, agricultural, and pharmaceutical pollution, which makes identification of the causal source in the environment difficult.

The effects of pharmaceutical exposure on host–parasite relationships was reviewed by Morley (2009), who pointed out that several thousand different drugs are detectable in the aquatic environment and one must consider the potential effects upon both the host and parasite. As with other contaminants, pharmaceuticals have been demonstrated to indirectly affect host–parasite disease dynamics by impairing the immune system of the host, making them more susceptible to infection or development of cancers (Box 6.1). However, others, such as antibiotics, antiparasitics, and animal hormones, can have direct impacts on non-target organisms and disease in the marine environment (Morley 2009) (Box 6.2).

6.3.3 Fossil fuels

Fossil fuels, such as oil, natural gas, and coal, threaten our atmosphere and oceans, primarily through climate change (including increasing sea surface temperatures), release of particulate matter, and CO_2 resulting in ocean acidification (Burge and Hershberger Chapter 5, this volume). Pollution with these natural products can also cause ecological destruction, primary through the accidental release of large quantities of oil that can span hundreds of kilometers and cause the mass mortality of birds, marine mammals, fish, invertebrates, plants, and planktonic life (Wan and Chen 2018).

Marine oil spills (or well blowouts) can also alter the structure of microbial communities due to selection for oil-tolerant bacteria, such as *Marinobacter* sp. (Raddadi et al. 2017), and other changes in marine microbial (prokaryotic and eukaryotic) assemblages (Bik et al. 2012). Microscale simulations of the effect of oil spills indicate significant changes to normal ecological processes that are associated with changes in microbial assemblages (Cappello et al. 2007), but these simulations have not explored linkages to disease. However, contamination from oil spills has been associated with physiological diseases of marine organisms. Early studies reported a correlation between oil exposure and stress-induced disease in the 1970s (Sindermann 1982), such as increased incidences of "fin-rot" among fish living in oil-contaminated estuarine environments and increased shell-disease in shrimp exposed to oil contamination (Gopalan and Young 1975; Minchew and Yarbrough 1977). Oil spills have been reported to cause seabird mortality but do not correlate with specific diseases; rather, mortality is a direct result of oil exposure (Newman et al. 2007). Further research on the bioaccumulation of fossil fuel contaminants in animals,

Box 6.1 Case study: Pollutants and cancers in marine turtles

Cancerous growth is often due to genomic/cellular aberrations triggered by senescence, infection (viral/bacterial), or introduction of a mutagen. Many of the chemicals present in industrial runoff are considered mutagenic and this has been associated with both increased incidences of infection, which result in cancer, and the prevalence of cancers as a result of chemical mutation (McAloose and Newton 2009). Several sea turtles suffer from fibriopapillomatosis, a viral disease that results in the systemic growth of cancerous tumors, primarily on the epidermal tissues (Figure 6.3). Higher than normal incidences of infection have been associated with marine environments heavily impacted by agricultural, industrial, or urban pollution (Herbst and Klein 1995; Foley et al. 2005).

Foley et al. (2005) examined US Sea Turtle Stranding and Salvage Network data for 4,328 green turtles (*Chelonia mydas*) found stranded in the eastern half of the USA (Massachusetts to Texas) between 1980 and 1998. Their data suggested that 22.6 percent of the turtles had tumors associated with fibriopapillomatosis, which was most prevalent in turtles found along the Gulf Coast of Florida (51.9 percent). Most infections corresponded with shallow coastal waters that had large areas of degraded and polluted habitat and low wave energy, providing likely circumstances for the pooling of pollutants. Direct confirmation using experimental procedures is required to corroborate these environmental observations.

Figure 6.3 A sea turtle showing advanced fibriopapillomatosis. (Photo courtesy of Inwater Research Group, 2017.)

such as bivalves, has determined that reproductive biology and increased disease susceptibility can be caused by the presence and uptake of lipophilic organic compounds, such as residual oil and other lipid-soluble contaminants (McDowell et al. 1999).

Oil spills are most often followed by remediation events consisting of oil clean-up efforts using chemicals that disperse the oil. The release of such chemicals into the environment is an additional concern and has been shown to adversely affect the settlement and survival of numerous marine fauna and flora (Goodbody-Gringley et al. 2013). Oil dispersants were also found to exacerbate "blue sac disease" (BSD) (a build-up of nitrogenous compounds and metabolic waste) in rainbow trout (*Oncorhynchus mykiss*) by increasing the bioavailability of toxic

Box 6.2 Case study: Estrogens and speedy parasite development

Estrogens are hormones that regulate several growth patterns across the Animalia, including both sexual and non-sexual characteristics. Estrogen compounds are often waste products in human urine and are associated with widely practiced hormone regulation treatments, and commonly make their way into the aquatic environment in unnatural quantities. Macnab et al. (2016) reported that fish pre-exposed to estrogens (specifically 17-β-oestradiol) during laboratory studies did not present increased susceptibility to infection by cestode parasites (*Schistocephalus solidus*) but instead the estrogen exposure caused accelerated parasite growth in male fish (Figure 6.4). The authors speculated several reasons for the host-sex-specific parasite growth, including immunosuppression, the pathological effects of infection, the parasite gaining nutritional benefit from cellular changes caused through endocrine disruption (e.g., vitellogenin expression), and parasite transregulation. These studies highlight the complexity faced when unravelling the multifactorial nature of multi-pollutant and multiple responses to host–parasite interactions. Environmental exposure could result in increased prevalence of the parasite due to shorter developmental times, reducing the overall health of susceptible host populations.

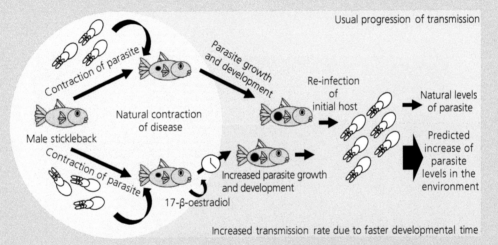

Figure 6.4 Diagrammatic representation of increased transmission rate of *Schistocephalus solidus* due to estrogen exposure. Fish that consume infected copepods acquire the infection, which develops faster in the host when it is exposed to estrogen. Over time, increased development rate could result in a greater prevalence of the parasite in copepod hosts, making transmission of the disease more likely to fish hosts.

fuels (Schein et al. 2009). Despite not being linked to disease agents, the effects of excess environmental oil levels have clear impacts on host physiology that may predispose them to disease.

6.3.4 Plastic pollution

In 1907, Leo Hendrik Baekeland developed a hardy material that could be moulded to fulfil a range of human needs (Laist 1987). Since development and wide-scale use, 12.7 million metric tons of the material known as "plastic" enters our oceans annually (Jambeck et al. 2015). Plastic pollution is now the most abundant form of human debris in the marine environment. The release and build-up of plastics in the ocean is of major concern due to its slow decomposition (Barnes et al. 2009). During its long decomposition period it will encounter wildlife, who mistake it for shelter or food. In addition to direct impacts upon wildlife, plastics (macro- and microplastics) in the marine environment can house microbes and have implications for disease susceptibility and transmission.

Macroplastics, such as discarded plastic bags, are now considered a habitat type within our oceans and can become colonized by a diverse microbial assemblage referred to as the "plastisphere" (Zettler et al. 2013). The plastisphere is a microbial community comprised of an assortment of heterotrophs, autotrophs, predatory microbes, and symbionts (Lobelle and Cunliffe 2011; Zettler et al. 2013). Zettler et al. (2013) utilized high-throughput rRNA gene sequencing methods to assess the microbial diversity present upon plastic debris and determined that a significant proportion aligned with bacteria from the genus *Vibrio*, a group commonly found to cause disease in humans and animals. Keswani et al. (2016) reviewed plastic colonization by harmful micro-organisms, which form biofilms that may exacerbate algal blooms, suggesting a concern for beach-goers among multiple other threats (Figure 6.5). In many cases, specific risks for disease transmission remain to be explicitly tested; however, bacteria identified from plastic debris may be a source of human and animal infections (Keswani et al. 2016).

Beyond risks to human health, plastics have also been linked with coral diseases in the Indian and Pacific oceans (Lamb et al. 2018). Corals have been

in global decline due to climate change and the unprecedented spread of disease. Lamb et al. (2018) assessed 159 reefs for the presence of disease and the relationship with plastic pollution, identifying that the likelihood of corals contracting disease increases from 4 percent to nearly 90 percent when plastic was in contact with coral. Morphologically complex corals that are more likely to become entangled in floating plastic were most at risk. Their study estimated that > 11 billion plastic items may be entangling corals across our oceans and this is projected to rise by 40 percent in the next 5–10 years, with implications for disease emergence, spread, and transmission.

Animals that ingest or become entangled in plastic can suffer from stress, which has been linked to a weakened immune system (Moloney et al. 2014). For large marine animals, such as sea turtles and marine mammals, ingestion of plastics can result in a loss of appetite, reduced energy levels, and chemical contamination (Nelms et al. 2016). This can increase susceptibility to disease, such as fibropapillomatosis in sea turtles (Santos et al. 2011). Similar increase in disease

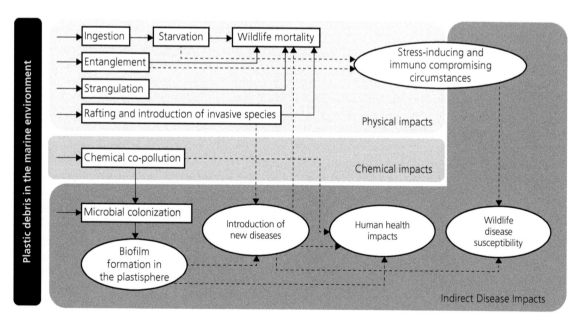

Figure 6.5 The relation ship between plastic debris in the ocean and potential risk for various ecological and anthropogenic sources, including the potential for altering disease dynamics indirectly within the marine environment. Solid lines express well-studied links between plastic pollution and wildlife impact, whereas dashed lines represent links between plastic pollution and disease introduction and susceptibility that require further understanding. (This diagram is adapted from a diagram included in Keswani et al. (2016).)

risk has been reported for dolphins (De Meirelles and Do Rego Barros 2007; Di Benedivo and Ramos 2014), whales (Jacobsen et al. 2010), and fish (Hoss and Settle 1990) that have ingested plastic. Many of these animals are apex predators whose loss to disease can have ecological consequences throughout the food web (McLaughlin et al. Chapter 2, this volume).

Macroplastics are eventually eroded away, giving rise to microplastics (Auta et al. 2017). Other sources of microplastic contamination include beauty products, such as exfoliation beads, and industrial waste materials (Fendall and Sewell 2009). Microplastics and their chemical constituents have been identified from the blood, organs, and surfaces of a wide range of animals (Watts et al. 2014; Auta et al. 2017; Collard et al. 2017) and these pollutants are entering the marine food web and human tissues (Carbery et al. 2018). Microplastics, and their chemical constituents, can potentially mimic cancer-causing compounds, reduce the host immune system, impair reproductive biology, and are able to spread quickly through the marine environment (Auta et al. 2017; Fang et al. 2018). For example, microplastics have been shown to have a significant effect on the immune system of the blue mussel *Mytilus edulis* (Von Moos et al. 2012). The microplastic particles (high-density polyethylene) were specifically targeted by the immune system and bioaccumulated over time, penetrating several tissue types. Von Moos et al. (2012) suggest a correlation with stress and increased risk of infection, primarily associated with the lysosomal membrane damage caused by microplastics. Similarly, microplastics have been shown to cause physiological degradation in the Norway lobster (*Nephrops norvegicus*) via a reduced feeding rate, lower body mass, reduced lipid storage, and depressed metabolism (Welden and Cowie 2016). Reduced physiological condition likely predisposes these lobsters to disease and may have impacts upon the fishery around Europe.

6.4 Physical pollution

Physical pollution is the impact of coastal or oceanic development, leisure activities, release of light and sound pollution, and marine debris that alters coastal hydrography or topology and the likelihood of wounding vulnerable wildlife. This section explores how physical disturbances in the marine environment can directly and indirectly affect disease dynamics and ecology.

6.4.1 Marine debris and discarded fishing equipment

In addition to the chemical effects of waste materials, the physical presence of marine debris (i.e., household garbage, discarded tools, and other human refuse) can predispose marine wildlife to wounding through interaction with, or ingestion of, harmful physical materials. Marine debris has also been associated with the movement of invasive species, which could harbor and introduce disease (Rech et al. 2016). The ingestion of macroplastics by various wildlife is explored in Section 6.3.4 dedicated to plastic pollution (and see Figure 6.5), but other materials pose a threat to internal and external damage and can increase the risk of infectious disease (Pagenkopp Lohan et al. Chapter 7, this volume; Behringer et al. Chapter 10, this volume).

Improperly discarded sharp objects (e.g., syringes, blades, glass) have the potential to cause external harm to marine life, opening wounds that may become infected. However, one of the primary sources of global ocean pollution is the 6.4 million tons of derelict fishing gear lost in the oceans annually (Wilcox et al. 2015). This gear can entangle and kill marine life, but more insidiously, derelict traps and nets can continue to "ghost fish" for many years after their loss or discard for various reasons (e.g., hurricanes, loss of the marker buoys).

In the Florida Keys (USA), the Caribbean spiny lobster (*Panulirus argus*) is a lucrative fishery but particularly prone to the effects of ghost fishing. It is predicted that > 10,000 traps are lost annually and may continue to fish, confine, and thereby kill lobsters and other marine animals, such as sharks, fishes, and crabs (Butler and Matthews 2015). The spiny lobster is a social species that attracts conspecifics through chemicals in its urine and it congregates for defense and mating (Anderson and Behringer 2013). This system has also been linked with pathogen avoidance, whereby healthy hosts can smell conspecifics infected with "Panulirus argus Virus 1," a viral disease of the hemolymph (Behringer et al. 2018). When healthy hosts smell infected conspecifics, they either force the infected animal out of the shelter or move away to avoid infection (Behringer et al. 2006). If trapped in a ghost trap, infected conspecifics will still be attracted to cues in the urine, but the healthy individuals will not be capable of escaping infection, thereby increasing the prevalence and possible impact of the virus in the vicinity (Behringer et al. 2012; Butler and Matthews 2015). The use of casitas, a method of open lobster fishing, seems preferential to avoid increasing the prevalence of natural disease levels in spiny lobster (Briones-Fourzán et al. 2012).

6.4.2 Marine disease in response to drought and altered freshwater flows

Freshwater input into marine environments is largely a natural process but can be exacerbated by extreme weather events (i.e., hurricanes) linked to global climate change (Burge and Hershberger Chapter 5, this volume). However, it can also be the result of human manipulation of freshwater and saltwater resources via riverine dams (Tonra et al. 2015) or changes to marine topology via the building of docks and creation of artificial beaches (Bulleri and Chapman 2010), which can affect salt concentration and nutrient levels in the marine environment. These changes can have diverse consequences for the movement and mixing of freshwater and saltwater, as well as nutrient availability, causing variation in environmental suitability, community assemblages, and the presence/absence of certain microbial species (Rubin et al. 2017).

Excessive freshwater input can drive motile species from affected areas, but for many sessile species, rapid environmental change may lead to mortality or a stress-induced increase in disease susceptibility. In corals, this has been observed on a global scale due to high levels of freshwater release in sewage and from storm drains, resulting in significant loss of nearby coral communities (Wear and Vega Thurber 2015). Conversely, freshwater is sometimes used as a treatment to remove disease from some aquaculture species (Powell et al. 2015) and could benefit some species that use it to avoid or remove disease naturally (Behringer et al. 2018).

Increased salinity can also result in increased stress for marine organisms, especially sessile species. Salinity changes can be particularly destructive for coastal plant life, where increased salinity has resulted in a greater prevalence and severity of wasting disease in eel grass (Jakobsson-Thor et al. 2018). *Labyrinthula zosterae*, a heterokont parasite, is sensitive to freshwater and saltwater concentrations and is found at higher prevalence in more saline conditions (Jakobsson-Thor et al. 2018). Further degradation of sea grass habitats due to exacerbated disease levels reduces the available area for settling larvae, food resources, and habitat for herbivorous and sheltering species.

6.4.3 Sound and light pollution

Excessive sound (audiopollution) and light (photopollution) are common pollutants of ocean ecosystems. At the environmental level, sound pollution has been found to affect habitat location (Radford et al. 2011), reproduction (Erbe et al. 2018), and communication (Putland et al. 2018) in marine animals. Light has been observed to alter animal navigation, predator–prey relationships, behavior, reproduction, and other traits (Preeti et al. 2018).

Audiopollution associated with vessel traffic is omnipresent in marine systems at multiple frequencies, where sound levels can exceed 111 dB (Bittencourt et al. 2014). Such pollution can affect communication (Lesage et al. 1999) and increase stress (Wright et al. 2011), particularly among whales and dolphins. The drastic effect of audiopollution on marine mammal communication and immunity/health is exemplified by a study measuring nerve and immune system health in whales (*Delphinapterus leucas*) and dolphins (*Tursiops truncatus*), finding that observable changes were present in the immune systems of dolphins exposed to prolonged and intense audiopollution (Romano et al. 2004). These changes include altered norepinephrine, epinephrine, and dopamine levels, which increased with increasing sound level. There was also a marked decrease in monocytes present in the circulatory system. This study determined that shorter bursts of sound exposure had smaller effects, but prolonged audiopollution could have significant impacts on the health of the marine mammals (Romano et al. 2004). Although studies have not directly linked stress from audiopollution to infectious disease, this area clearly warrants further investigation.

Artificial light (photopollution) has been found to disrupt the natural light cycles of a wide range of animals and plants (Gaston et al. 2014). In the marine environment, photopollution is linked mainly to coastal towns, shipping operations, and offshore structures. The effects of photopollution have resulted in issues with animal navigation (Longcore and Rich 2004) and can increase stress in many species, including humans (Gaston et al. 2014). Photopollution has been identified as a potential factor in animal health and disease emergence in the marine environment, but the area lacks in-depth assessment (Gaston et al. 2014). Some parasites alter the phototactic response of their host, so excessive photopollution combined with extended hunting periods for predatory animals may increase the incidence of host-manipulating parasites, such as *Gammarus insensibilis* infected with cerebral metacercariae (Gates et al. 2018). Light pollution may therefore be a driver for increased parasitism in the ocean.

Links between audiopollution, photopollution, and marine disease remain limited and require greater exploration.

6.5 Summary

- The human impact upon marine ecosystems is far reaching and many key studies suggest that disease dynamics can be substantially affected by biological, chemical, and physical pollution.
- Many factors remain underexplored and would strongly benefit from the union of environmental, experimental, molecular, and pathological approaches. Data from adequately replicated and designed (e.g., Basic Access Control Interface (BACI) designs) studies are needed to validate these findings from the "cocktail" of pollutive agents we see in our oceans (Wear and Vega Thurber 2015; Wear 2019).
- Biological pollution (e.g., sewage) stands as the most influential perturbation studied to date, with several key examples of marine diseases being linked to terrestrial environments, and, in some cases, ending up at our dinner tables.
- Chemical contamination in our oceans is unnaturally high and the link between the presence of contaminating pharmaceuticals and metals to

specific diseases (including their dynamics) requires detailed exploration and experimental work to provide baseline data.
- Links between physical pollution in marine systems and disease are the least well understood, but evidence suggests such links probably exist.

References

Anderson, J.R., and Behringer, D.C. 2013. Spatial dynamics in the social lobster *Panulirus argus* in response to diseased conspecifics. *Marine Ecology Progress Series*, 474, pp.191–200

Anderson, R.S., Unger, M.A., and Burreson, E.M. 1996. Enhancement of *Perkinsus marinus* disease progression in TBT-exposed oysters (*Crassostrea virginica*). *Marine Environmental Research*, 42(1–4), pp.177–180

Auta, H.S., Emenike, C.U., and Fauziah, S.H. 2017. Distribution and importance of microplastics in the marine environment: a review of the sources, fate, effects, and potential solutions. *Environment International*, 102, pp.165–176

Barnes, D.K., Galgani, F., Thompson, R.C., and Barlaz, M. 2009. Accumulation and fragmentation of plastic debris in global environments. *Philosophical Transactions of the Royal Society of London B: Biological Sciences*, 364(1526), pp.1985–1998

Behringer, D., Butler IV, M., and Shields, J. 2006. Ecology: avoidance of disease by social lobsters. *Nature*, 441, pp.421–421

Behringer, D.C., Butler, M.J., Moss, J., and Shields, J.D. 2012. PaV1 infection in the Florida spiny lobster (*Panulirus argus*) fishery and its effects on trap function and disease transmission. *Canadian Journal of Fisheries and Aquatic Sciences*, 69, pp.136–144

Behringer, D.C., Karvonen, A., and Bojko, J. 2018. Parasite avoidance behaviours in aquatic environments. *Philosophical Transactions of the Royal Society of London B: Biological Sciences* 373(1751), 20170202

Bellou, M., Kokkinos, P., and Vantarakis, A. 2013. Shellfish-borne viral outbreaks: a systematic review. *Food and Environmental Virology*, 5(1), pp.13–23

Bertness, M.D., Bruno, J.F., Silliman, B.R., and Stachowicz, J.J. 2014. A short history of marine community ecology. In: M.D. Bertness, J.F. Bruno, B.R. Sulliman, and J.J. Stachowicz (eds) *Marine Community Ecology and Conservation* (Chapter 1). Sinauer Associates, Sunderland, MA, pp.2–8

Bettarel, Y., Sime-Ngando, T., Amblard, C., and Laveran, H. 2000. A comparison of methods for counting viruses in aquatic systems. *Applied and Environmental Microbiology*, 66(6), pp.2283–2289

Bik, H.M., Halanych, K.M., Sharma, J., and Thomas, W.K. 2012. Dramatic shifts in benthic microbial eukaryote communities following the Deepwater Horizon oil spill. *PLoS One*, 7(6), e38550

Bittencourt, L., Carvalho, R.R., Lailson-Brito, J., and Azevedo, A.F. 2014. Underwater noise pollution in a coastal tropical environment. *Marine Pollution Bulletin*, 83(1), 331–336

Blanar, C.A., Munkittrick, K.R., Houlahan, J., MacLatchy, D.L., and Marcogliese, D.J. 2009. Pollution and parasitism in aquatic animals: a meta-analysis of effect size. *Aquatic Toxicology*, 93(1), pp.18–28

Boxall, A.B.A., Fogg, L.A., Blackwell, P.A., Blackwell, P., Kay, P., Pemberton, E.J., and Croxford, A. 2004. Veterinary medicines in the environment. In: A. Boxall, K. Barrett, and M. Crane (eds) *Reviews of Environmental Contamination and Toxicology* (pp. 1–91). Springer, New York.

Boyce, N.P., and Yamada, S.B. 1977. Effects of a parasite, *Eubothrium salvelini* (Cestoda: Pseudophyllidea), on the resistance of juvenile sockeye salmon, *Oncorhynchus nerka*, to zinc. *Journal of the Fisheries Board of Canada*, 34(5), pp.706–709

Briones-Fourzán, P., Candia-Zulbarán, R.I., Negrete-Soto, F., Barradas-Ortiz, C., Huchin-Mian, J.P., and Lozano-Álvarez, E. 2012. Influence of local habitat features on disease avoidance by Caribbean spiny lobsters in a casita-enhanced bay. *Diseases of Aquatic Organisms*, 100, pp.135–148

Bruno, J.F., Petes, L.E., Drew Harvell, C., and Hettinger, A. 2003. Nutrient enrichment can increase the severity of coral diseases. *Ecology Letters*, 6, pp.1056–1061

Bulleri, F., and Chapman, M.G. 2010. The introduction of coastal infrastructure as a driver of change in marine environments. *Journal of Applied Ecology*, 47(1), pp.26–35

Butler, C.B., and Matthews, T.R. 2015. Effects of ghost fishing lobster traps in the Florida Keys. *ICES Journal of Marine Science*, 72(1), pp.185–198

Cappello, S., Caruso, G., Zampino, D., Monticelli, L.S., Maimone, G., Denaro, R., et al. 2007. Microbial community dynamics during assays of harbour oil spill bioremediation: a microscale simulation study. *Journal of Applied Microbiology*, 102(1), pp.184–194

Carbery, M., O'Connor, W., and Palanisami, T. 2018. Trophic transfer of microplastics and mixed contaminants in the marine food web and implications for human health. *Environment International*, 115, pp.400–409

Champ, M.A. 2000. A review of organotin regulatory strategies, pending actions, related costs and benefits. *Science of the Total Environment*, 258(1–2), pp.21–71

Civitello, D.J., Allman, B.E., Morozumi, C., and Rohr, J.R. 2018. Assessing the direct and indirect effects of food provisioning and nutrient enrichment on wildlife infectious disease dynamics. *Philosophical Transactions of the Royal Society B: Biological Sciences*, 373, 20170101

Collard, F., Gilbert, B., Compère, P., Eppe, G., Das, K., Jauniaux, T., and Parmentier, E. 2017. Microplastics in livers of European anchovies (*Engraulis encrasicolus*, L.). *Environmental Pollution*, 229, pp.1000–1005

Connor, R., Renata, A., Ortigara, C., Koncagül, E., Uhlenbrook, S., Lamizana-Diallo, B. M., et al. 2017. *The United Nations World Water Development Report 2017. Wastewater: The Untapped Resource*. UNESCO, Paris.

Conrad, P.A., Miller, M.A., Kreuder, C., James, E.R., Mazet, J., Dabritz, H., et al. 2005. Transmission of *Toxoplasma*: clues from the study of sea otters as sentinels of *Toxoplasma gondii* flow into the marine environment. *International Journal for Parasitology*, 35(11–12), pp.1155–1168

Cressler, C.E., Nelson, W.A., Day, T., and McCauley, E. 2013. Disentangling the interaction among host resources, the immune system and pathogens. *Ecology Letters*, 17, pp.284–293

De Meirelles, A.C.O., and Do Rego Barros, H.M.D. 2007. Plastic debris ingested by a rough-toothed dolphin, *Steno bredanensis*, stranded alive in northeastern Brazil. *Biotemas*, 20(1), pp.127–131

Di Beneditto, A.P.M., and Ramos, R.M.A. 2014. Marine debris ingestion by coastal dolphins: what drives differences between sympatric species? *Marine Pollution Bulletin*, 83(1), pp.298–301

Eliasson, J. 2015. The rising pressure of global water shortages. *Nature News*, 517(7532), p.6

Erbe, C., Dunlop, R., and Dolman, S. 2018. Effects of noise on marine mammals. In: H. Slabbekoorn, R.J. Dooling, A.N. Popper, and R.R. Fay (eds) *Effects of Anthropogenic Noise on Animals* (pp. 277–309). Springer, New York.

Ewert, D.L., and Paynter, M.J.B. 1980. Enumeration of bacteriophages and host bacteria in sewage and the activated-sludge treatment process. *Applied and Environmental Microbiology*, 39(3), pp.576–583

Fang, C., Zheng, R., Zhang, Y., Hong, F., Mu, J., Chen, M., et al. 2018. Microplastic contamination in benthic organisms from the Arctic and sub-Arctic regions. *Chemosphere*, 209, pp.298–306

FAO. 2017. *World Fertilizer Trends and Outlook to 2020 Summary Report*. Food and Agricultural Organization of the United Nations, Rome.

Fendall, L.S., and Sewell. M.A. 2009. Contributing to marine pollution by washing your face: microplastics in facial cleansers. *Marine Pollution Bulletin*, 58, pp.1225–1228

Foley, A.M., Schroeder, B.A., Redlow, A.E., Fick-Child, K.J., and Teas, W.G. 2005. Fibropapillomatosis in stranded green turtles (*Chelonia mydas*) from the eastern United States (1980–98): trends and associations with environmental factors. *Journal of Wildlife Diseases*, 41(1), pp.29–41

Ford, A.T., Fernandes, T.F., Robinson, C.D., Davies, I.M., and Read, P.A. 2006. Can industrial pollution cause intersexuality in the amphipod, *Echinogammarus marinus? Marine Pollution Bulletin*, 53, pp.100–106

Froelich, B., Ayrapetyan, M., and Oliver, J.D. 2013. Integration of *Vibrio vulnificus* into marine aggregates and its subsequent uptake by *Crassostrea virginica* oysters. *Applied and Environmental Microbiology*, 79(5), pp.1454–1458

Galloway, J.N., Dentener, F.J., Capone, D.G., Boyer, E.W., Howarth, R.W., Seitzinger, S.P., Asner, G.P., et al. 2004. Nitrogen cycles: past, present, and future. *Biogeochemistry*, 70, pp.153–226

Galloway, J.N., Townsend, A.R., Erisman, J.W., Bekunda, M., Cai, Z., Freney, J.R., Martinelli, L.A., et al. 2008. Transformation of the nitrogen cycle: recent trends, questions, and potential solutions. *Science*, 320, pp.889–892

Galloway, T.S., and Depledge, M.H. 2001. Immunotoxicity in invertebrates: measurement and ecotoxicological relevance. *Ecotoxicology*, 10, pp.5–23

Gaston, K.J., Duffy, J.P., Gaston, S., Bennie, J., and Davies, T.W. 2014. Human alteration of natural light cycles: causes and ecological consequences. *Oecologia*, 176(4), pp.917–931

Gates, A.R., Sheader, M., Williams, J.A., and Hawkins, L.E. 2018. Infection with cerebral metacercariae of microphallid trematode parasites reduces reproductive output in the gammarid amphipod *Gammarus insensibilis* (Stock 1966) in UK saline lagoons. *Journal of the Marine Biological Association of the United Kingdom*, 98(6), pp.1391–1400

Gauthier, D.T. 2015. Bacterial zoonoses of fishes: a review and appraisal of evidence for linkages between fish and human infections. *The Veterinary Journal*, 203(1), pp.27–35

Gentry, J., Vinjé, J., Guadagnoli, D., and Lipp, E.K. 2009. Norovirus distribution within an estuarine environment. *Applied Environmental Microbiology*, 75(17), pp.5474–5480

Gheorgiu, C., Marcogliese, D.J., and Scott, M. 2006. Concentration-dependent effects of waterborne zinc on population dynamics of *Gyrodactylus turnbulli* (Monogenea) on isolated guppies (*Poecilia reticulata*). *Parasitology*, 132(2), pp.225–232

Goodbody-Gringley, G., Wetzel, D.L., Gillon, D., Pulster, E., Miller, A., and Ritchie, K.B. 2013. Toxicity of Deepwater Horizon source oil and the chemical dispersant, Corexit® 9500, to coral larvae. *PLoS One*, 8(1), e45574

Gopalan, U.K., and Young, J.S. 1975. Incidence of shell disease in shrimp in the New York Bight. *Marine Pollution Bulletin*, 6(10), pp.149–153

Grodzki, M., Schaeffer, J., Piquet, J.C., Le Saux, J.C., Chevé, J., Ollivier, J., et al. 2014. Bioaccumulation efficiency, tissue distribution and environmental occurrence of hepatitis E virus in bivalve shellfish from France. *Applied and Environmental Microbiology*, 80(14), pp.4269–4276

Herbst, L.H., and Klein, P.A. 1995. Green turtle fibropapillomatosis: challenges to assessing the role of environmental cofactors. *Environmental Health Perspectives*, 103, pp.27–30

Hoss, D.E., and Settle, L.R. 1990. Ingestion of plastics by teleost fishes. *Proceedings of the Second International Conference on Marine Debris. NOAA Technical Memorandum.* NOAA-TM-NMFS-SWFSC-154, Miami, FL, pp.693–709

Hughes, R.G., Potouroglou, M., Ziauddin, Z., and Nicholls, J.C. 2018. Seagrass wasting disease: nitrate enrichment and exposure to a herbicide (Diuron) increases susceptibility of *Zostera marina* to infection. *Marine Pollution Bulletin*, 134, pp.94–98

Islam, M.S., and Tanaka, M. 2004. Impacts of pollution on coastal and marine ecosystems including coastal and marine fisheries and approach for management: a review and synthesis. *Marine Pollution Bulletin*, 48, pp.624–649

Jacobsen, J.K., Massey, L., and Gulland, F. 2010. Fatal ingestion of floating net debris by two sperm whales (*Physeter macrocephalus*). *Marine Pollution Bulletin*, 60(5), pp.765–767

Jacobson, T., Holmström, K., Yang, G., Ford, A.T., Berger, U., and Sundelin, B. 2010. Perfluorooctane sulfonate accumulation and parasite infestation in a field population of the amphipod *Monoporeia affinis* after microcosm exposure. *Aquatic Toxicology*, 98(1), pp.99–106

Jacobson, T., Sundelin, B., Yang, G., and Ford, A.T. 2011. Low dose TBT exposure decreases amphipod immunocompetence and reproductive fitness. *Aquatic Toxicology*, 101(1), pp.72–77

Jakobsson-Thor, S., Toth, G.B., Brakel, J., Bockelmann, A.C., and Pavia, H. 2018. Seagrass wasting disease varies with salinity and depth in natural *Zostera marina* populations. *Marine Ecology Progress Series*, 587, pp.105–115

Jambeck, J.R., Geyer, R., Wilcox, C., Siegler, T.R., Perryman, M., Andrady, A., Narayan, R., and Law, K.L. 2015. Plastic waste inputs from land into the ocean. *Science*, 347(6223), pp.768–771

Johnson, P. T. J., and Carpenter, S.R. 2008. Influence of eutrophication on disease in aquatic ecosystems: patterns, processes, and predictions, pp.71–99. In: R. Ostfeld, F. Keesing, and V. Eviner (eds) *Infectious Disease Ecology: Effects of Ecosystems on Disease and of Disease on Ecosystems*. Princeton University Press, NJ

Johnson, P.T.J., Townsend, A.R., Cleveland, C.C., Glibert, P.M., Howarth, R.W., McKenzie, V.J., Rejmankova, E., et al. 2010. Linking environmental nutrient enrichment and disease emergence in humans and wildlife. *Ecological Applications*, 20, pp.16–29

Kazama, S., Masago, Y., Tohma, K., Souma, N., Imagawa, T., Suzuki, A., Liu, X., Saito, M., Oshitani, H., and Omura, T. 2016. Temporal dynamics of norovirus determined through monitoring of municipal wastewater by pyrosequencing and virological surveillance of gastroenteritis cases. *Water Research*, 92, pp.244–253

Keswani, A., Oliver, D.M., Gutierrez, T., and Quilliam, R.S. 2016. Microbial hitchhikers on marine plastic debris: human exposure risks at bathing waters and beach environments. *Marine Environmental Research*, 118, pp.10–19

Ketchum, B.H. 1970. Biological implications of global marine pollution. In: S.F. Singer (ed.) *Global Effects of Environmental Pollution* (pp. 190–194). Springer, Dordrecht.

Khan, R.A., and Thulin, J. 1991. Influence of pollution on parasites of aquatic animals. *Advances in Parasitology*, 30, pp.201–238

Kim, K., and Harvell, C.D. 2004. The rise and fall of a six-year coral–fungal epizootic. *American Naturalist*, 164, pp.552–563

Lafferty, K.D. 1997. Environmental parasitology: what can parasites tell us about human impacts on the environment? *Parasitology Today*, 13(7), pp.251–255

Laist, D.W. 1987. Overview of the biological effects of lost and discarded plastic debris in the marine environment. *Marine Pollution Bulletin*, 18(6), pp.319–326

Lamb, J.B., Willis, B.L., Fiorenza, E.A., Couch, C.S., Howard, R., Rader, D.N., et al. 2018. Plastic waste associated with disease on coral reefs. *Science*, 359(6374), pp.460–462

Le Guyader, S., Atmar, R., Maalouf, H., and Le Pendu, J. 2013. Shellfish contamination by norovirus: strain selection based on ligand expression? Rinsho to uirusu. *Clinical Virology*, 41(1), pp.3–18

Lesage, V., Barrette, C., Kingsley, M.C., and Sjare, B. 1999. The effect of vessel noise on the vocal behavior of belugas in the St. Lawrence River estuary, Canada. *Marine Mammal Science*, 15(1), pp.65–84

Lobelle, D., and Cunliffe, M. 2011. Early microbial biofilm formation on marine plastic debris. *Marine Pollution Bulletin*, 62(1), pp.197–200

Longcore, T., and Rich, C. 2004. Ecological light pollution. *Frontiers in Ecology and the Environment*, 2(4), pp.191–198

MacKenzie, K. 1999. Parasites as pollution indicators in marine ecosystems: a proposed early warning system. *Marine Pollution Bulletin*, 38(11), pp.955–959

Macnab, V., Katsiadaki, I., Tilley, C.A., and Barber, I. 2016. Oestrogenic pollutants promote the growth of a parasite in male sticklebacks. *Aquatic Toxicology*, 174, pp.92–100

Marcogliese, D.J., and Pietrock, M. 2011. Combined effects of parasites and contaminants on animal health: parasites do matter. *Trends in Parasitology*, 27(3), pp.123–130

McAloose, D., and Newton, A.L. 2009. Wildlife cancer: a conservation perspective. *Nature Reviews Cancer*, 9(7), p.517

McBride, G.B., Stott, R., Miller, W., Bambic, D., and Wuertz, S. 2013. Discharge-based QMRA for estimation of public health risks from exposure to stormwater-borne pathogens in recreational waters in the United States. *Water Research*, 47(14), pp.5282–5297

McDowell, J.E., Lancaster, B.A., Leavitt, D.F., Rantamaki, P., and Ripley, B. 1999. The effects of lipophilic organic contaminants on reproductive physiology and disease processes in marine bivalve molluscs. *Limnology and Oceanography*, 44(3), pp.903–909

Miller, M.A., Gardner, I.A., Kreuder, C., Paradies, D.M., Worcester, K.R., Jessup, D.A., et al. 2002. Coastal freshwater runoff is a risk factor for *Toxoplasma gondii* infection of southern sea otters (*Enhydra lutris nereis*). *International Journal for Parasitology*, 32(8), pp.997–1006

Minchew, D.C., and Yarbrough, J.D. 1977. The occurrence of fin rot in mullet (*Mugil cephalus*) associated with crude oil contamination of an estuarine pond-ecosystem. *Journal of Fish Biology*, 10(4), pp.319–323

Mohanty, B.R., and Sahoo, P.K. 2007. Edwardsiellosis in fish: a brief review. *Journal of Biosciences*, 32(3), pp.1331–1344

Moloney, R.D., Desbonnet, L., Clarke, G., Dinan, T.G., and Cryan, J.F. 2014. The microbiome: stress, health and disease. *Mammalian Genome*, 25(1–2), pp.49–74

Morley, N.J. 2009. Environmental risk and toxicology of human and veterinary waste pharmaceutical exposure to wild aquatic host–parasite relationships. *Environmental Toxicology and Pharmacology*, 27(2), pp.161–175

Morley, N.J. 2012. The effects of radioactive pollution on the dynamics of infectious diseases in wildlife. *Journal of Environmental Radioactivity*, 106, pp.81–97

Murphree, R.L., and Tamplin, M.L. 1995. Uptake and retention of *Vibrio cholerae* O1 in the Eastern oyster, *Crassostrea virginica*. *Applied and Environmental Microbiology*, 61(10), pp.3656–3660

Nachev, M., and Sures, B. 2016. Environmental parasitology: parasites as accumulation bioindicators in the marine environment. *Journal of Sea Research*, 113, pp.45–50

National Research Council. 1993. *Managing Wastewater in Coastal Urban Areas*. National Academies Press, Washington, DC.

Nelms, S.E., Duncan, E.M., Broderick, A.C., Galloway, T.S., Godfrey, M.H., Hamann, M., et al. 2016. Plastic and marine turtles: a review and call for research. *ICES Journal of Marine Science*, 73(2), pp.165–181

Newman, S.H., Chmura, A., Converse, K., Kilpatrick, A.M., Patel, N., Lammers, E., and Daszak, P. 2007. Aquatic bird disease and mortality as an indicator of changing ecosystem health. *Marine Ecology Progress Series*, 352, pp.299–309

Orth, R J., Carruthers, T.J.B., Dennison, W.C., Duarte, C.M., Fourqurean, J.W., Heck, K.L., Hughes, A.R., et al. 2006. A global crisis for seagrass ecosystems. *BioScience*, 56, pp.987–996

Pandolfi, J. M., Jackson, J.B.C., Baron, N., Bradbury, R.H., Guzman, H.M., Hughes, T.P., Kappel, C.V., et al. 2005. Are U.S. coral reefs on the slippery slope to slime? *Science*, 307, 1725

Powell, M.D., Reynolds, P., and Kristensen, T. 2015. Freshwater treatment of amoebic gill disease and sea-lice in seawater salmon production: considerations of water chemistry and fish welfare in Norway. *Aquaculture*, 448, pp.18–28

Powles, H., Bradford, M.J., Bradford, R.G., Doubleday, W.G., Innes, S., and Levings, C.D. 2000. Assessing and protecting endangered marine species. *ICES Journal of Marine Science*, 57(3), pp.669–676

Preeti, J.K.R., Thakur, M., Suman, M., and Kumar, R. 2018. Consequences of pollution in wildlife: a review. *Pharma Innovation Journal*, 7, pp.94–102

Preston, D.L., Mischler, J.A., Townsend, A.R., and Johnson, P.T. 2016. Disease ecology meets ecosystem science. *Ecosystems*, 19(4), pp.737–748

Putland, R.L., Merchant, N.D., Farcas, A., and Radford, C.A. 2018. Vessel noise cuts down communication space for vocalizing fish and marine mammals. *Global Change Biology*, 24(4), pp.1708–1721

Raddadi, N., Giacomucci, L., Totaro, G., and Fava, F. 2017. *Marinobacter* sp. from marine sediments produce highly stable surface-active agents for combatting marine oil spills. *Microbial Cell Factories*, 16(1), p.186

Radford, C.A., Stanley, J.A., Simpson, S.D., and Jeffs, A.G. 2011. Juvenile coral reef fish use sound to locate habitats. *Coral Reefs*, 30(2), pp.295–305

Rech, S., Borrell, Y., and García-Vazquez, E. (2016). Marine litter as a vector for non-native species: what we need to know. *Marine Pollution Bulletin*, 113(1–2), 40–43

Romano, T.A., Keogh, M.J., Kelly, C., Feng, P., Berk, L., Schlundt, C.E., et al. 2004. Anthropogenic sound and marine mammal health: measures of the nervous and immune systems before and after intense sound exposure. *Canadian Journal of Fisheries and Aquatic Sciences*, 61(7), pp.1124–1134

Rubin, S.P., Miller, I.M., Foley, M.M., Berry, H.D., Duda, J.J., Hudson, B., et al. 2017. Increased sediment load during a large-scale dam removal changes nearshore subtidal communities. *PLoS One*, 12(12), e0187742

Santos, L.H., Araújo, A.N., Fachini, A., Pena, A., Delerue-Matos, C., and Montenegro, M.C.B.S.M. 2010. Ecotoxicological aspects related to the presence of pharmaceuticals in the aquatic environment. *Journal of Hazardous Materials*, 175(1–3), pp.45–95

Santos, R.G., Martins, A.S., da Nobrega Farias, J., Horta, P.A., Pinheiro, H.T., Torezani, E., et al. 2011. Coastal habitat degradation and green sea turtle diets in southeastern Brazil. *Marine Pollution Bulletin*, 62(6), pp.1297–1302

Sazykina, T.G., and Kryshev, A.I. 2003. EPIC database on the effects of chronic radiation in fish: Russian/FSU data. *Journal of Environmental Radioactivity*, 68(1), pp.65–87

Schein, A., Scott, J.A., Mos, L., and Hodson, P.V. 2009. Oil dispersion increases the apparent bioavailability and toxicity of diesel to rainbow trout (*Oncorhynchus mykiss*). *Environmental Toxicology and Chemistry*, 28(3), pp.595–602

Shapiro, K., Silver, M.W., Largier, J.L., Conrad, P.A., and Mazet, J.A. 2012. Association of *Toxoplasma gondii* oocysts with fresh, estuarine, and marine macroaggregates. *Limnology and Oceanography*, 57(2), pp.449–456

Sharafi, K., Fazlzadehdavil, M., Pirsaheb, M., Derayat, J., and Hazrati, S. 2012. The comparison of parasite eggs and protozoan cysts of urban raw wastewater and efficiency of various wastewater treatment systems to remove them. *Ecological Engineering*, 44, pp.244–248

Sindermann, C.J. 1982. Implications of oil pollution in production of disease in marine organisms. *Philosophical Transactions of the Royal Society B*, 297(1087), pp.385–399

Smith, F.W. 1941. Sponge disease in British Honduras, and its transmission by water currents. *Ecology*, 22(4), pp.415–421

Stentiford, G.D., Neil, D.M., Peeler, E.J., Shields, J.D., Small, H.J., Flegel, T.W., et al. 2012. Disease will limit future food supply from the global crustacean fishery and aquaculture sectors. *Journal of Invertebrate Pathology*, 110(2), pp.141–157

Sueiro, M.C., Bagnato, E., and Palacios, M.G. 2017. Parasite infection and immune and health-state in wild fish exposed to marine pollution. *Marine Pollution Bulletin*, 119(1), pp.320–324

Sures, B. 2004. Environmental parasitology: relevancy of parasites in monitoring environmental pollution. *Trends in Parasitology*, 20(4), pp.170–177

Sutherland, K.P., Porter, J.W., Turner, J.W., Thomas, B.J., Looney, E.E., Luna, T.P., et al. 2010. Human sewage identified as likely source of white pox disease of the threatened Caribbean elkhorn coral, *Acropora palmata*. *Environmental Microbiology*, 12(5), pp.1122–1131

Sutherland, K.P., Berry, B., Park, A., Kemp, D.W., Kemp, K.M., Lipp, E.K., and Porter, J.W. 2016. Shifting white pox aetiologies affecting *Acropora palmata* in the Florida Keys, 1994–2014. *Philosophical Transactions of the Royal Society B: Biological Sciences*, 371(1689), 20150205

Timoney, J.F., and Abston, A. 1984. Accumulation and elimination of *Escherichia coli* and *Salmonella typhimurium* by hard clams in an in vitro system. *Applied and Environmental Microbiology*, 47(5), pp.986–988

Tonra, C.M., Sager-Fradkin, K., Morley, S.A., Duda, J.J., and Marra, P.P. 2015. The rapid return of marine-derived nutrients to a freshwater food web following dam removal. *Biological Conservation*, 192, pp.130–134

Vega Thurber, R.L., Burkepile, D.E., Fuchs, C., Shantz, A.A., McMinds, R., and Zaneveld, J.R. 2013. Chronic nutrient enrichment increases prevalence and severity of coral disease and bleaching. *Global Change Biology*, 20, pp.544–554

Vidal-Martinez, V.M., Pech, D., Sures, B., Purucker, S.T., and Poulin, R. 2010. Can parasites really reveal environmental impact? *Trends in Parasitology*, 26(1), pp.44–51

Vitousek, P.M., Aber, J.D., Howarth, R.W., Likens, G.E., Matson, P.A., Schindler, D.W., Schlesinger, W.H., et al. 1997. Human alteration of the global nitrogen cycle: sources and consequences. *Ecological Applications*, 7, pp.737–750

Von Moos, N., Burkhardt-Holm, P., and Köhler, A. 2012. Uptake and effects of microplastics on cells and tissue of the blue mussel *Mytilus edulis* L. after an experimental exposure. *Environmental Science and Technology*, 46(20), pp.11327–11335

Wan, Z., and Chen, J. 2018. Human errors are behind most oil-tanker spills. *Nature*, 560, pp.161–163

Ward, J.R., and Lafferty, K.D. 2004. The elusive baseline of marine disease: are diseases in ocean ecosystems increasing? *PLoS Biology*, 2(4), e120

Warner, E.B., and Oliver, J.D. 2008. Population structures of two genotypes of *Vibrio vulnificus* in oysters (*Crassostrea virginica*) and seawater. *Applied Environmental Microbiology*, 74, pp.80–85

Watts, A.J., Lewis, C., Goodhead, R.M., Beckett, S.J., Moger, J., Tyler, C.R., and Galloway, T.S. 2014. Uptake and retention of microplastics by the shore crab *Carcinus maenas*. *Environmental Science & Technology*, 48(15), pp.8823–8830

Wear, S.L. 2019. Battling a common enemy: joining forces in the fight against sewage pollution. *BioScience*, 69, pp.360–367

Wear, S.L., and Vega Thurber, R.V. 2015. Sewage pollution: mitigation is key for coral reef stewardship. *Annals of the New York Academy of Sciences*, 1355(1), pp. 15–30

Welden, N.A., and Cowie, P.R. 2016. Long-term microplastic retention causes reduced body condition in the langoustine, *Nephrops norvegicus*. *Environmental Pollution*, 218, pp.895–900

Wilcox, C., Heathcote, G., Goldberg, J., Gunn, R., Peel, D., and Hardesty, B.D. 2015. Understanding the sources and effects of abandoned, lost, and discarded fishing gear on marine turtles in northern Australia. *Conservation Biology*, 29(1), pp.198–206

Williams, H.H., and MacKenzie, K. 2003. Marine parasites as pollution indicators: an update. *Parasitology*, 126(7), pp.527–541

Won, E.J., Dahms, H.U., Kumar, K.S., Shin, K.H., and Lee, J.S. 2015. An integrated view of gamma radiation effects on marine fauna: from molecules to ecosystems. *Environmental Science and Pollution Research*, 22(22), pp.17443–17452

Wright, A.J., Deak, T., and Parsons, E.C.M. 2011. Size matters: management of stress responses and chronic stress in beaked whales and other marine mammals may require larger exclusion zones. *Marine Pollution Bulletin*, 63, pp.5–9

Zaneveld, J.R., Burkepile, D.E., Shantz, A.A., Pritchard, C.E., McMinds, R., Payet, J.P., Welsh, R., et al. 2016. Overfishing and nutrient pollution interact with temperature to disrupt coral reefs down to microbial scales. *Nature Communications*, 7, 11833

Zettler, E.R., Mincer, T.J., and Amaral-Zettler, L.A. 2013. Life in the "plastisphere": microbial communities on plastic marine debris. *Environmental Science & Technology*, 47(13), pp.7137–7146

ZoBell, C.E., and Feltham, C.B. 1942. The bacterial flora of a marine mud flat as an ecological factor. *Ecology*, 23(1), pp.69–78

CHAPTER 7

Invasions can drive marine disease dynamics

Katrina M. Pagenkopp Lohan, Gregory M. Ruiz, and Mark E. Torchin

7.1 Introduction

In a progressively interconnected world, the global-scale movement of people, plants, animals, and cargo has grown exponentially, increasing the potential for new invasions. Over the past century, this rise in global trade is associated with increases in the documented rate of invasions by non-native species, particularly in countries with the highest trade volumes (Hulme 2009). To date, most research on biological invasions has focused on free-living macro-organisms, rather than marine parasites (Poulin 2017) or microbes (Wyatt and Carlton 2002), primarily since they are difficult to detect and identify and often lack information on their native range. However, modern molecular techniques have aided detecting and identifying parasites and microscopic organisms. In this chapter, we examine parasite invasions and highlight the potential importance of invasions by microbial *parasites* (all italicized words in the text are defined in the Glossary in Box 7.1). More specifically, we explore the particular traits of microbial parasites that can influence invasion success, the scale and history of anthropogenic vectors contributing to their spread, and the potential ecological and evolutionary implications of these invasions. Our analysis underscores the dearth of knowledge regarding the scale and impact of *biological invasions* of microbial parasites, making this an important area for future research.

Marine parasites occur in all three domains (Eukarya, Bacteria, Archaea), throughout the world's oceans (Rohde 2005), and have evolved independently across most lineages (Weinstein and Kuris 2016). Approximately 40 percent of biodiversity on Earth is estimated to be parasitic (Rohde 1982, Poulin and Morand 2000), though high species discovery rates indicate these estimates are probably low (Poulin 2014). Additionally, while researchers have attempted total biodiversity estimates of helminths, there is still insufficient knowledge to extrapolate similar estimates for the extant number of pathogenic viruses, bacteria, *protists*, and fungi (Dobson et al. 2008). Across these diverse taxa, marine parasites differ in form, function, specificity, transmission modes, and pathogenicity (Rohde 2005), all characteristics that can affect the probability of dispersal and establishment. For example, directly transmitted parasites require only a single *host* species to establish a new population, whereas those with complex life cycles require multiple suitable host species, any one of which many be limiting or absent.

Most introduced parasites are thought to invade new areas with their hosts (Torchin and Kuris 2005), but it is also possible for parasites to arrive independent of hosts, as either a dispersive or resting/dormant stage (Ruiz et al. 2000a, Pagenkopp Lohan et al. 2016, Solarz and Najberek 2017), utilizing a variety of natural and anthropogenic dispersal

Pagenkopp Lohan, K.M., Ruiz, G.M., and Torchin, M.E., *Invasions can drive marine disease dynamics* In: *Marine Disease Ecology*. Edited by: Donald C. Behringer, Kevin D. Lafferty, and Brian. R. Silliman, Oxford University Press (2020). © Oxford University Press. DOI: 10.1093/oso/9780198821632.003.0007

Box 7.1 Glossary

Ballast water = water used by commercial ships for trim and stability during voyages. Usually pumped from adjacent coastal or ocean water, along with a taxonomically diverse range of organisms (from bacteria to fish), and discharged at subsequent ports of call, creating an efficient transfer mechanism for marine and freshwater organisms

Biofouling organisms = organisms associated with hard substrate, including the hulls of vessels, which transfer biofouling organisms throughout the world. These organisms include both sessile organisms (such as barnacles and mussels) and mobile organisms

Biological invasion = establishment of a self-sustaining population beyond its historical biogeographic range, usually as a result of human-mediated transfer across dispersal barriers such as oceans and continents

Disease = a disorder of structure or function in an organism that produces specific signs that are not caused by physical injury alone

Facultative parasite = an organism that can survive using a parasitic or free-living lifestyle. A host is not required for the completion of its life cycle, but this does not prevent the organism from acquiring a host, allowing it to further propagate

Flux = the number of organisms moved (transferred) between locations over some unit of time. This may refer to number of organisms or number of species. We are referring specifically to organisms moved by human activities. This is related to propagule pressure, which characterizes organisms released to a location that are capable of reproduction

Host = an organism on or in which a parasite or commensal organism lives

Host specificity = the taxonomic breadth of hosts that a single parasite species is able to exploit. Those with high host specificity may only infect a single species, while those with low host specificity may infect many hosts from different classes or phyla

Metabarcoding = a genetic method that uses general primers to mass-amplify a single region of DNA from a collection of organisms or from an environmental sample

Metagenomics = a genetic method that generates sequence data from all the genetic material (i.e., genomes) from a collection of organisms or from an environmental sample

Metatranscriptomics = a genetic method that generates sequence data from all the genes that are expressed within cells (i.e., transcriptome) from a collection of organisms or from an environmental sample

Native = an organism or species within its historical and evolutionary range, where its presence is due solely to natural dispersal processes

Non-native = an organism or species that is outside its historical and evolutionary range, where its presence is due usually to human-mediated dispersal

Parasite = an organism that lives at the expense of another (the host), causing some level of harm, generally living in or on the host, and is structurally adapted to this way of life

Propagule pressure = the number of individuals released to an area over some period of time. Often considered the number of a single species, it also refers to the total number of multiple species. Implicit is that the organisms are viable and capable of reproduction

Protist = a general term to refer to "eukaryotic organisms with unicellular, colonial, filamentous, or parenchymatous organization that lack vegetative tissue differentiation, except for reproduction" (Adl et al. 2007)

Sapronotic agent = an organism that can survive equally as well on abiotic substrate at ambient temperatures as on biotic substrate. These organisms can only complete their life cycles on abiotic substrate. Once inside a host, they cannot complete their life cycle and thus do not contribute to population growth

Spill-back = when a native parasite reproduces in an introduced host and subsequently transmits back to the native host population (Figure 1 in Kelly et al. 2009)

Spill-over = when co-introduced parasites are transmitted from non-native to local native hosts (Prenter et al. 2004, Figure 1 in Kelly et al. 2009)

Vector = an abiotic or biotic vessel for dispersing and/or transmitting a parasite. Biotic vectors often refer to insects that transmit diseases from one host to another, generally through biting, but can also include the use of a living organism for parasite dispersal, such as a parasite attached to the mucosal layer of a fish, but not infecting the fish itself. Abiotic vectors generally refer to human-mediated dispersal mechanisms, such as ships, planes, or cars, where parasites can be associated with or attached to these vehicles and be transported to new locations

Zoonotic = can refer to a parasite or disease that normally inhabits or causes disease in wild or domestic animals, but can also infect or cause disease in humans

mechanisms (or *vectors*). For example, larval and juvenile stages of hosts are common in ships' *ballast water* and thought to have few parasites upon arrival (Lafferty and Kuris 1996, Torchin and Lafferty 2009; see also section 7.3.2), since parasite prevalence and intensity is often size- and age-dependent. Still, results from new genetic methods (e.g., *metabarcoding*) suggest that infected planktonic adult organisms (e.g., copepods, dinoflagellates) and free-living or resting stages of parasites may be abundant on commercial ships (Aguirre-Macedo et al. 2008, Kim et al. 2015, Pagenkopp Lohan et al. 2016, 2017), but the extent to which these parasites successfully establish is uncertain (Torchin and Kuris 2005, Torchin and Lafferty 2009).

Whether during or after delivery, detection of *non-native* parasites and other small-bodied organisms is generally much more difficult (and less likely) compared to most free-living macro-organisms (Box 7.2) (Wyatt and Carlton 2002). Parasite invasions that cause mass mortalities are sometimes an exception (Litchman 2010). Nonetheless, many parasite taxa are well known for their indirect effects (Hatcher et al. 2012), which may go unnoticed for long periods of time, while still causing substantial ecological changes in populations and communities. Importantly, the biogeography of potentially invasive etiological agents can be elusive, as the taxa are sometimes new to science or lack historical records of geographical range. As a result, there is

---

Box 7.2 Consequences of small size on invasion ecology baselines

While it is no surprise that large, conspicuous organisms such as fish, crabs, and shelled molluscs are more easily detected and identified than smaller organisms, this simple phenomenon has important broad-scale consequences for existing baseline measures in invasion ecology that are especially relevant to marine parasites. Some key consequences are outlined below as general size-dependent relationships across taxonomic groups, although we recognize that exceptions exist and increased use of molecular tools may shift existing relationships going forward.

The extent of both taxonomic resolution and biogeographic knowledge increases with body size for organisms across taxonomic groups. One consequence of this relationship is that the likelihood of recognizing a species as non-native is inversely related to body size, when organisms are detected. Referred to as the "smalls rule of invasion ecology" (Carlton 2009), such resolution bias is one possible explanation for the few small-bodied species of marine invaders recognized to date (Ruiz et al. 2000b).

The level of effort required to detect organisms and characterize communities (species composition) is also largely size-dependent, as magnitude of effort (e.g., cumulative number of samples collected, locations sampled, and percent of taxa characterized) undoubtedly increases with body size. This is historically the case in marine systems, contributing partly to the poor taxonomic and biogeographic baselines noted above, and continues to be true today. Although some groups of small organisms (e.g., phytoplankton) have received significant sampling effort, and molecular inventories are being developed in some areas for micro-organisms,

these are exceptions. We surmise that local-to-regional inventories for small organisms are often extremely limited compared to macrobiota (e.g., fish, plants, invertebrates, algae), documenting only a minor fraction of extant biodiversity.

The ecological and social effects of most species around the world have not been evaluated, but existing data on species impacts have focused most on relatively large, conspicuous organisms. There are certainly documented examples of strong effects of small organisms in marine ecosystems—such as toxic phytoplankton and bacteria to a diverse range of parasites from molluscs, fish, and plants—but we suggest nonetheless that a strong body size bias exists in the frequency of impact studies, resulting in less knowledge about effects of small organisms compared to large organisms, especially at population, community, and ecosystem levels (Ruiz et al. 1999).

Taken together, these consequences of small body size suggest the number and impact of invasions by small-bodied species may be grossly underestimated relative to large-bodied species in marine systems. The likelihood of detection, taxonomic identification, and biogeographic knowledge are all lower for small organisms. As a result, the correct classification as a non-native species in any region suffers because it is the product of all three components. Further, for those small species known to be introduced, impacts may be much less explored than those of larger organisms. Novel techniques (i.e., metabarcoding, metagenomics) are now being more frequently used to overcome "smalls rule," particularly for detecting early invasions (see Box 7.5 for more detail).

often uncertainty about whether *disease* outbreaks result from new introductions or simply a change in environmental conditions (Harvell et al. 1999, Torchin and Kuris 2005). This issue is exemplified by the oyster parasite *Haplosporidium nelsoni*, the etiological agent of MSX (multinucleated unknown or multinuclear sphere X) disease and cause of mass mortalities in the northwestern Atlantic, which was not recognized as an introduced species from Asia until 50 years post-arrival (Burreson et al. 2000).

Here, we evaluate invasions by marine parasites from three perspectives. First, we explore how traits influence invasion success of microbial parasites in marine environments. Second, we examine the scale and opportunity for marine parasites to spread by anthropogenic vectors and evaluate their current invasion record. Third, in light of these factors, we consider the ecological and evolutionary implications of parasite invasions. We focus primarily on microbial parasites, which have received less attention in invasion ecology than metazoan parasites, despite being transferred in large numbers across the world by human activities. Overall, we show that knowledge about the extent and impact of marine parasite invasions is surprisingly limited, especially for microbial taxa, which have vast potential for colonization and diverse impacts. We surmise that

microbial invasions may be frequent but often undetected to date, relative to larger-bodied organisms, and represent a significant knowledge gap in invasion ecology and parasite biogeography.

7.2 Life-history traits that influence parasite invasions

Many studies identify specific traits of free-living species that facilitate successful invasion of new areas (Jeschke and Strayer 2006, van Kleunen et al. 2010), with parasites and especially *zoonotic* parasites only recently receiving much attention (Cleaveland et al. 2001, Rideout et al. 2017). The primary traits considered to facilitate parasite invasions include (1) the presence of at least one suitable host in the new location; (2) low *host specificity*; (3) a simple life cycle (direct and/or asexual reproduction); and (4) a physical environment that is conducive to the survival of the free-living stage (Taraschewski 2006, Rideout et al. 2017). We expand this discussion to include life-history characteristics that likely influence the ability of a parasite to survive transport and colonize a new geographic area (Table 7.1). We then consider how these traits are distributed across microbial parasite taxa. While much of this discussion is limited by a lack of detailed life-history

Table 7.1 Microbial parasitic taxa in marine systems, including particular life-history traits. Taxonomic assignments are based on Adl et al. (2005, 2007, 2012). Those characteristics predicted to lead to invasion success are in green and those that will impede invasion are in red. We consider host specificity as the range of higher taxonomic groupings of hosts used.

Supergroup	Taxa	Life cycle	Longevity?	Host specificity	Facultative?	Resistant stage?	References
Opisthokonta	Amoebozoa	Direct	Long	Unknown	Yes	Some	Rohde 2005
	Fungi	Direct	Long	Low	Yes	No	Fisher et al. 2012
	Microsporidia	Direct or complex	Variable	High	No	No	Rohde 2005, Keeling and Fast 2002
Rhizaria	Haplosporidia	Direct or complex	Unknown	High	No	Unknown	Rohde 2005
	Paramyxeans	Direct or complex	Short	Low	No	No	Berthe et al. 2004
	Paradinians	Unknown	Unknown	High	No	Unknown	Skovgaard and Daugbjerg 2008, Ward et al. 2018
	Mikrocytids	Direct	Unknown	High	No	Unknown	Abbott and Meyer 2014
	Phagomyxida	Direct	Unknown	Low	No	Known for one species	Neuhauser et al. 2011, 2014

Supergroup	Taxa	Life cycle	Longevity?	Host specificity	Facultative?	Resistant stage?	References
Chromalveolata	Apicomplexa	Direct or complex	Variable	Variable	No	Some	Rohde 2005
	Labyrinthomorpha	Direct	Long	Low	Most	Unknown	Rohde 2005
	Ciliophora	Direct	Unknown	Low	Many	No	Rohde 2005
	Perkinsozoa	Direct	Long	Low	No	Yes	Villalba et al. 2004
	Dinophyta	Direct	Unknown	Low	No	Unknown	Coats 1999, Rohde 2005

information across many parasite taxa, our goal is to provide a useful framework to spur future research.

7.2.1 Host specificity

Finding a suitable host is critical for parasite survival, and thus low host specificity can facilitate invasion success (Cleaveland et al. 2001, Woolhouse et al. 2001, Taraschewski 2006) and allow parasites to invade communities in a new location. Further, parasites with low host specificity are more likely to switch hosts (Holt et al. 2003) and be geographically widespread (Krasnov et al. 2005), potentially increasing the likelihood of transport and survival in a new location. Independent or interacting with host specificity, prior host invasion and establishment can facilitate future invasions of host-specific parasites. For example, many species of molluscs and crustaceans are successful invaders globally (e.g., Asian oyster *Crassostrea gigas*, European green crab *Carcinus maenus*), and thus a suitable host is already present for the coevolved parasites from their *native* range (Torchin et al. 2005, Miura et al. 2006).

A few microbial taxa appear to be host-specific, such as haplosporidians (Rohde 2005), microsporidians (Rohde 2005), and some papillomaviruses which infect cetaceans (Van Bressem et al. 1999). For example, *Haplosporidium scopli, H. nemertis,* and *H. louisiana* are all known from a single host species (Burreson and Ford 2004). Many more microbial parasites have moderate to low host specificity (Table 7.1), including the ability to infect hosts across families and orders, such as viruses that infect fishes (Crane and Hyatt 2011), cetaceans (Van Bressem et al. 1999, 2009), bivalves (Renault and Novoa 2004),

and shrimp (Lightner 2011). For example, tattoo skin disease (TSD) poxviruses infect cetaceans across four orders (Delphinidae, Phocoenidae, Ziphiidae, and Balaenidae) (Van Bressem et al. 2009). Multiple bacteria that infect fishes (Toranzo et al. 2005), bivalves (Paillard et al. 2004), and marine mammals (Van Bressem et al. 2009) have low to moderate host specificity. Among these, *Listonella anguillarum* is a bacterium capable of infecting fishes across six orders (Toranzo et al. 2005). Finally, some protistan taxa, such as particular apicomplexans, labyrinthulids, perkinsids, sydinids, and ciliates, have low host specificity (Table 7.1), as illustrated by *Halioticida noduliformans*, an oomycete parasite that is capable of infecting abalone and lobsters (Holt et al. 2018). It is important to acknowledge that host specificity of many taxa may change, with future biodiversity surveys uncovering new host records and genetic methods (i.e., direct sequencing, metabarcoding, phylogenetics) continuing to redefine species complexes.

7.2.2 Life cycle complexity

The likelihood of parasite colonization should be inversely proportional to life cycle complexity (Kennedy 1994), when controlling for other factors (e.g., host specificity and abundance), as parasites with complex life cycles must find more than one compatible host, often sequentially (Torchin and Lafferty 2009). Still, there are many examples of introduced parasites with complex life cycles (e.g., the trematode *Fasciolodes magna*; Malcicka et al. 2015) that demonstrate these barriers can be overcome (Agosta et al. 2010, Malcicka et al. 2015).

There are conspicuous patterns of life cycle complexity across parasitic microbial taxa (Table 7.1). Although sometimes not the case in terrestrial ecosystems (e.g., Weinert et al. 2009, Sharma and Lal 2017), viruses, bacteria, and many protistan taxa (e.g., ciliates and perkinsids) in marine systems appear to have direct life cycles, only requiring a single host. However, a few microbial groups have parasites with more complex life cycles that require multiple hosts. For example, *H. nelsoni* cannot be directly transmitted among its oyster hosts and thus is believed to require an unidentified secondary host (Burreson and Ford 2004). *Martelia refringens*, a major pathogen of the oyster *Ostrea edulis*, uses a copepod (*Paracartia grani*) as a secondary host in its life cycle (Audemard et al. 2002). Additionally, most microbial parasites can or solely reproduce asexually. These organisms can have short generation times, which may greatly increase mutation rates and lead to a greater likelihood of adapting to new hosts. This is particularly true for viruses, which evolve faster than other parasites, and specifically RNA viruses (Drake and Holland 1999), making them major culprits of emerging disease (Cleaveland et al. 2001).

7.2.3 Longevity and durability of free-living stages

Another key attribute affecting the likelihood of invasion is whether the free-living or transmissive stage can survive the environmental conditions during transport and arrival to new locations. A long-lived transmissive stage, the ability to create a resistant cyst, or high tolerance to a wide range of physical–chemical environments would all increase the likelihood of surviving transit (Kennedy 1994), especially if the parasite is not transported with its host (Solarz and Najberek 2017). However, these issues can be readily overcome if the parasite is introduced to a habitat ecologically similar to its native one (Kennedy 1994).

Many microbial parasites have evolved mechanisms for long-term survival outside of their hosts (Table 7.1). Some viruses are extremely tolerant, capable of surviving desiccation, freezing, long periods of time in the water column, or even passage through the digestive system of a non-host (Huchin-Mian et al. 2009, Crane and Hyatt 2011, Lightner 2011). For example, Taura syndrome virus (TSV), which infects shrimp, can survive and remain infectious after passage through a bird's digestive track (Lightner 2011). Aggregating in biofilms can provide pathogenic bacteria protection from biotic and abiotic threats, including predation, UV light, toxins, dehydration, salinity, antibiotics, and other antimicrobial chemicals and processes (Hall-Stoodley et al. 2004). Bacteria can also invade protists, such as amoebae, and be protected when the protist encysts (Barker and Brown 1994). The longevity of protistan parasites is highly variable, with some having tolerant free-living stages that can survive long durations (e.g., weeks to months) without their hosts, such as the prezoosporangia of *Perkinsus* spp. (Villalba et al. 2004), oocysts of *Toxoplasma gondii* (Dubey 2004), and the spores of microsporidians (Gajadhar and Allen 2004).

7.2.4 Facultative and sapronotic parasites

Not all disease-causing organisms are obligate parasites or require a host for survival. Some are referred to as *facultative parasite*s, which may have fewer barriers to invasion compared to obligate parasites. There are many examples of facultative parasites among microbes. Bacterial pathogens can be facultative, including bacterial pathogens of bivalves (Paillard et al. 2004), crustaceans (Vogan et al. 2008), corals (Lesser et al. 2007), and cetaceans (Van Bressem et al. 2009). For example, a number of bacteria (*Alteromonas* spp., *Vibrio* spp.) that cause epizootic shell disease in crustaceans occur without any hosts in the aquatic environment (Vogan et al. 2008). A number of protistan parasites are also facultative, including members of the Labyrinthulomycetes (Burge et al. 2013, Sullivan et al. 2013), ciliate parasites of crustaceans (Morado and Small 1994, Small et al. 2013), fungi that infect corals (Burge et al. 2013), and amoebae (Schuster et al. 2004). For example, *Aspergillus sydowii* is a terrestrial fungus that has infiltrated Caribbean waters and causes opportunistic infections and disease in corals (Burge et al. 2013).

Sapronotic agents can also survive without a host but, unlike facultative parasites, cannot complete their life cycles once inside a host (Hubálek 2003). While the initial description specified only sapronoses in humans, Kuris et al. (2014) expanded this to include any organism that is typically free-living but can parasitize and cause disease if presented the opportunity. As sapronotic parasites are not reproductively dependent on their hosts, they do not adapt or evolve virulence to a host (Kuris et al. 2014). Examples of sapronotic agents are known for bacteria, fungi, and protists (Kuris et al. 2014). Thus, we hypothesize that both facultative and sapronotic parasites have the same potential to invade new regions as free-living microbes.

7.2.5 Recipe for invasion success

A number of traits in combination likely contribute to the successful introduction and establishment of parasites, including (1) low host specificity or no host requirement, (2) a high tolerance for variation in environmental conditions, including the existence of resting or dormant stages, and (3) a direct life cycle with asexual reproduction. This combination of traits is often encountered in microbial parasites (Table 7.1) (Cleaveland et al. 2001, Rideout et al. 2017), and their importance is further highlighted by the World Organisation for Animal Health (OIE) priorities, where biosecurity threats are focused on microbial parasites (Box 7.3).

Box 7.3 OIE reportable diseases in the marine realm

The World Organisation on Animal Health has generated a list of diseases of aquatic organisms to "ensure transparency in and enhance knowledge of the worldwide animal health situation" (http://www.oie.in). The overall goal is to prevent the future spread of these parasites to novel locations and hosts. Upon examining the twenty-six diseases of aquatic animals currently on the list, 96 percent of the etiological agents are microbial, whereas only one metazoan is present. Within the microbial taxa, 60 percent are viruses, 28 percent are protists, and 12 percent are bacteria. While this list is primarily based on the likelihood of the parasite causing mass mortalities if established in a new area, rather than the overall likelihood of invasion success or establishment, it highlights a discrepancy between invasion research, where primarily larger organisms are detected (Box 7.2) and the biosecurity threat clearly posed by microbial parasites, mostly those important to aquaculture (Chapter 10—Behringer et al.). Additionally, as discussed in the text, many of these microbial taxa likely have characteristics increasing their chances of surviving introduction and establishment in new areas.

Host taxa	Disease name	Disease-causing agent	Pathogen type
Fish	Epizootic hematopoietic necrosis disease	Epizootic hematopoietic necrosis virus (EHNV)	Virus
	Epizootic ulcerative syndrome	*Aphanomyces invadans*	Fungus
	Infection with *Gyrodactylus salaris*	*Gyrodactylus salaris*	Metazoan
	Infection with highly polymorphic region (HPR0) infectious salmon anemia virus	HPR0 infectious salmon anemia virus	Virus
	Infection with salmonid alphavirus	Salmonid alphavirus	Virus
	Infectious hematopoietic necrosis	Infectious hematopoietic necrosis virus (IHNV)	Virus
	Koi herpesvirus disease	Koi herpesvirus	Virus
	Red sea bream iridoviral disease	Red sea bream iridovirus	Virus
	Spring viremia of carp	Spring viremia of carp virus	Virus
	Viral hemorrhagic septicemia	Viral hemorrhagic septicemia virus	Virus

continued

Box 7.3 *Continued*

Host taxa	Disease name	Disease-causing agent	Pathogen type
Mollusc	Infection with abalone herpesvirus	Abalone herpesvirus	Virus
	Infection with *Bonamia exitiosa*	*Bonamia exitiosa*	Protist
	Infection with *Bonamia ostreae*	*Bonamia ostreae*	Protist
	Infection with *Marteilia refringens*	*Martelia refringens*	Protist
	Infection with *Perkinsus marinus*	*Perkinsus marinus*	Protist
	Infection with *Perkinsus olseni*	*Perkinsus olseni*	Protist
	Infection with *Xenohaliotis californiensis*	*Xenohaliotis californiensis*	Bacteria
Crustacean	Acute hepatopancreatic necrosis disease	Unidentified bacteria	Bacteria
	Crayfish plague	*Aphanomyces astaci*	Fungus
	Necrotizing hepatopancreatitis	*Hepatobacter penaei*	Bacteria
	Infection with infectious hypodermal and hematopoietic necrosis virus	Hypodermal and hematopoietic necrosis virus	Virus
	Infection with infectious myonecrosis virus	Myonecrosis virus	Virus
	White tail disease	*Macrobrachium rosenbergii* nodavirus	Virus
	Infection with Taura syndrome virus	Taura syndrome virus	Virus
	Infection with white spot syndrome virus	White spot syndrome virus	Virus
	Infection with yellow head virus genotype 1	Yellow head virus genotype 1	Virus

7.3 Assessing marine parasite species transfers and invasions

Multiple approaches have been used to evaluate invasion dynamics in marine ecosystems around the globe, focusing on either (1) the delivery process or (2) patterns of invasion. Studies of species delivery examine the *flux* of organisms or *propagule pressure* associated with human activities, such as shipping or trade, which move species across oceans and from local-to-regional scales. Just as living vectors transmit parasites, invasion ecology considers human-mediated transfer mechanisms as vectors, evaluating associated species composition and abundance, the magnitude or size of transfers, and differences in space or time of vector operation (e.g., changing trade patterns or tempo, regulations, management) that affect biotic exchange (Muirhead et al. 2015, Fowler et al. 2016, Carney et al. 2017).

In contrast, studies of invasion patterns rely on the invasion record, examining detections of non-native species to evaluate spatial, temporal, and taxonomic patterns of invasion as well as the relative contribution of particular vectors to the number of known invasions (Ruiz et al. 2000b, Wonham and Carlton 2005, Katsanevakis et al. 2013). Throughout this chapter, vectors refer to anthropogenic vectors (not biological vectors). We review briefly the current state of knowledge about marine parasite delivery and invasions in sections 7.3.2 and 7.3.3.

7.3.1 Temporal changes in vector operation and species transfer

Although quantifying the total flux of organisms per unit time within and across vectors remains challenging, some coarse comparisons have been advanced (Williams et al. 2013) and these suggest invasions may be increasing over time. Figure 7.1 expands on earlier portrayals of vector dynamics for those known to contribute significantly to either marine species transfers or invasions and outlines both the likely temporal changes per vector in total flux of all marine organisms and the current relative flux of parasites (comparing small to large parasite taxa).

The estimated rate of organism delivery for most individual vectors has increased greatly over time (Figure 7.1B), in terms of both organism abundance and species richness. These increases are driven by a combination of increases in the magnitude of trade (number of inoculation events per year) and changes in per capita organism release per inoculum, with the latter resulting from changes in size of inoculum, management or operation, and speed of transfer. We estimate temporal increases in the number and frequency of species transfer events per year for all vectors except stocking (Figure 7.1B1), which declined strongly due to regulatory and policy changes. Directional changes in the per capita organism delivery (Figure 7.1B2) are more variable

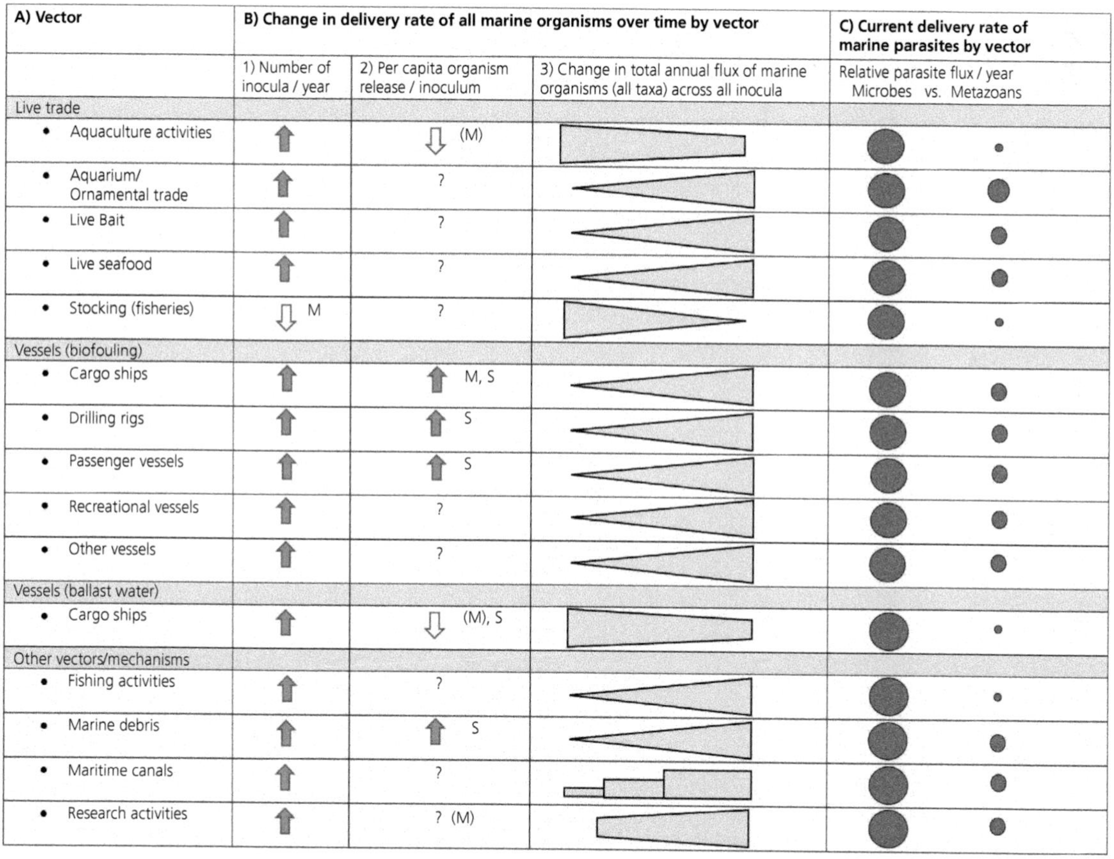

Figure 7.1 Marine organism flux associated with different vectors. Estimates of organism flux represent both abundance and species richness. Shown by vector (A) are changes through time in the delivery rate of marine organisms for all taxa combined (B), including free-living, parasitic, and symbiotic species. Temporal changes in delivery rate (B3) result from the product of components the number of inocula per year (B1) and per capita release of organisms/inoculum (B2). Estimated temporal changes (increase, decrease, or uncertain (?)) in these components are shown for each vector, along with outcome (increase or decrease) and shape function of total annual flux. Some temporal changes in components are driven by shifts in management (practice and/or regulation) or size of inocula, shown respectively as M and S (with those resulting in decreases in organisms shown in parentheses). Also shown by vector is a comparison of the relative flux of small versus large parasite taxa at the present time (C). (Note that all comparisons (B and C) are within vector only and do not represent comparisons across vectors.)

and often difficult to characterize, due to lack of robust measures for many vectors (Williams et al. 2013). For most of these vectors, the speed of transfer has increased over time, increasing survivorship per delivery event. Also, the global reach and connectivity has increased through time, expanding the number of source regions and total species pool involved. Nonetheless, changes in per capita delivery attributes are often uncertain, except for aquaculture, vessel biofouling, ballast water, and marine debris.

While the scale of aquaculture has continued to expand (Ottinger et al. 2016), the mode of operation has shifted in at least three ways that have reduced the unintentional transfers of associated non-target organisms over time (Grosholz et al. 2015). First, shipment of aquaculture stocks has moved from older individuals collected from the field to laboratory-raised larvae or young post-larvae. This has reduced the abundance and species richness of hitch-hiking organisms and parasites, which were a major source of invasions in some regions (e.g., Cohen and Carlton 1997). Second, protocols for screening and quarantine have been adopted and are widely used to reduce transfer of specific infectious agents (see OIE, http://www.oie.in; Box 7.3).

Third, many countries have increased regulations and oversight of species transplants for aquaculture purposes. Nonetheless, local or regional movement of aquaculture species (e.g., oysters or mussels moved among locations during production) may remain common in some areas, facilitating secondary spread of non-native species.

There have been multiple changes in organism delivery by vessel biofouling and ballast water over the past decades (Figure 7.1B2). Per capita delivery rates of *biofouling organisms* by cargo ships, drilling rigs, and passenger vessels have probably increased, due partly to increases in vessel size and speed, which increases the underwater surface area available for colonization and the survivorship of colonizers, respectively, as well as limited management or regulation (Box 7.4). Temporal trends in per capita delivery rates of biofouling for recreational boats and other vessels (e.g., tugs and barges) are less clear, because operations and hull husbandry practices have received less attention. Unlike vessel biofouling, per capita delivery rates for ships' ballast water have declined, due to extensive regulations to reduce associated species transfers and invasions in coastal ecosystems (Box 7.4). Representing a major

---

Box 7.4 Changing management of biota by commercial ships

Commercial ships, a major source of invasions in coastal ecosystems, transfer marine taxa associated primarily with biofouling and ballast water communities. Management practices have changed over time for each, affecting the biotic assemblages delivered by ships.

To date, biofouling management has focused primarily on vessel performance (e.g., reducing vessel drag and fuel consumption) and there are still very few restrictions or regulations on biota associated with ships' hulls and underwater surfaces, although this issue is now receiving serious attention at regional-to-international scales for large macro-organisms (Davidson et al. 2016). Nonetheless, the management of hull biofouling has undergone a major shift since the 1970s, with an international ban of tributyl tin in antifouling paints (due to environmental impacts), being replaced with other substances that are likely less effective at reducing biofouling accumulation.

In contrast, delivery of organisms in ships' ballast water has been the focus of significant management and regulation (Carney et al. 2017, IMO 2017), advancing a stepwise series of requirements now being phased in to reduce concentrations of organisms transferred among ports. The current requirement of international, national, and regional regulations is to meet specific discharge standards for ballast water that are based on organism size. For example, management of ships' ballast water aims to reduce concentrations of large (> 50 µm) organisms to below 10/m³, a reduction of several orders of magnitude compared to untreated water (Minton et al. 2015). In contrast, the target for small (10–50 µm) organisms may not represent any reduction, and that for the smallest (< 10 µm) organisms is restricted to only three taxa (including the bacteria *Vibrio cholerae*, *Escherichia coli*, and intestinal *Enterococci*), despite the large concentrations and diversity of microbes in ships' ballast, including parasites (Drake et al. 2001, Pagenkopp Lohan et al. 2017; Box 7.5).

behavioral change of the global shipping fleet, we note that these management strategies focus on large-bodied organisms and may have little or no effect on microbial taxa.

Less appreciated until recently is the growing importance of marine debris for long-distance dispersal of marine organisms, including their parasites. While there has been considerable focus on microplastics and growing accumulations of larger material in the open ocean (Jambeck et al. 2015), the potential consequence for transoceanic dispersal has come into recent focus following the 2011 earthquake and tsunami in Japan that launched boats, docks, floats, and other material into the sea and resulted in the arrival of hundreds of Asian coastal species to the shores of Hawaii and North America (Carlton et al. 2017). This event underscores the growing volume and size of debris available for species transport among coastal regions by ocean currents and its interaction with events, including storms and hurricanes that are increasing in frequency and severity due to climate change.

Overall, we surmise large, global-scale increases in propagule pressure are occurring for most individual vectors, in terms of both organism abundance and species richness transferred by human activities (Figure 7.1B3). Importantly, the cumulative species richness being transferred across vectors is also increasing, as global trade expands to include more transfers, source regions, and destinations. While per capita changes are uncertain for many vectors, the large increases in transfer numbers (Figure 7.1B1) are likely driving increases in total flux. In the case of ships' ballast water, a large increase in trade appears to be partially compensating for the reduced per capita delivery due to heightened management (Carney et al. 2017). In the case of major shipping canals, such as the Panama and Suez Canals, expansion has occurred multiple times, resulting in punctuated increases in size and number of transiting vessels.

7.3.2 Flux of marine parasites

While Figure 7.1B estimates temporal changes in flux within each vector for all organism types, we also suggest the general patterns are robust and apply equally to free-living, symbiotic, and parasitic organisms. In general, we expect transfer of microbes to be greater than metazoans, reflecting the inverse relationship in abundance and species richness with body size among taxonomic groups (Gaston and Blackburn 2007). Considering parasites specifically, we hypothesize that the current flux is greater for microbes versus metazoans across all vectors (Figure 7.1C). Any such difference in propagule pressure has implications for invasions, since the probability of successful establishment increases with density, abundance, and richness (Ruiz et al. 2000b). Thus, the combined effects of life-history traits (section 7.2) and propagule supply suggest that microbial parasite invasions may be especially frequent relative to metazoan parasites.

We also hypothesize that the relative flux of microbial versus metazoan parasites is greatest within those vectors that transfer primarily larval, young, and small organisms, since prevalence and intensity of infection by metazoan parasites is often size- and age-dependent within host species. For example, the relative flux of microbial parasites is likely greatest for aquaculture, stocking, and ballast water (Figure 7.1C), due to management and regulations that reduce the transfer of older, larger individuals, compared to that for (1) the aquarium trade, which generally focuses on large individuals for sale, and (2) other vectors that include a wide range of organism sizes and ages.

More broadly, the flux of parasites across various vectors is largely unappreciated and poorly resolved, especially for microbes. There are protocols to prevent the transfer of parasites with aquaculture activities (Box 7.3) that often focus on screening for a subset of known taxa, but the transfer of small organisms receives little attention compared to larger macrofauna for many vectors. This focus on larger organisms is partially an outgrowth of the limited baseline knowledge about the extent and impact of invasions by small-bodied organisms (Box 7.2).

Finally, the current scale of transfer opportunity for parasites is illustrated by commercial ships, a dominant vector for marine species transfers. Most of the world's goods are moved by commercial ships, with a global fleet in excess of 120,000 active vessels (Moser et al. 2015). Approximately 100,000 commercial ship arrivals occur each year in US waters alone, discharging > 250 million m³ of ballast water in 2018 alone (according to the National Ballast Information Clearinghouse (NBIC), http://invasions.si.edu/nbic/). Regulations require vessels to meet ballast water discharge standards (Box 7.4), which will not be fully implemented for several years. Assuming all ships currently met these discharge standards, ballast-mediated transfer of large (>50 μm) organisms to US waters would be constrained to 2.5 billion individuals/year (10 organisms/m³), and that for small (10–50 μm) organisms would be 2.5 × 10¹⁵ individuals/year (10 organisms/ml). Bacteria and viruses arrive in concentrations of roughly 10⁹ bacteria/ml and 10¹⁰ virus-like particles/ml (Drake et al. 2001), suggesting total current delivery in the order of 10²³ bacteria/year and 10²⁴ viruses/year.

A similar approach illustrates the potential scale of transfer of marine biota on the hulls of vessels. The mean annual flux of underwater surface area for commercial ships arriving to the USA is estimated to be 510 km², or roughly three times the area of Washington, DC (Miller et al. 2018). While less is known about the delivery and release of biota associated with vessel hulls, it is clear that this potent vector transfers a wide diversity of micro-organisms, invertebrates, and macroalgae (Zabin et al. 2014, Davidson et al. 2018; see also section 7.3.3), with few restrictions on the transfer of biota (Box 7.4).

7.3.3 Invasion history of marine parasites

While there are conspicuous examples of well-documented parasite invasions in marine ecosystems, parasites are generally underrepresented in marine invasion records relative to free-living species. For example, of the 520 non-native species known currently to have established populations in marine and estuarine waters of North America through 2018 (according to the National Exotic Marine and Estuarine Species Information System (NEMESIS), https://invasions.si.edu/nemesis,), only 11 percent are parasites or other symbionts (Figure 7.2A). The latter are dominated by helminths (40 percent of total), protists (24 percent), or crustaceans (19 percent), which are associated primarily with fishes, crustaceans, and molluscan hosts (Figure 7.2B and C).

The current record of invasion history underestimates the full extent of invasions across all taxa, as many species go undetected or are unrecognized as non-native (Carlton 2009), and this underestimate is likely most pronounced for small-bodied organisms (Box 7.2). Furthermore, parasites may be among the least likely to be added to the invasion record, because parasite biogeography is poorly understood and parasites are generally difficult to identify to low taxonomic levels, which is compounded by the issue of cryptic species (Miura et al. 2006, Leung et al. 2009). It is also noteworthy that most non-native parasites have been documented for host taxa with larger bodies (Figure 7.2C), reflecting a size bias in the range of hosts studied.

An alternative hypothesis is that invasions by small-bodied species and parasites are disproportionately rare compared to larger free-living species. In general, marine species richness and abundance decrease with increasing body size in natural communities (Gaston and Blackburn 2007), suggesting that more small species exist and are available for transfer. Further, current data on vectors suggest that transfers of small organisms are greater than for large organisms (see section 7.3.2). Thus, the alternative hypothesis requiring lower establishment rates for small species and parasites is contrary to these patterns. For free-living microbes, some have suggested they are more widespread geographically than large organisms, lacking biogeography and therefore limiting their invasions, despite extensive human-mediated transfers of small organisms (e.g., Dobbs and Rogerson 2005, Carlton 2009). This size paradox in invasion ecology remains unresolved, requiring further analysis of biogeography for both free-living and parasitic organisms, and efforts/technology to detect small-bodied invaders and their parasites.

While most species are host to a diverse parasite fauna in their native range, hosts often leave behind many parasites when they invade new regions (see section 7.4). This suggests some limitation exists, due to either transfer opportunity or colonization success, in cases examined to date. It is, however, noteworthy that most marine species have not been surveyed for introduced parasites, suggesting that many may be undetected. Moreover, most of the existing work has focused on relatively large-bodied

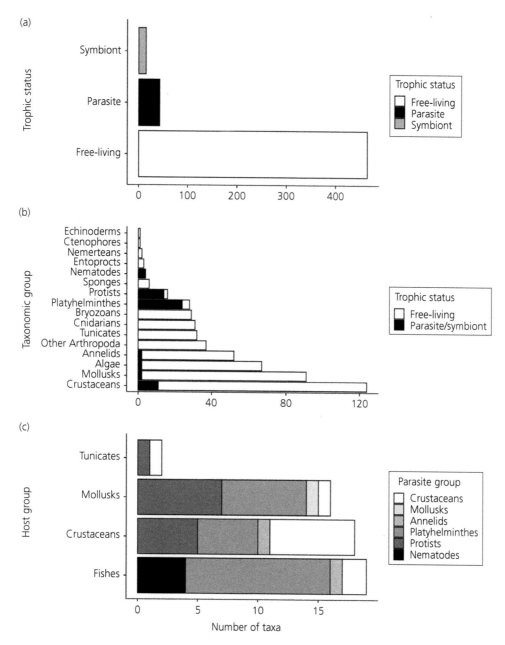

Figure 7.2 Invasion baseline of parasites, symbionts, and free-living species for coastal marine habitats in North America. Shown are the number of established species documented in marine and estuarine waters of the continental USA and Canada for invertebrates, algae, and micro-organisms (excluding vertebrates and vascular plants; updating Ruiz et al. 2000), highlighting (A) the relative contribution of parasitic and symbiotic taxa to the total species pool, (B) number of reported non-native parasite and symbiont species (combined) across taxonomic groups compared to free-living taxa, and (C) reported distribution (number) or parasite and symbiont species (by taxonomic group) across host taxonomic groups. Invasion record data presented are summarized and available from NEMESIS (https://invasions.si.edu/nemesis), including the invasion record through 2017.

hosts and metazoan parasites (Figure 7.2B and C), with 76 percent of known introduced marine parasites in North America being metazoans. Yet, life-history requirements (e.g., host specificity, complex life cycles) may differentially limit invasion success of metazoan compared to microbial parasites, as demonstrated by *C. maenus* escaping macroparasites but not microparasites when invading new areas (Bojko et al. 2018). We hypothesize therefore that marine invasions by microbes are much more common than metazoans, due to the sheer number of organisms transferred and life-history characteristics that facilitate invasions, which suggests a pronounced disparity in detection for microbial parasites to date.

7.4 Ecological and evolutionary consequences of parasite invasion

Relatively little is known about the ecological consequences of introduced parasites in marine systems, particularly microbial parasites. However, reviews by Blakeslee et al. (2013) and Goedknegt et al. (2016) demonstrate some of the complex ways that invasions and diseases interact, albeit most of the available examples relate to metazoan parasites. Extensive evidence indicates that introduced marine hosts escape most of their native parasites (see reviews by Torchin et al. 2002, Torchin and Mitchell 2004, Torchin and Kuris 2005, Blakeslee et al. 2013, Goedknegt et al. 2016), but bring a subset of infectious agents (Torchin et al. 2002, 2003, Blakeslee et al. 2013, Goedknegt et al. 2016). These can have devastating consequences through disease outbreaks in native communities that result in population declines. For example, the monogenean *Nitzchia suronis* was introduced with its host, *Acipenser stellatus*, to the Aral Sea and subsequently infected and caused mass mortalities of the native sturgeon, *Acipenser nudiventris* (Osmanov 1971, Zholdasova 1997). Parasites can also have ecosystem-wide effects by altering competition, predation, and herbivory (Wood et al. 2007, Hatcher et al. 2012) and food webs (Chapter 2—McLaughlin et al.). This was demonstrated by the pilchard herpes-virus introduced to Australia in the mid-1990s (Hyatt et al. 1997, McCallum et al. 2004). Imported with baitfish from Japan and the Americas, the herpes-virus

rapidly spread to native fishes, notably the pilchard (*Sardinopsis sagax*), causing widespread mortalities and population declines with wide-ranging effects, including impacting the seabird populations that rely on the fishes. Further, the indirect effects of parasites and pathogens can be substantial (reviewed by Dunn et al. 2012). Parasites can alter host behavior/activity (Kuris 1997, Hughes et al. 2012) and immune responses (Lee and Klasing 2004), which all have the potential to influence invasion consequences. However, for most introduced parasites, we lack vital information about the community-level consequences or indirect impacts.

The introduction of infectious agents and potential hosts can lead to several outcomes. Parasites are sometimes introduced with their hosts, so if newly introduced parasites have broad enough host specificity, they can *"spill-over"* and infect native hosts (Prenter et al. 2004, Figure 1 in Kelly et al. 2009). Some of the best examples in marine systems come from aquaculture co-introductions, such as the sturgeon example above and oyster introductions (Behringer et al. Chapter 10, this volume). Historically, entire oyster reefs were moved to new locations, introducing oysters and associated free-living organisms (Ruesink et al. 2005, Blakeslee et al. 2013). While we still do not know the whole scope of parasite introductions associated with oyster movements, several have spread and caused disease in native molluscs (Friedman 1996, Minchin 1996, Burreson and Ford 2004, Blakeslee et al. 2013, Goedknegt et al. 2016). One example is the haplosporidian *Bonamia ostreae*, which is thought to have been introduced from the West Coast of North America to Europe via the movement of European oysters (*Ostrea edulis*) from California back to Europe where it caused mass mortalities in the native oysters (Goedknegt et al. 2016).

Goedknegt et al. (2016) provide a comprehensive overview of marine parasites co-introduced with invading hosts. Eighty percent of these introductions were presumably a result of historical aquaculture and fisheries introductions. Interestingly, of the thirty-five marine parasites reported to have spilled over to native hosts, only 37 percent are microparasites and 63 percent are macroparasites. However, as noted in section 7.1, introduced microparasites are likely underreported compared to macroparasites. Fortunately, most modern aquaculture practices have increased biosecurity measures and comply with quarantine measures to screen for parasites and diseases (Torchin and Kuris 2005; Box 7.3).

Introduced species may accumulate novel parasites and pathogens from the native range (Torchin et al. 2002, 2003), and if the introduced host is competent, it may amplify transmission and *"spill-back"* to native hosts (Kelly et al. 2009). One example is the amplification and spread of the native "ich" disease by introduced shad to native salmon in the Pacific Northwest (Hershberger et al. 2010). Here, the introduced host is considered responsible for moving the marine parasite into the Columbia River system; however, the impacts and transmission of this native parasite in freshwater remain uncertain (Goedknegt et al. 2016). Several generalist parasites have been reported to infect introduced hosts, but the extent to which this causes spill-back and increased transmission rates to native hosts is less clear and difficult to demonstrate. Microorganisms, particularly viruses (Faillace et al. 2017), may further impact invasion dynamics through high mutation rates allowing rapid evolution and potential for switching and adapting to novel hosts. This might be the case in an emerging fishery for European green crabs in North America, where the non-native crabs are used as bait for the lobster (*Homarus americanus*) fishery. Bojko et al. (2018) have recently identified several pathogens that could be transmitted to the native lobsters as well as other crustaceans.

We still know relatively little about the ecological and evolutionary consequences of diseases associated with marine compared to terrestrial invasions (Torchin and Kuris 2005, Torchin and Lafferty 2009, Blakeslee et al. 2013, Goedknegt et al. 2016) and how both natural and anthropogenic vectors might interact, resulting in new disease outbreaks (Miura et al. 2006). This is particularly true for microbial parasites which are generally more difficult to detect (Litchman 2010). Nonetheless, the unique nature of the marine environment combined with increasing dispersal through shipping may facilitate the rapid spread of both introduced free-living species (Grosholz 2002) and new parasites and diseases (McCallum et al. 2004).

7.5 Looking forward: integrating parasite, disease, and invasion ecology

As human populations and global trade continue to increase, opportunities for worldwide dispersal of marine parasites will continue to expand. Although a large proportion of species on Earth are parasitic (de Meeus and Renaud 2002, Dobson et al. 2008) and many cause diseases, knowledge of biogeography (Poulin 2014) and the extent of invasions is relatively limited for parasites, particularly in the ocean (Harvell et al. 1999). Increasing evidence suggests extensive transfers of micro-organisms to new locations by commercial ships (Box 7.5) and other vectors, including resistant free-living infective stages and sapronotic agents (Ruiz et al. 2000a, Drake et al. 2001,

---

Box 7.5 Protistan parasites in ballast water

Previous research showed very low resolution, but high taxonomic breadth of microbial taxa reported in ballast water (Drake et al. 2001), suggesting that many parasites were likely present and transported in ballast water. Thus, we re-examined a dataset from Pagenkopp Lohan et al. (2017) to identify and characterize the protistan and metazoan parasites present in ballast water samples (*n* = 61) from thirty-nine vessels entering ports across three US coasts. We used metabarcoding to identify 783 parasite operational

taxonomic units (OTUs) either to species level or to a group where all organisms are parasitic (e.g., Apicomplexa, Syndinidae). Only two of these OTUs were identified as metazoan parasites (i.e., myxosporeans); all other parasite OTUs were protists. To further assess the accuracy of the genus- or species-level identifications, we chose three parasite taxa for additional analyses: *Syndinium turbo* and *Hematodinium* spp. (Figure 7.3) and *Perkinsus* spp. (Figure 7.4). For these OTUs, we created alignments using

continued

Box 7.5 *Continued*

the MAFFT (Katoh et al. 2002) plugin in Geneious v11.0.4 (Biomatters Ltd) using the representative sequences from those OTUs and additional sequences of those taxa from the PR2 database (Guillou et al. 2013). We then created neighbor-joining trees in Geneious using the Kimura-2 model and 1,000 bootstrap replicates. Raw sequence data can be found under NCBI BioProject PRJNA371431.

These trees highlight the diversity of just some of the parasites found in ballast water. OTU_977 is positioned well within the clade of other *Syndinium turbo* sequences, confirming its identification; however, OTU_1364 and OTU_2818 are not as

easily confirmed (Figure 7.3). Both appear to be syndinids, with OTU_1364 potentially being an undescribed species of *Hematodinium* and OTU_2818 grouping with another sequence only identified as Dino Group IV. For the eleven OTUs identified as *Perkinsus* spp., none of these sequences forms clades with known *Perkinsus* or *Parvilucifera* spp. (Figure 7.4). Consistent with previous studies (Guillou et al. 2008, Poulin 2014), this suggests that there are many unidentified parasites in the world's oceans and that these parasites are being transported by ships across vast distances, providing opportunities for novel parasite invasions.

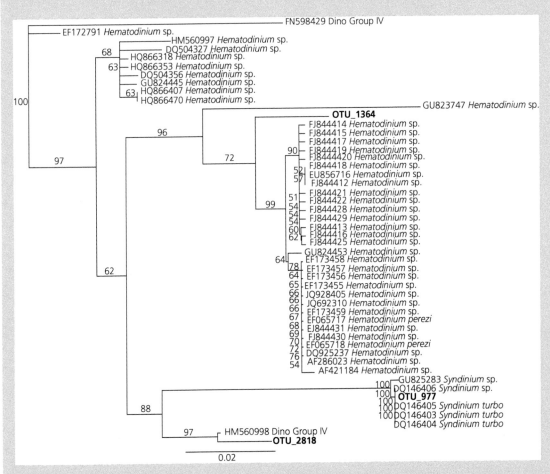

Figure 7.3 Neighbor-joining tree generated in Geneious v11.0.4 to further characterize OTUs identified from ballast water samples as *Syndinium turbo* or *Hematodinium* spp.

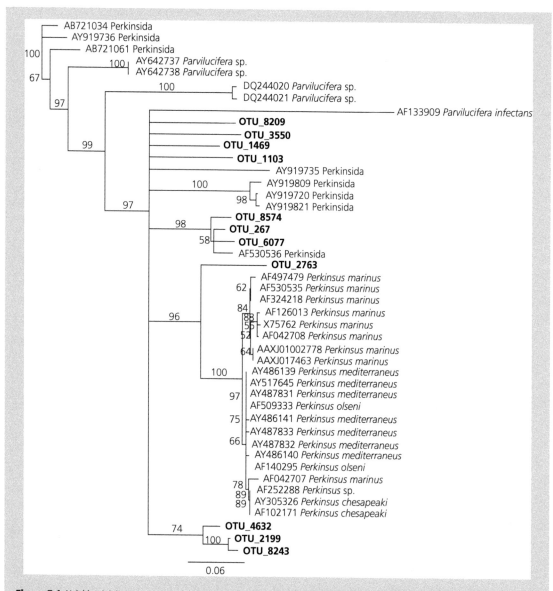

Figure 7.4 Neighbor-joining tree generated in Geneious v11.0.4 to further characterize OTUs identified from ballast water samples as *Perkinsus* spp.

Pagenkopp Lohan et al. 2016, 2017). While examples of marine parasite invasions are well known, the full diversity of parasites transferred, their fate, and their impacts on populations and communities remain unquantified. We hypothesize that the magnitude and impact of parasite invasions by microbial marine taxa is vastly underestimated. The scale of human-mediated transfer of these organisms creates diverse opportunities for invasions, and their life-history traits make them highly likely to survive and colonize novel hosts in new areas.

Understanding the full scope and consequences of marine parasite invasions requires multi-disciplinary research that integrates genomics, biogeography, parasitology, and disease ecology, focusing especially on microbial parasite diversity, invasion dynamics, and disease-related impacts across multiple scales. All of this information is critical to evaluate what factors determine invasion success. Recent advances in molecular approaches (e.g., metabarcoding, *metagenomics*, *metatranscriptomics*) can uncover regional parasite biodiversity (Bass et al. 2015; Frasca et al. Chapter 11, this volume), especially for microbial parasites easily overlooked in morphological surveys (Box 7.2). Following up with histopathology (Bateman et al. Chapter 1, this volume) and ecological surveys of host individuals, populations, and communities will provide greater understanding of their impacts (Burge et al. 2016). Additionally, examination of museum specimens could allow for differentiating how environmental changes, human impacts, and new outbreaks have interacted to shape diseases in today's oceans and elucidate historical biogeography of some groups (Harmon et al. 2019). For example, researchers used museum specimens to demonstrate that the virus responsible for sea star wasting disease was present in these outbreak locations long before these outbreaks were first reported (as far back as 1942 (Hewson et al. 2014)).

We also call for detailed analyses of parasite communities associated with key anthropogenic vectors (e.g., marine debris, ballast water, ships' hulls) to further elucidate the flux of microbial parasites to coastal ecosystems by human activities (e.g., Williams et al. 2013). Ideally, these measures would be combined with (1) challenge experiments, to evaluate the potential for transit survival and colonization in recipient locations, and (2) surveys of local biota to detect whether invasions have occurred and how host individuals, populations, and communities are being directly and indirectly impacted by the invasive parasites.

Finally, independent of parasite delivery, it is important to recognize that coastal ecosystems are undergoing rapid changes, which may affect their susceptibility to invasions (Ruiz et al. 1999, 2000). For example, changes in community structure and species composition through disturbances, biodiversity loss, and other factors will affect the opportunity for invasions. In the case of parasites, the presence or abundance of particular host species may change through time. Moreover, hosts experience various types of stress (e.g., temperature, eutrophication, acidification), which may affect their immune response and susceptibility. Such a shift in stressors may operate in a similar fashion to disturbances, which are known to affect invasion of free-living organisms (Hobbs 1989, Horvitz 1997). Thus, temporal changes in host abundance, composition, and condition may also affect invasion outcome for marine parasites. The effects of host composition and stress on invasion outcome is largely unexplored but deserves serious attention, particularly as the global climate continues to change, with many potential consequences for marine ecosystems, their inhabitants, and future invaders.

7.6 Summary

- Microbial parasites have life-history characteristics that facilitate invasions.
- Anthropogenic activities result in large transfers of marine parasites across the globe, creating many opportunities for invasions in diverse marine environments.
- While the invasion potential of marine parasites and other microbes is vast, due to propagule pressure and life-history traits, remarkably few microbial parasite invasions are documented from coastal ecosystems to date.
- We hypothesize that parasite invasions are common but mostly unreported, especially for microbial taxa, due to size bias in detection.
- Although the evolutionary and ecological consequences of marine parasite invasions can be potent, these are largely unexplored, even where invasions have been detected.
- To understand the actual scale and impact of marine parasites, research that integrates measures of diversity, biogeography, invasion ecology, and disease-related impacts are urgently needed.

Acknowledgments

We thank our many colleagues who have indulged us with endless conversations brainstorming about parasites and invasions over the years, particularly Kristina Hill-Spanik, Armand Kuris, and Kevin Lafferty. We also thank Brian Steves and Paul Fofonoff for help in generating figures from NEMESIS data.

References

Abbott, C. L. and Meyer, G. R. 2014. Review of *Mikrocytos* microcell parasites at the dawn of a new age of scientific discovery,. *Diseases of Aquatic Organisms* 110: 25–32.

Adl, S.M., Simpson, A. G. B., Farmer, M. A., Andersen, R. A., Anderson, O. R., Barta, J. R., Bowser, S. S., Brugerolle, G., Fensome, R. A., Fredericq, S., James, T. Y., Karpov, S., Kugrens, P., Krug, J., Lane, C. E., Lewis, L. A., Lodge, J., Lynn, D. H., Mann, D. G., McCourt, R. M., Mendoza, L., Moestrup, O., Mozley-Standridge, S. E., Nerad, T. A., Shearer, C. A., Smirnov, A. V., Spiegel, F. W., and Taylor, M. F. J. R. 2005. The new higher level classification of eukaryotes with emphasis on the taxonomy of protists,. *Journal of Eukaryotic Microbiology* 52: 299–451.

Adl, S. M., Leander, B. S., Simpson, A. G. B., Archibald, J. M., Anderson, O. R., Bass, D., Bowser, S. S., Brugerolle, G., Farmer, M. A., Karpov, S., Kolisko, M., Lane, C. E., Lodge, D. J., Mann, D. G., Meisterfeld, R., Mendoza, L., Moestrup, O., Mozley-Standridge, S. E., Smirnov, A. V., and Spiegel, F. 2007. Diversity, nomenclature, and taxonomy of protists. *Systematic Biology* 56: 684–89.

Adl, S. M., Simpson, A. G. B., Lane, C. E., Lukes, J., Bass, D., Bowser, S. S., Brown, M. W., Burki, F., Dunthorn, M., Hampl, V., Heiss, A., Hoppenrath, M., Lara, E., le Gall, L., Lynn, D. H., McManus, H., Mitchell, E. A. D., Mozley-Stanridge, S. E., Parfrey, L. W., Pawlowski, J., Rueckert, S., Shadwick, L., Schoch, C. L., Smirnov, A., and Spiegel, F. W. 2012. The revised classification of eukaryotes,. *Journal of Eukaryotic Microbiology* 59:429–93.

Agosta, S. J., Janz, N., and Brooks, D. R. 2010. How specialists can be generalists: resolving the "parasite paradox" and implications for emerging infectious disease. *Zoologia* 27: 151–62.

Aguirre-Macedo, M. L., Vidal-Martinez, V. M., Herrera-Silveira, J. A., Valdes-Lozano, D. S., Herrera-Rodriguez, M., and Olvera-Novoa, M. A. 2008. Ballast water as a vector of coral pathogens in the Gulf of Mexico: the case of the Cayo Arcas coral reef. *Marine Pollution Bulletin* 56: 1570–7.

Audemard, C., Le Roux, F., Barnaud, A., Collins, C., Sautour, B., Sauriau, P. G., de Montaudouin, X., Coustau, C., Combes, C., and Berthe, F. 2002. Needle in a haystack: Involvement of the copepod *Paracartia grani*, in the life-cycle of the oyster pathogen *Marteilia refringens*. *Parasitology* 124: 315–23.

Barker, J. and Brown, M. R. W. 1994. Trojan-horses of the microbial world—protozoa and the survival of bacterial pathogens in the environment. *Microbiology* 140: 1253–9.

Bass, D., Stentiford, G. D., Littlewood, D. T. J., and Hartikainen, H. 2015. Diverse applications of environmental DNA methods in parasitology. *Trends in Parasitology* 31: 499–513.

Berthe, F. C. J., Le Roux, F., Adlard R. D., and Figueras, A. 2004. Marteiliosis in molluscs: A review,. *Aquatic Living Resources* 17: 433–48.

Blakeslee, A. M., Altman, I., Miller, A. W., Byers, J. E., Hamer, C. E., and Ruiz, G. M. 2012. Parasites and invasions: a biogeographic examination of parasites and hosts in native and introduced ranges. *Journal of Biogeography* 39: 609–22.

Blakeslee, A. M. H., Fowler, A. E., and Keogh, C. L. 2013. Marine invasions and parasite escape: updates and new perspectives. *Advances in Marine Biology,* 66: 87–169.

Bojko, J., Stebbing, P. D., Dunn, A. M., Bateman, K. S., Clark, F., Kerr, R. C., Stewart-Clark, S., Johannesen, A., and Stentiford, G. D. 2018. Green crab *Carcinus maenas* symbiont profiles along a North Atlantic invasion route. *Diseases of Aquatic Organisms* 128: 147–68.

Burge, C. A., Kim, C. J., Lyles, J. M., and Harvell, C. D. 2013. Special issue oceans and humans health: the ecology of marine opportunists. *Microbial Ecology* 65: 869–79.

Burge, C. A., Friedman, C. S., Getchell, R., House, M., Lafferty, K. D., Mydlarz, L. D., Prager, K. C., Sutherland, K. P., Renault, T., Kiryu, I., and Vega-Thurber, R. 2016. Complementary approaches to diagnosing marine diseases: A union of the modern and the classic. *Philosophical Transactions of the Royal Society B—Biological Sciences* 371: 20150207.

Burreson, E. M. and Ford, S. E. 2004. A review of recent information on the Haplosporidia, with special reference to *Haplosporidium nelsoni* (MSX disease). *Aquatic Living Resources* 17: 499–517.

Burreson, E. M., Stokes, N. A., and Friedman, C. S. 2000. Increased virulence in an introduced pathogen: *Haplosporidium nelsoni* (MSX) in the eastern oyster *Crassostrea virginica*. *Journal of Aquatic Animal Health* 12: 1–8.

Carlton, J. T. 2009. Deep invasion ecology and the assembly of communities in historical time. In: J. A. Crooks and G. Rilov (eds) *Biological Invasions in Marine Ecosystems*, pp. 13–56. Berlin: Springer.

Carlton, J. T., Chapman, J. W., Geller, J. B., Miller, J. A., Carlton, D. A., McCuller, M. I., Treneman, N. C., Steves, B. P., Ruiz, G. M. 2017. Tsunami-driven rafting: Transoceanic species dispersal and implications for marine biogeography. *Science* 357: 1402–5.

Carney, K. J., Minton, M. S., Holzer, K. K., Miller, A. W., McCann, L. D., and Ruiz, G. M. 2017. Evaluating the combined effects of ballast water management and trade dynamics on transfers of marine organisms by ships. *PLoS One* 12 (3).

Cleaveland, S., Laurenson, M. K., and Taylor, L. H. 2001. Diseases of humans and their domestic mammals: pathogen characteristics, host range, and the risk of emergence. *Philosophical Transactions of the Royal Society of London Series B—Biological Sciences* 356: 991–9.

Coats, D. W. 1999. Parasitic life styles of marine dinoflagellates,. *Journal of Eukaryotic Microbiology* 46: 402–9.

Cohen, A. N. and Carlton, J. T. 1997. Transoceanic transport mechanisms: Introduction of the Chinese mitten crab, *Eriocheir sinensis*, to California. *Pacific Science* 51: 1–11.

Crane, M. and Hyatt, A. 2011. Viruses of fish: An overview of significant pathogens. *Viruses* 3: 2025–46.

Davidson, I., Scianni, C., Hewitt, C., Everett, R., Holm, E., Tamburri, M., and Ruiz, G. 2016. Mini-review: Assessing the drivers of ship biofouling management—aligning industry and biosecurity goals. *Biofouling* 32: 411–28.

Davidson, I. C., Scianni, C., Minton, M. S., Ruiz, G. M., and Vamosi, S. 2018. A history of ship specialization and consequences for marine invasions, management and policy. *Journal of Applied Ecology* 55: 1799–811.

Dobbs, F. C. and Rogerson, A. 2005. Ridding ships' ballast water of microorganisms. *Environmental Science & Technology* 39: 259a–64a.

Dobson, A., Lafferty, K., Kuris, A., Hechinger, R. F., and Jetz, W. 2008. Homage to Linnaeus: How many parasites? How many hosts? *Proceedings of the National Academy of Sciences* 105: 11482–89.

Drake, J. W. and Holland, J. J. 1999. Mutation rates among RNA viruses. *Proceedings of the National Academy of Sciences* 96: 13910–13.

Drake, L. A., Choi, K. H., Ruiz, G. M., and Dobbs, F. C. 2001. Global redistribution of bacterioplankton and virioplankton communities. *Biological Invasions* 3: 193–9.

Dubey, J. P. 2004. Toxoplasmosis—a waterborne zoonosis. *Veterinary Parasitology* 126: 57–72.

Dunn, A. M., Torchin, M. E., Hatcher, M. J., Kotanen, P. M., Blumenthal, D. M., Byers, J. E., Coon, C. A. C., Frankel, V. M., Holt, R. D., Hufbauer, R. A., Kanarek, A. R., Schierenbeck, K. A., Wolfe, L. M., and Perkins, S. E. 2012. Indirect effects of parasites in invasions. *Functional Ecology* 26:1262–74.

Faillace, C. A., Lorusso, N. S., and Duffy, S. 2017. Overlooking the smallest matter: Viruses impact biological invasions. *Ecology Letters* 20: 524–38.

Fisher, M. C., Henk, D. A., Briggs, C. J., Brownstein, J. S., Madoff, L. C., McCraw, S. L., and Gurr, S. J. 2012. Emerging fungal threats to animal, plant, and ecosystem health,. *Nature* 484: 186–94.

Fowler, A. E., Blakeslee, A. M. H., Canning-Clode, J., Repetto, M. F., Phillip, A. M., Carlton, J. T., Moser, F. C.,

Ruiz, G. M., Miller, A. W. 2016. Opening Pandora's bait box: A potent vector for biological invasions of live marine species. *Diversity and Distributions* 22: 30–42.

Friedman, C. S. 1996. Haplosporidian infections of the Pacific oyster, *Crassostrea gigas* (Thunberg), in California and Japan. *Journal of Shellfish Research* 15: 597–600.

Gajadhar, A. A. and Allen, J. R. 2004. Factors contributing to the public health and economic importance of waterborne zoonotic parasites. *Veterinary Parasitology* 126: 3–14.

Goedknegt, M. A., Feis, M. E., Wegner, K. M., Luttikhuizen, P. C., Buschbaum, C., Camphuysen, K., van der Meer, J., and Thieltges, D. W. 2016. Parasites and marine invasions: Ecological and evolutionary perspectives. *Journal of Sea Research* 113: 11–27.

Grosholz, E. D. 2002. Ecological and evolutionary consequences of coastal invasions. *Trends in Ecology & Evolution* 17: 22–7.

Grosholz, E. D., Crafton, R. E., Fontana, R. E., Pasari, J. R., Williams, S. L., and Zabin, C. J. 2015. Aquaculture as a vector for marine invasions in California. *Biological Invasions* 17: 1471–84.

Guillou, L., Bachar, D., Audic, S., Bass, D., Berney, C., Bittner, L., Boutte, C., Burgaud, G., de Vargas, C., Decelle, J., del Campo, J., Dolan, J. R., Dunthorn, M., Edvardsen, B., Holzmann, M., Kooistra, W. H. C. F., Lara, E., Le Bescot, N., Logares, R., Mahe, F., Massana, R., Montresor, M., Morard, R., Not, F., Pawlowski, J., Probert, I., Sauvadet, A. L., Siano, R., Stoeck, T., Vaulot, D., Zimmermann, P., and Christen, R. 2013. The Protist Ribosomal Reference database (PR2): A catalog of unicellular eukaryote small sub-unit rRNA sequences with curated taxonomy. *Nucleic Acids Research* 41: D597–604.

Guillou, L., Viprey, M., Chambouvet, A., Welsh, R. M., Kirkham, A. R., Massana, R., Scanlan, D. J., and Worden, A. Z. 2008. Widespread occurrence and genetic diversity of marine parasitoids belonging to Syndiniales (Alveolata). *Environmental Microbiology* 10: 3349–65.

Hall-Stoodley, L., Costerton, J. W., and Stoodley, P. 2004. Bacterial biofilms: From the natural environment to infectious diseases. *Nature Reviews Microbiology* 2: 95–108.

Harmon, A., Littlewood, D.T.J., and Wood C.L. 2019. Parasites lost: Using natural history collections to track disease change across deep time. *Frontiers in Ecology and the Environment* 17: 157–66.

Harvell, C. D., Kim, K., Burkholder, J. M., Colwell, R. R., Epstein, P. R., Grimes, D. J., Hofmann, E. E., Lipp, E. K., Osterhaus, A. D. M. E., Overstreet, R. M., Porter, J. W., Smith, G. W., and Vasta, G. R. 1999. Emerging marine diseases—Climate links and anthropogenic factors. *Science* 285: 1505–10.

Hatcher, M. J., Dick, J. T. A., and Dunn, A. M. 2012. Diverse effects of parasites in ecosystems: Linking interdependent processes. *Frontiers in Ecology and the Environment* 10: 186–94.

Hershberger, P. K., van der Leeuw, B. K., Gregg, J. L., Grady, C. A., Lujan, K. M., Gutenberger, S. K., Purcell, M. K., Woodson, J. C., Winton, J. R., Parsley, M. J. 2010. Amplification and transport of an endemic fish disease by an introduced species. *Biological Invasions* 12: 3665–75.

Hewson, I., Button, J. B., Gudenkauf, B. M., Miner, B., Newton, A. L., Gaydos, J. K., Wynne, J., Groves, C. L., Hendler, G., Murray, M., Fradkin, S., Breitbart, M., Fahsbender, E., Lafferty, K. D., Kilpatrick, A. M., Miner, C. M., Raimondi, P., Lahner, L., Friedman, C. S., Daniels, S., Haulena, M., Marliave, J., Burge, C. A., Eisenlord, M. E., and Harvell, C. D. 2014. Densovirus associated with sea-star wasting disease and mass mortality. *Proceedings of the National Academy of Sciences* 111: 17278–83.

Hobbs, R. J. 1989. The nature and effects of disturbance relative to invasions. In: H. A. Mooney and J. A. Drake (eds) *Biological Invasions: A Global Perspective*, pp. 389–405. New York: John Wiley and Sons.

Holt, R. D., Dobson, A. P., Begon, M., Bowers, R. G., and Schauber, E.M. 2003. Parasite establishment in host communities. *Ecology Letters* 6: 837–42.

Holt, C., Foster, R., Daniels, C. L., van der Giezen, M., Feist, S. W., Stentiford, G. D., and Bass, D. 2018. *Halioticida noduliformans* infection in eggs of lobster (*Homarus gammarus*) reveals its generalist parasitic strategy in marine invertebrates. *Journal of Invertebrate Pathology* 154: 109–16.

Horvitz, C. C. 1997. The impact of natural disturbances. In: D. Simberloff, D. C. Schmitz, and T. C. Brown (eds) *Strangers in Paradise: Impact and Management of Non-Indigenous Species in Florida*, pp. 63–74. Washington, DC: Island Press.

Hubálek, Z. 2003. Emerging human infectious diseases: Anthroponoses, zoonoses,and sapronoses. *Emerging Infectious Diseases* 9: 403–4.

Huchin-Mian, J. P., Briones-Fourzan, P., Sima-Alvarez, R., Cruz-Quintana, Y., Perez-Vega, J. A., Lozano-Alvarez, E., Pascual-Jimenez, C., Rodriguez-Canul, R. 2009. Detection of *Panulirus argus* virus 1 (PaV1) in exported frozen tails of subadult–adult Caribbean spiny lobsters *Panulirus argus*. *Diseases of Aquatic Organisms* 86: 159–62.

Hughes, D., Brodeur, J., and Thomas, F. 2012. *Host Manipulation by Parasites*. Oxford: Oxford University Press.

Hulme, P. E. 2009. Trade, transport and trouble: managing invasive species pathways in an era of globalization. *Journal of Applied Ecology* 46: 10–18.

Hyatt, A. D., Hine, P. M., Jones, J. B., Whittington, R. J., Kearns, C., Wise, T. G., Crane, M. S., and Williams, L. M. 1997. Epizootic mortality in the pilchard *Sardinops sagax*

neopilchardus in Australia and New Zealand in 1995. Identification of a herpesvirus within the gill epithelium. *Diseases of Aquatic Organisms* 28: 17–29.

IMO. 2017. *International Convention for the Control and Management of Ships' Ballast Water and Sediments (BWM)*. London: International Maritime Organization.

Jambeck, J. R., Geyer, R., Wilcox, C., Siegler, T. R., Perryman, M., Andrady, A., Narayan, R., and Law, K. L. 2015. Plastic waste inputs from land into the ocean. *Science* 347: 768–71.

Jeschke, J. M. and Strayer, D. L. 2006. Determinants of vertebrate invasion success in Europe and North America. *Global Change Biology* 12: 1608–19.

Katoh, K., Misawa, K., Kei-ichi, K., and Miyata, T. 2002. MAFFT: A novel method for rapid multiple sequence alignment based on fast Fourier transform. *Nucleic Acids Research* 30: 3059–66.

Katsanevakis, S., Zenetos, A., Belchior, C., and Cardoso, A. C. 2013. Invading European seas: Assessing pathways of introduction of marine aliens. *Ocean Coast Management* 76: 64–74.

Keeling, P. J. and Fast, N. M. 2002. Microsporidia: Biology and evolution of highly reduced intracellular parasites,. *Annual Review of Microbiology* 56: 93–116.

Kelly, D. W., Paterson, R. A., Townsend, C. R., Poulin, R., and Tompkins, D. M. 2009. Parasite spillback: A neglected concept in invasion ecology? *Ecology* 90: 2047–56.

Kennedy, C. R. 1994. The distribution and abundance of the nematode *Anguillicola australiensis* in eels *Anguilla reinhardtii* in Queensland, Australia. *Folia Parasitologica* 41: 279–85.

Kim, Y., Aw, T. G., Teal, T. K., and Rose, J. B. 2015. Metagenomic investigation of viral communities in ballast water. *Environmental Science and Technology* 49: 8396–407.

Krasnov, B. R., Shenbrot, G. I., Mouillot, D., Khokhlova, I. S., and Poulin, R. 2005. Spatial variation in species diversity and composition of flea assemblages in small mammalian hosts: Geographical distance or faunal similarity? *Journal of Biogeography* 32: 633–44.

Kuris A. M. 1997. Host behavior modification: an evolutionary perspective. In: N. E. Beckage (eds) *Parasites and Pathogens*. Boston: Springer.

Kuris, A. M., Lafferty, K. D., and Sokolow, S. H. 2014. Sapronosis: A distinctive type of infectious agent. *Trends in Parasitology* 30: 386–93.

Lafferty, K. D. and Kuris, A. M. 1996. Biological control of marine pests. *Ecology* 77: 1989–2000.

Lee, K. A. and Klasing, K. C. 2004. A role for immunology in invasion biology. *Trends in Ecology and Evolution* 19: 523–9.

Lesser, M. P., Bythell, J. C., Gates, R. D., Johnstone, R. W., and Hoegh-Guldberg, O. 2007. Are infectious diseases really killing corals? Alternative interpretations of the experimental and ecological data. *Journal of Experimental Marine Biology and Ecology* 346: 36–44.

Leung, T. L. F., Keeney, D. B., and Poulin, R. 2009. Cryptic species complexes in manipulative echinostomatid trematodes: When two become six. *Parasitology* 136: 241–52.

Lightner, D. V. 2011. Virus diseases of farmed shrimp in the western hemisphere (the Americas): A review. *Journal of Invertebrate Pathology* 106: 110–30.

Litchman, E. 2010. Invisible invaders: Non-pathogenic invasive microbes in aquatic and terrestrial ecosystems. *Ecology Letters* 13: 1560–72.

Malcicka, M., Agosta, S. J., and Harvey, J. A. 2015. Multi-level ecological fitting: indirect life cycles are not a barrier to host switching and invasion. *Global Change Biology* 21: 3210–18.

McCallum, H., Kuris, A., Harvell, C., Lafferty, K., Smith, G., and Porter, J. 2004. Does terrestrial epidemiology apply to marine systems? *Trends in Ecology & Evolution* 19: 585–91.

Miller, A. W., Davidson, I. C., Minton, M. S., Steves, B., Moser, C. S., Drake, L. A., and Ruiz, G. M. 2018. Evaluation of wetted surface area of commercial ships as biofouling habitat flux to the United States. *Biological Invasions* 20: 1977–90.

Minchin, D. 1996. Management of the introduction and transfer of marine mollusks. *Aquatic Conservation— Marine and Freshwater Ecosystems* 6: 229–44.

Minton, M. S., Verling, E., Miller, A. W., and Ruiz, G. M. 2015. Reducing propagule supply and coastal invasions via ships: Effects of emerging strategies. *Frontiers in Ecology and the Environment* 3: 304–8.

Miura, O., Torchin, M. E., Kuris, A. M., Hechinger, R. F., and Chiba, S. 2006. Introduced cryptic species of parasites exhibit different invasion pathways. *Proceedings of the National Academy of Sciences* 103: 19818–23.

Morado, J. F. and Small, E. B. 1994. Morphology and stomatogenesis of *Mesanophrys pugettensis* (Scuticociliatida, Orchitophryidae), a facultative parasitic ciliate of the Dungeness crab, *Cancer magister* (Crustacea, Decapoda). *Transactions of the American Microscopical Society* 113: 343–64.

Moser, C. S., Wier, T. P., Grant, J. F., First, M. R., Tamburri, M. N., Ruiz, G. M., Miller, A. W., and Drake, L. A. 2015. Quantifying the total wetted surface area of the world fleet: A first step in determining the potential extent of ships' biofouling. *Biological Invasions* 18: 265–77.

Muirhead, J. R., Minton, M. S., Miller, W. A., and Ruiz, G. M. 2015. Projected effects of the Panama Canal expansion on shipping traffic and biological invasions. *Diversity and Distributions* 21: 75–87.

Neuhauser, S., Kirchmair, M., and Gleason, F.H. 2011. Ecological roles of the parasitic phytomyxids (plasmodiophorids) in marine ecosystems—A review. *Marine and Freshwater Research*, 62: 365–71.

Neuhauser, S., Kirchmair, M., Bulman, S., and Bass, D. 2014. Cross-kingdom host shifts of phytomyxid parasites. *BMC Evolutionary Biology*, 14: 33.

Osmanov, S. D. 1971. *Parazity Ryb Uzbekistana* (Parasites of Fishes of Uzbekistan). Tashkent: Fan.

Ottinger, M., Clauss, K., and Kuenzer, C. 2016. Aquaculture: Relevance, distribution, impacts and spatial assessments—A review. *Ocean & Coastal Management* 119: 244–66.

Pagenkopp Lohan, K. M., Fleischer, R. C., Carney, K. J., Holzer, K. K., and Ruiz, G. M. 2016. Amplicon-based pyrosequencing reveals high diversity of protistan parasites in ships' ballast water: Implications for biogeography and infectious diseases. *Microbial Ecology* 71: 530–42.

Pagenkopp Lohan, K. M., Fleischer, R. C., Carney, K. J., Holzer, K. K., and Ruiz, G. M. 2017. Molecular characterisation of protistan species and communities in ships' ballast water across three U.S. coasts. *Diversity and Distributions* 23: 680–91.

Paillard, C., Le Roux, F., and Borrego, J. J. 2004. Bacterial disease in marine bivalves, a review of recent studies. *Aquatic Living Resources* 17: 477–98.

Poulin, R. 2014. Parasite biodiversity revisited: frontiers and constraints. *International Journal of Parasitology* 44: 581–9.

Poulin, R. 2017. Invasion ecology meets parasitology: Advances and challenges. *International Journal for Parasitology* 6: 361–3.

Poulin, R. and Morand, S. 2000. The diversity of parasites. *Quarterly Review of Biology* 75: 277–93.

Prenter, J., MacNeil, C., Dick, J. T. A., and Dunn, A. M. 2004. Roles of parasites in animal invasions. *Trends in Ecology and Evolution* 19: 385–90.

Renault, T. and Novoa, B. 2004. Viruses infecting bivalve mollusks. *Aquatic Living Resources* 17: 397–409.

Rideout, B. A., Sainsbury, A. W., and Hudson, P. J. 2017. Which parasites should we be most concerned about in wildlife translocations? *Ecohealth* 14: 42–6.

Rohde, K. 1982. *Ecology of Marine Parasites*, p. 245. St Lucia: University of Queensland Press.

Rohde, K. 2005. *Marine Parasitology*. Clayton, Australia: CSIRO Publishing.

Ruesink, J. L., Lenihan, H. S., Trimble, A. C., Heiman, K. W., Micheli, F., Byers, J. E., and Kay, M. C. 2005. Introduction of non-native oysters: Ecosystem effects and restoration implications. *Annual Review of Ecology Evolution and Systematics* 36: 643–89.

Ruiz, G. M., Fofonoff, P., Hines, A. H., and Grosholz, E. D. 1999 Non-indigenous species as stressors in estuarine and marine communities: Assessing invasion impacts and interactions. *Limnology and Oceanography* 44: 950–72.

Ruiz, G. M., Rawlings, T. K., Dobbs, F. C., Drake, L. A., Mullady, T., Huq, A., and Colwell, R. R. 2000a. Global spread of microorganisms by ships—Ballast water discharged from vessels harbours a cocktail of potential pathogens. *Nature* 408: 49–50.

Ruiz, G. M., Fofonoff, P. W., Carlton, J. T., Wonham, M. J., and Hines, A. H. 2000b. Invasion of coastal marine communities in North America: Apparent patterns, processes, and biases. *Annual Review of Ecology and Systematics* 31: 481–531.

Sharma, A. and Lal, S. K. 2017. Zika virus: Transmission, detection, control, and prevention. *Frontiers in Microbiology* 8: https://doi.org/10.3389/fmicb.2017.00110.

Skovgaard, A. and Daugbjerg, N. 2008. Identity and systematic position of *Paradinium poucheti* and other *Paradinium*-like parasites of marine copepods based on morphology and nuclear-encoded SSU rDNA. *Protist* 159: 401–13.

Small, H. J., Miller, T. L., Coffey, A. H., Delaney, K. L., Schott, E., and Shields, J. D. 2013. Discovery of an opportunistic starfish pathogen, *Orchitophrya stellarum*, in captive blue crabs, *Callinectes sapidus*. *Journal of Invertebrate Pathology* 114: 178–85.

Solarz, W. and Najberek, K. 2017. Alien parasites may survive even if their original hosts do not. *Ecohealth* 14: 3–4.

Sullivan, B. K., Sherman, T. D., Damare, V. S., Lilje, O., and Gleason, F. H. 2013. Potential roles of *Labyrinthula* spp. in global seagrass population declines. *Fungal Ecology* 6: 328–38.

Taraschewski, H. 2006. Hosts and parasites as aliens. *Journal of Helminthology* 80: 99–128.

Toranzo, A. E., Magariños, B., and Romalde, J. L. 2005. A review of the main bacterial fish diseases in mariculture systems. *Aquaculture* 246: 37–61.

Torchin, M. E., Byers, J. E., and Huspeni, T. C. 2005. Differential parasitism of native and introduced snails: replacement of a parasite fauna. *Biological Invasions* 7:885–94.

Torchin, M. E. and Kuris, A. 2005. Introduced marine parasites. In: K. Rohde (ed.) *Marine Parasitology*, pp. 358–66. Clayton, Australia: CSIRO Publishing.

Torchin, M. E. and Lafferty, K. 2009. Escape from parasites. In: G. Rilov and J. A. Crooks (eds) *Biological Invasions in Marine Ecosystems*, pp. 203–14. Berlin: Springer.

Torchin, M. E., Lafferty, K. D., Dobson, A. P., McKenzie, V. J., and Kuris, A. M. 2003. Introduced species and their missing parasites. *Nature* 421: 628–30.

Torchin, M. E., Lafferty, K. D., and Kuris, A. M. 2002. Parasites and marine invasions. *Parasitology* 124: S137–51.

Torchin, M. E. and Mitchell, C. E. 2004. Parasites, pathogens, and invasions by plants and animals. *Frontiers in Ecology and the Environment* 2: 183–90.

Van Bressem, M. F., Raga, J. A., Di Guardo, G., Jepson, P. D., Duignan, P. J., Siebert, U., Barrett, T., Santos, M. C., Moreno, I. B., Siciliano, S., Aguilar, A., and Van Waerebeek, K. 2009. Emerging infectious diseases in cetaceans worldwide and the possible role of environmental stressors. *Diseases of Aquatic Organisms* 86: 143–57.

Van Bressem, M. F., Van Waerebeek, K., and Raga, J. A. 1999. A review of virus infections of cetaceans and the potential impact of morbilliviruses, poxviruses and papillomaviruses on host population dynamics. *Diseases of Aquatic Organisms* 38: 53–65.

van Kleunen, M., Dawson, W., Schlaepfer, D., Jeschke, J. M., and Fischer, M. 2010. Are invaders different? A conceptual framework of comparative approaches for assessing determinants of invasiveness. *Ecology Letters* 13: 947–58.

Villalba, A., Reece, K. S., Camino Ordás, M., Casas, S. M., and Figueras, A. 2004. Perkinsosis in mollusks: A review. *Aquatic Living Resources* 17: 411–32.

Vogan, C. L., Powell, A., and Rowley, A. F. 2008. Shell disease in crustaceans—just chitin recycling gone wrong? *Environmental Microbiology* 10: 826–35.

Ward, G. M., Neuhauser, S., Groben, R., Ciaghi, S., Berney, C., Romac, S., and Bass, D. 2018. Environmental sequencing fills the gap between parasitic haplosporidians and free-living giant amoebae. *Journal of Eukaryotic Microbiology* 65: 574–86.

Weinert, L. A., Werren, J. H., Aebi, A., Stone, G. N., and Jiggins, F. M. 2009. Evolution and diversity of Rickettsia bacteria. *BMC Biology* 7: 6.

Weinstein, S. B. and Kuris, A. M. 2016. Independent origins of parasitism in Animalia. *Biology Letters* 12: 7.

Williams, S. L., Davidson, I. C., Pasari, J. R., Ashton, G. V., Carlton, J. T., Crafton, R. E., Fontana, R. E., Grosholz, E. D., Miller, A. W., Ruiz, G. M., and Zabin, C. J. 2013. Managing multiple vectors for marine invasions in an increasingly connected world. *BioScience* 63: 952–66.

Wonham, M. J. and Carlton, J. T. 2005. Trends in marine biological invasions at local and regional scales: The Northeast Pacific Ocean as a model system. *Biological Invasions* 7: 369–92.

Wood, C., Byers, J., Cottingham, K., Altman, I., Donahue, M., and Blakeslee, A. 2007. Parasites alter community structure. *Proceedings of the National Academy of Sciences* 104(22): 9335–9.

Woolhouse, M. E. J., Taylor, L. H., and Haydon, D. T. 2001. Population biology of multihost pathogens. *Science* 292: 1109–12.

Wyatt, T. and Carlton, J. T. 2002. Phytoplankton introductions in European coastal waters: Why are so few invasions reported? *CIESM Workshop Monographs* 20: 41–6.

Zabin, C. J., Ashton, G. V., Brown, C. W., Davidson, I. C., Sytsma, M. D., and Ruiz, G. M. 2014. Small boats provide connectivity for nonindigenous marine species between a highly invaded international port and nearby coastal harbors. *Management of Biological Invasions* 5: 97–112.

Zholdasova, I. 1997. Sturgeons and the Aral Sea ecological catastrophe. *Environmental Biology of Fishes* 48: 373–80.

Disease Problems and their Management

CHAPTER 8

Disease outbreaks can threaten marine biodiversity

C. Drew Harvell and Joleah B. Lamb

8.1 Introduction

Host–pathogen theory predicts that host-specific pathogens can regulate host population dynamics, whereas multi-host pathogens can cause extreme population impacts, including extinction of susceptible species if they are continuously infected from reservoir species (McCallum and Dobson 1995, McCallum 2012). In this chapter, we focus on global disease outbreaks that have impacted marine communities (Harvell et al. 2002, Burge et al. 2014, Harvell 2019) and pose risks to human health and livelihoods (Daszak et al. 2000). Mass mortality (loss of more than 10 percent of a population) from infectious disease has recently impacted wild marine taxa and habitat-forming taxa such as coccolithophores (Frada et al. 2008), seagrasses (Martin et al. 2016), corals (Harvell et al. 2007), abalone (Crosson and Friedman 2018), sea stars (Montecino-Latorre et al. 2016), urchins (Clemente et al. 2014), marine mammals (Rubio-Guerri et al. 2013), and turtles (Flint et al. 2010).

Since 2013, sea star wasting disease (SSWD), linked to a multi-host sea star-associated densovirus (SSaDV, family Parvoviridae), has caused massive, ongoing mortality, from Mexico to Alaska (Hewson et al. 2014). Over twenty asteroid species have been affected in what is currently the largest documented epizootic of a non-commercial marine taxon (Hewson et al. 2014). Since 1984, an outbreak of withering foot syndrome caused by a rickettsial bacterium has contributed to listing three species of California abalone on the endangered species list (Crosson and Friedman 2018). Beginning in the 1930s, episodes of eelgrass wasting disease, caused by *Labyrinthula zosterae*, have devastated eelgrass beds in the continental USA and Europe (Sullivan et al. 2013, Martin et al. 2016). Reef-building corals around the world are increasingly impacted by infectious diseases caused by a combination of dysbiosis and infectious pathogens.

8.2 Disease outbreaks of threatened foundation, keystone, and ecological engineering species

Single species often have far-reaching impacts on ecosystem structure and functioning, and thereby biodiversity. The most influential species include two kinds. *Foundation species* dominate abundance and biomass within a system, modify the physical environment, and create habitat for many other species (Bruno et al. 2003). *Keystone species* are strong interactors, and thus exert strong impacts disproportionate to their low abundance (Paine 1969). In this chapter, we describe the impacts of disease outbreaks on threatened, foundation, keystone, and ecological engineering species (Figure 8.1), using specific case studies of reef-building corals (Section 8.2.1), sea stars (Section 8.2.2), seagrass (Section 8.2.3), and abalone (Section 8.2.4).

Harvell, C.D., and Lamb, J.B., *Disease outbreaks can threaten marine biodiversity* In: *Marine Disease Ecology*. Edited by: Donald C. Behringer, Kevin D. Lafferty, and Brian. R. Silliman, Oxford University Press (2020). © Oxford University Press.
DOI: 10.1093/oso/9780198821632.003.0008

Figure 8.1 Marine disease threatens biodiversity and key species with regard to ecosystem function, including reef-building corals (top left), seagrass meadows (top right), abalone (bottom left), and sea stars (bottom right). (Photos: M. Primivani, B. Tissot, J. Lamb, and C. Harvell.)

8.2.1 Foundation species: *reef-building corals*

As foundation species, corals have been estimated to support up to nine million marine species, representing one of the most biodiverse ecosystems in the world (Roberts et al. 2002). Since the late 1980s, partly driven by infectious disease, coral cover has decreased by 50–75 percent, jeopardizing associated marine species and the US $375 billion in goods and services coral reefs provide to people each year through fisheries, tourism, and coastal protection (Burke et al. 2011).

When

Outbreaks of coral disease emerged in the 1970s as a significant driver of global coral reef degradation (reviewed in Harvell et al. 2007). Evidence from paleontological monitoring suggests that coral

disease epizootics were not as common in the past (Aronson et al. 2003). Lafferty et al. (2004) reviewed modern changes in coral disease over time and found that although bleaching reports did increase over several decades, peaks in infectious disease reports instead corresponded to El Niño events.

Signs and cause

Disease outbreaks have been described across all major ocean basins, with three-quarters of reported diseases estimated globally affecting species in the Caribbean, which is known as a disease "hot spot" (Randall et al. 2014; Harvell 2019). Yet, despite this extensive monitoring and exploration, classical culturing approaches to determine causative agents have been applied to only a few described diseases (Frasca et al. Chapter 11, this volume) that can repeatedly initiate a consistent disease phenotype, including aspergillosis (causative fungi, *Aspergillus sydowii*), white pox disease (causative bacterium, *Serratia marcescens*), and bacterial bleaching (*Vibrio shiloi*). However, these traditional approaches are arguably challenging when applied to corals (Mera and Bourne 2018).

Documented coral infectious diseases are often multi-host syndromes, enabling them to be unusually destructive. For example, aspergillosis infects six octocoral species (Weil 2004), whereas black band disease (a polymicrobial consortia) has been reported on forty-two coral species in the Caribbean, and on an additional forty coral species from twenty-one genera in the Indo-Pacific (Green and Bruckner 2000) (Figure 8.2).

Infectious agents may spread rapidly in the ocean (McCallum et al. 2003). For example, the coral disease white plague (WP) spread along the coast of Florida at rates of approximating 200 km per year (Richardson et al. 1998). Despite reports of disease significantly impacting corals worldwide, knowledge underlying the distributions, causative agents, and environmental drivers is lacking. Outbreaks of coral diseases stand out as being driven largely by opportunistic agents and a changing environment

(Harvell 2019). The dynamics of infectious wildlife diseases are known to be influenced by shifting interactions among the host, pathogen, and other members of the microbiome (Wobeser 2006). This is also a common case for corals (Ainsworth et al. 2010).

Several common Caribbean coral diseases, including yellow-band disease (*Vibrio* consortium and virus-like particles), white pox disease, WP type II (*Aurantimonas coralicida*), and dark-spot syndrome (undetermined causative agent as of 2019), do not display transmission dynamics characteristic of contagious diseases (Muller and van Woesik 2012, 2014, Mera and Bourne 2018), suggesting that intrinsic properties of the holobiont may play a large role in disease initiation and progression. As in some human diseases, it may be that heterogeneous communities of micro-organisms are responsible and act to disrupt microbiome homeostasis (Lamont and Hajishengallis 2015). In these cases, it is important to examine host organismal traits that affect disease susceptibility and environmental thresholds that serve as tipping points for disruption of microbiome homeostasis and disease induction, in addition to focusing on transmission dynamics of pathogens (Burge and Hershberger Chapter 5, this

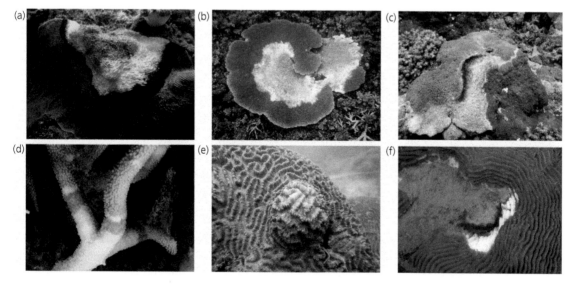

Figure 8.2 Visual characteristics of six coral syndromes commonly affecting reef corals in the Indo-Pacific: (a) black band disease, (b) white syndrome, (c) skeletal eroding band, (d) brown band disease, (e) growth anomaly, and (f) atramentous necrosis. (Photos: J. Lamb and B. Willis.)

volume). Coral host–pathogen work is plagued by a changing and polymicrobial series of pathogenic bacteria, prompting an increasing focus and need to better understand the interplay between the coral microbiome and the environment and the potential role of dysbiosis in triggering outbreaks (Zaneveld et al. 2017).

Among wildlife, outbreaks of coral diseases stand out as being driven largely by a changing environment (Raymundo et al. Chapter 9, this volume), particularly warming temperatures (Harvell et al. 2002) but also chronic exposure to pollutants like sewage (Lamb et al. 2017), sediment (Pollock et al. 2014), nutrients (Vega Thurber et al. 2014), and aquaculture (Lafferty et al. 2015). These outbreaks have contributed to losses of coral species and are driving a regime shift and, in some locations, collapse of the coral ecosystem.

Coral injury increases the likelihood of disease development by disrupting immune system function during wound healing processes and providing an entry site for opportunist pathogens (Mydlarz et al. 2006). This has been shown following tissue damage from feeding reef organisms (Aeby and Santavy 2006, Nicolet et al. 2013); the passage of cyclones (Brandt et al. 2013); and anthropogenic activities including high-intensity tourism (Lamb and Willis 2011, Lamb et al. 2014), destructive fishing methods and gear (Lamb et al. 2015, 2016), entanglement with plastic debris (Lamb et al. 2018), and ship groundings (Raymundo et al. 2018).

Impact

An outbreak of white band disease virtually eliminated two dominant reef-building corals, *Acropora cervicornis* and *A. palmata*, on many Caribbean reefs and these corals are now listed on the US Endangered Species List (reviewed in Sutherland et al. 2016). Successive disease outbreaks decreased populations of these two significant reef-building acroporid corals by 95 percent and contributed substantially to observed ecological phase shifts from coral- to algal-dominated reefs (Weil 2004).

Infectious disease is demonstrably pivotal in changing the composition, structure, and function of coral reef communities (Morton et al. Chapter 3, this volume)—for example, by opening up substratum for colonization by less competitive species or triggering a phase shift to algal domination (Aronson et al. 2003). Disease-induced reductions in coral cover may also limit direct competitive interactions between neighboring colonies (Bruno et al. 2007). The loss of coral reef structural complexity associated with anthropogenically driven coral diseases (Lamb et al. 2018) has the ability to reduce the economic productivity of fisheries by up to three-fold (Rogers et al. 2014).

8.2.2 Ecosystem engineering species: *seagrass*

Ecosystem engineers modify environments to create unique habitats, which provide better habitat for biodiversity and improved ecological functions, with services for humans. Globally, seagrass cover has declined by 29 percent from 1879 to 2009, with the rate of loss on the rise at 7 percent per year since 1990 (Waycott et al. 2009). These declines are attributed to anthropogenic and environmental stressors, many of which are also synergistic with disease susceptibility, such as terrestrial runoff, physical disturbance, fisheries activities, algal blooms, and ocean warming (Orth et al. 2006). Eelgrass (*Zostera marina*), a seagrass species which forms extensive coastal meadows throughout temperate waters, is a key ecosystem engineer and foundation species. Here, we review historical and current studies documenting large impacts of disease on seagrass, with huge biodiversity repercussions, given the importance of these meadows for habitat.

When

Outbreaks of eelgrass wasting disease (EGWD) historically caused catastrophic losses of eelgrass beds, including documented losses of > 90 percent along the US and European Atlantic coasts in the 1930s (Short et al. 1988) and in Florida in the 1980s (Martin et al. 2016). EGWD currently affects eelgrass populations in Chesapeake Bay (Short et al. 1988), along the US West Coast (Groner et al. 2014, 2016b, 2018, Dawkins et al. 2018), Florida (Martin et al. 2016), and Europe (Godet et al. 2008). Recent work shows rising levels of EGWD in Washington state, in well-monitored meadows in the San Juan Islands (Groner et al. 2014, 2016a, 2016b, 2018, Dawkins et al. 2018), and in Puget Sound.

Signs and cause

EGWD causes sharp, dark-edged, necrotic lesions on the plant, associated with fungus-like protists in the genus *Labyrinthula*; in eelgrass, the causative agent of EGWD is *Labyrinthula zosterae* (Figure 8.3). Other eelgrass pathogens that are less well studied but of emerging importance are *Phytophora* spp. and *Halophytophora* spp. (reviewed in Sullivan et al. 2018). The presence of *L. zosterae* is confirmed with histology, which visualizes the *L. zosterae* spindle-shaped cells within host tissues and culture (Figure 8.3).

A quantitative polymerase chain reaction (qPCR) assay was designed to quantify *L. zosterae* within plant tissues (using the *L. zosterae* ITS region). Multiple virulent strains of *L. zosterae* exist globally, including both pathogenic and non-pathogenic strains (Martin et al. 2016). Strains of *L. zosterae* with varying virulence have been isolated from the US West Coast (Muehlstein et al. 1991, Dawkins et al. 2018) and more globally (Martin et al. 2016). Cross-infection experiments reveal that *L. zosterae* strains are equally pathogenic to eelgrass hosts on both the US Atlantic and Pacific Coast (Martin et al. 2016). Some *L. zosterae* strains are suggested to have low host specificity and infect multiple seagrass species, including *Posidonia* and *Thalassia* (Martin et al. 2016), and may infect marine algae and other potential reservoirs (reviewed in Bockelmann et al. 2013).

As a widespread opportunist throughout the range of eelgrass (Sullivan et al. 2013, Martin et al. 2016), *L. zosterae* can be present but causes few or no signs of disease (Bockelmann et al. 2013, Martin et al. 2016). Outbreaks of lesions with changes in environmental drivers like temperature and nutrients, and vulnerability of older blades to *L. zosterae* prevalence have been identified in some regions (Groner et al. 2014, 2016a, Jakobsson-Thor et al. 2018). The triggers and biological interactions that lead to epidemics and seagrass wasting in complex field ecosystems remain poorly understood (Martin et al. 2016).

Outbreaks of EGWD in temperate regions are associated with warming temperatures, with highest prevalences within sites during the warmest months of the year (Bockelmann et al. 2013). Temperate *L. zosterae* strains appear to grow optimally at temperatures from 14 to 24 °C, with most catastrophic field outbreaks associated with elevated temperatures in this range (Sullivan et al. 2013). Dosage-controlled laboratory experiments show that *L. zosterae in vitro* cultures grow faster at 18 than 11 °C, and *in vivo*, making larger lesions at 18 than 11 °C (Dawkins et al. 2018). A modelling study simulated observed reductions in eelgrass beds during warm periods and mediated by *L. zosterae* and EGWD (Bull et al. 2012). Thus, warmer water temperatures will likely lead

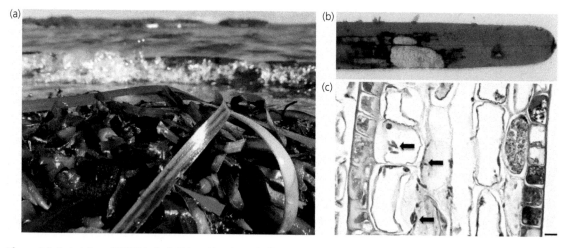

Figure 8.3 Typical signs of EGWD in the field (a and b, red arrow) affecting *Z. marina*. (c) Histological characterization of *Labyrinthula* cells (black arrows). Scale bar = 10 μm. (Photos in a and b: O. Graham and P. Dawkins. Histological section in c from Groner et al. (2014).)

to increased *L. zosterae* transmission and damage to eelgrass health in a warming ocean.

Impact

EGWD has been globally a major factor in the dynamics of eelgrass over the last century (Short et al. 1987, Muehlstein et al. 1991, Sullivan et al. 2013). During the 1930s, eelgrass populations declined drastically and suddenly on both sides of the North Atlantic and in parts of the US Pacific Northwest, giving way to bare sediments or ephemeral algae. *Labyrinthula zosterae* was isolated from diseased plants at that time and suspected as the culprit, along with unusually warm water temperatures in Europe (reported in Sullivan et al. 2013). The 1930s' outbreak reportedly wiped out over 90 percent of eelgrass along the East Coast of North America in 1931–1932, impacting numerous waterfowl species that use eelgrass as habitat and food (reviewed in Sullivan et al. 2013) and effectively killing the fishery for bay scallops. In the 1980s, a similar die-off began in New England, traced to *L. zosterae* (Short et al. 1987, Muehlstein et al. 1991).

In both *Z. marina* (eelgrass) and *Thalassia testudinum* (turtle grass), wasting disease has caused mass die-offs at regional to pandemic scales (reviewed in Sullivan et al. 2013). Although Puget Sound eelgrass beds are recorded as stable overall (Shelton et al. 2017), some monitored beds recently disappeared (Wyllie-Echeverria and Harvell, pers. obs.), while other beds declined between 2013 and 2017 (Eisenlord and Harvell, unpublished data). Changes in eelgrass bed size and density likely have several causes; however, our recent data from the San Juan Islands suggest that high EGWD prevalence may be causing declines in bed density (Groner et al. 2016a).

8.2.3 Keystone species: *sea stars*

Keystone species are defined as those that play a disproportionate role in their community relative to their abundance (Paine 1969, Power et al. 1996). In this section, we report on the large impacts to ecologically pivotal sea stars and eighteen other species from a multi-host disease epidemic, and the resulting cascading ecosystem changes.

When

Large numbers of *Asterias forbesi* sea stars died on the northeast coast of the USA beginning in 2011. By 2013, sea stars in over twenty species died catastrophically along the US West Coast, first noticed in British Columbia and the Olympic peninsula. After 2017, SSWD was observed in Asia (Hewson et al. 2018).

The first reports on the West Coast were for ochre sea stars (*Pisaster ochraceus*) from Washington state in June 2013 and sunflower sea stars (*Pycnopodia helianthoides*) from British Columbia, Canada, in August 2013. By late fall 2013, reports of dying sea stars with lesions in upwards of fifteen species were reported from parts of California, Washington, and British Columbia. Summer 2014 was a time of huge mortality across the range from southern California to Alaska (Miner et al. 2018). Previous outbreaks of SSWD have been reported in the northeast Pacific, California, and northern New England since the 1970s (Eisenlord et al. 2016); however, these earlier SSWD outbreaks involved single species in localized areas. In addition to a virus-sized fraction, the 2013–2015 SSWD epidemic has been linked to an increase (Eisenlord et al. 2016, Miner et al. 2018, Harvell et al. 2019) or decrease (Menge et al. 2016) in sea temperature, while host immunity (Fuess et al. 2015), genetics (Wares and Schiebelhut 2016), and a suite of other host or environmental factors are likely involved.

Signs and cause

Infected sea stars develop lesions in the dermis that increase in depth and diameter, dissolving tissue from the outside in (Figure 8.4). One or all arms then detach from the central disc as individuals die, often leaving only white piles of ossicles and disconnected limbs. SSWD signs included lesions through the dermal wall into the coelom on the star dorsal surface, gonads and other internal organs spilling out, and arms detaching. These signs differ from previous reports of sea star wasting in causing catastrophic mortality of top-condition, reproductive sea stars and not a gradual wasting. Vast populations died simultaneously, causing intertidal and subtidal areas to be strewn with dead and dying sea stars and detached, moving arms, sometimes containing ripe gonads.

Figure 8.4 Images of (a) healthy and abundant populations of ochre sea star (*Pisater ochraceus*) before an outbreak of SSWD. Healthy sea stars develop (b) initial curling of arms, (c) development of dermal lesions, (d) loss of arms from central disk, and (e) extensive tissue necrosis and death. (Figure from Eisenlord et al. (2016).)

Sea star mortalities during this event were linked to a sea star-associated densovirus (SSaDV, family Parvoviridae), based on evidence provided by experimental challenge studies with the sunflower sea star and a metagenomic analysis of field and laboratory samples (Hewson et al. 2014). Inoculation experiments confirmed a viral-sized fraction as causative for at least one species, the sunflower sea star, in Washington state, and metagenomics suggested the densovirus as the potential virus, present in *Pycnopodia helianthoides*, *Pisaster ochraceus*, and *Evasterias troscheli* along its geographic range (Salish Sea to southern California) when surveyed in 2013–2014 (Hewson et al. 2014). (Bucci et al. 2017). Inoculation trials were not successful in producing disease signs in any species (Hewson et al. 2018).

Redesigned qPCR primers support an association between asteroid densovirus load (as measured by the WAaDs primer and probe set) and signs of SSWD only in *Pycnopodia helianthoides*, but not in either *Pisaster ochraceus* or *E. troscheli*. Surveys in 2016 showed highest loads of WAaDs in *Pycnopodia helianthoides* and *Pisaster brevispinus* and presence of lower levels of WAaDs in non-symptomatic species of the genera *Crossaster*, *Pteraster*, and *Henricia* (Hewson et al. 2018).

Impact

The outbreak caused strong declines in at least five species (*Pisaster ochraceus*, *Pisaster brevispinus*, *E. troscheli*, *Pycnopodia helianthoides*, and *Solaster stimpsoni*), with impacts on the remainder of poorly studied subtidal sea stars being unclear (Eisenlord et al. 2016, Montecino-Latorre et al. 2016, Harvell et al. 2019). The intertidal *Pisaster ochraceus* and *E. troschelli* declined precipitously in the summer of 2014 at all sites from California to Alaska. The impacts of this disease outbreak reduced populations of multiple species and extirpated at least two (*Pisaster ochraceus* and *Pycnopodia helianthoides*) from some geographic regions in the southern part of their range (Miner et al. 2018, Harvell et al. 2019). In Washington state, *Pisaster ochraceus* (Figure 8.4) declined by over 70 percent between 2014 and 2015, and *Pycnopodia helianthoides* declined over 90 percent for much of their range south of Alaska (Harvell et al. 2019). The epidemic continued in fall 2018, with lesioned sea stars in multiple species still present, although at a very reduced level (Harvell 2019).

Ochre sea stars are keystone species, and sunflower sea stars are pivotal predators (Burt et al. 2018, Harvell et al. 2019) and are likely keystones in some portions of their range. Ochre sea stars are capable of controlling populations of mussels and clams; sunflower sea stars control populations of green and purple sea urchins and clams in some parts of their range. The largest impacts of sea star removal observed are large outbreaks in populations of sea urchins in central California and southern British Columbia, which in turn are devastating kelp beds. Recent surveys document massive declines in California and British Columbia kelp beds (Schultz et al. 2016), driven by urchin increases and overgrazing. In 2017, there was some recovery of ochre sea stars and bouts of significant recruitment at multiple northern sites (Eisenlord, Winningham, Harvell, pers. com.). Data for populations of all subtidal sea stars are poor, except for the most common, the sunflower sea star (Figure 8.5). Citizen science diver data and National Oceanic and Atmospheric Administration (NOAA) trawl data show that sunflower sea star populations had not recovered in 2017 (Harvell et al. 2019). Early observations suggested the sunflower sea star was the most susceptible of the Asteroid species. This wide host range pathogen is currently endangering southern populations of this species, with well-studied confirmation of little or no recovery observed from southern California through British Columbia.

The SSWD outbreak that started in 2013 is considered the largest disease outbreak of marine wildlife, affecting well over twenty species and with an initial range from Mexico to Alaska and recent possible spread to Asia (Hewson et al. 2018), which would qualify it as a global pandemic. Taura syndrome of shrimp is a geographically larger and longer running outbreak, but since it affects predominantly farmed species, it is not considered a disease of wildlife (Lafferty et al. 2015).

The coelomocytes of sunflower sea stars inoculated with a viral-sized fraction from sick sea stars mounted an impressive immune response, showing potential for immune capability (Fuess et al. 2015). Ochre sea star populations show some imprint of selection, in possible association with SSWD, with the frequency of heterozygous EF1α in *Pisaster ochraceus* and transcriptomic analyses of EF1α mutants suggesting that these individuals have a greater cellular response to temperature stress (Chandler and Wares 2017). Moreover, differential survival and genetically based resistance of surviving adults and new recruits follows the epidemic (Schiebelhut et al. 2018).

Warm temperature anomalies between 2014 and 2016 likely fueled larger impacts from this epidemic. In laboratory experiments, ochre sea stars died at a faster rate, and in field populations, they had a higher risk of disease at warmer temperatures (Eisenlord et al. 2016). Similarly, populations of ochre sea stars in warmer parts of the range died sooner and have failed to return, compared to more northern locations

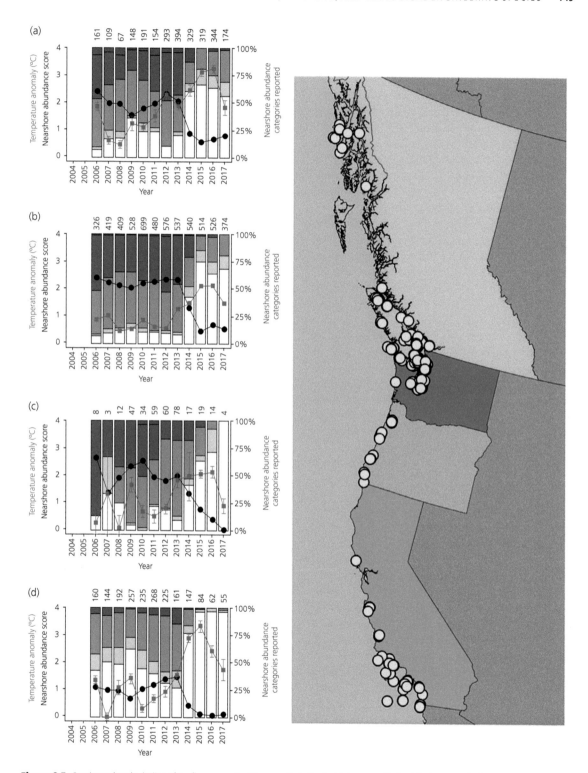

Figure 8.5 Continental-scale decline of sunflower sea star (*Pycnopodia helianthoides*) reported in Reef Environmental Education Foundation (REEF) roving diver surveys from 2006 to 2017. Yellow circles depict locations of surveys; black lines depict annual abundance surveys; red lines depict mean of the maximum temperature anomaly 60 days before each survey in (A) British Columbia, (B) Washington, (C) Oregon, and (D) California. Number of surveys per jurisdiction per year is shown above each plot. (Figure from Harvell et al. (2019).)

(Miner et al. 2018). Timing and extent of mortality of the sunflower sea star was linked with warm temperature anomalies (Harvell et al. 2019). The extent of the 2015–2016 marine heat wave in the Pacific northwest is best measured as a metric called temperature anomaly, which shows the site-specific increase in temperature from a 20-year mean. Analyses of the sunflower sea star decline with both absolute temperature and temperature anomaly show a better fit to the temperature anomaly (Harvell et al. 2019). In one case, higher declines in ochre sea stars were associated with cooler temperatures (Menge et al. 2016). The outbreak has, for now, significantly changed the seascape in ways that can be seen in intertidal and shallow subtidal rocky reefs, with high densities of new sea urchins and widespread urchin barrens (areas of denuded kelp beds). The outbreak has likely changed less-studied, deeper ocean regions due to the widespread disappearance of the sunflower sea star from waters as deep at 1,100 m (Harvell et al. 2019).

This outbreak stands with the abalone withering syndrome (Crosson and Friedman 2018) and amphibian chytrid fungus (Lips et al. 2006) as a stark reminder of the impact a novel, wide-host-range pathogen can have on groups of related species in wild populations (Harvell 2019).

8.2.4 One pathogen, one disease: abalone

Abalone are herbivorous marine gastropods and support fisheries around the world. Before 1980, there were seven species of abalone in California waters and a thriving fishery. In the mid-1980s, a combination of overharvesting, a warm-water El Niño, and a disease epidemic drove declines in California abalone species. Half the species are now under varying levels of endangerment and one is on the verge of extinction. In this section, we review the role of a well-studied multi-host rickettsial bacterium in pushing some of these species to endangerment.

When

Timing, level of impact, and geographic patterns in mortality are best studied for the intertidal black abalone (*Haliotis cracherodii*). Massive die-offs of intertidal black abalone due to withering syndrome

were first noted on the Channel Islands off the Californian coast in 1986 (Tissot 1995), and this was thought to be related to starvation due to El Niño. However, the subsequent spread over time suggested an infectious agent (Lafferty and Kuris 1993). By 1992, withering syndrome was observed near Point Conception on the mainland (Altstatt et al. 1996). By 1998, mass mortalities of black abalone due to withering syndrome had occurred throughout southern California (> 90 percent decline in numbers in all size classes) and there was a clear pattern of decline from south to north over time (Raimondi et al. 2002). Mortality rate was later linked to warm temperatures. Abalone continue to die from disease (Crosson and Friedman 2018).

Signs and cause

Withering syndrome is a fatal disease of abalones characterized by a severely shrunken foot. The signs develop slowly and result in the external loss of muscle tone and ability to grip the substrate (Figure 8.6). Histology reveals large inclusions and atrophy of gastro-intestinal tissue and the digestive gland (Crosson and Friedman 2018). Withering syndrome (WS) was eventually confirmed to be caused by infection with a Rickettsiales-like organism (RLO); the causative agent was described and provisionally named "*Candidatus Xenohaliotis californiensis*" (WS-RLO) (Friedman et al. 2000). The WS-RLO is an obligate, intracellular bacterium that infects abalone gastro-intestinal epithelia and disrupts the digestive gland (Figure 8.6). The bacterium is transmitted horizontally via a fecal–oral route, with initial infections located in the posterior esophagus tissue and, to a lesser extent, the intestine of host abalone (Friedman et al. 2002).

Impact

Currently, populations of five of the seven California abalone species are declining and receive varying levels of federal protection, ranging from "Species of Concern," including pinto (*Haliotis kamtschatkana*), green (*H. fulgens*), and pink (*H. corrugata*) abalones, to "Endangered," including white (*H. sorenseni*) and black (*H. cracherodii*) abalones (reviewed in Crosson et al. 2014).

Once a thriving wild and farmed fishery, black and white abalone are endangered in California.

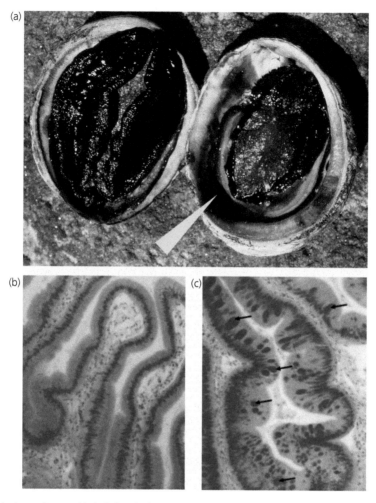

Figure 8.6 Effects of withering syndrome on black abalone (*Haliotis cracherodii*) (a, right, yellow arrow) compared to an uninfected abalone (a, left). (Photo: Jenny Dugan.) Light micrographs of abalone tissues stained with H&E of a normal post-esophagus (b) and a WS-RLO-infected post-esophagus (c), with arrows indicating WS-RLO cytoplasmic inclusions (bacterial colonies). Magnification × 200. (Micrographs from Crosson et al. (2014).)

The white abalone was listed as endangered under the Endangered Species Act in 2001. In the most recent surveys by the National Marine Fishery Service, white abalone populations had declined from historic levels of millions to less than 2,500. White abalone are on the verge of extinction. Due to precipitous declines, the black abalone were listed as endangered in 2009. Black abalone are considered locally extinct at most mainland sites south of Point Conception, California. The continued presence in California waters of the rickettsia that causes withering syndrome impedes recovery of these

abalone species. The National Marine Fisheries Service is preparing a 5-year review of black abalone and white abalone to determine if anything can be done to save them from extinction. A rickettsia-infecting phage and the existence of resistant individuals provide some hope for the black abalone (Friedman et al. 2014; Little et al. Chapter 4, this volume).

Pinto abalone are the only abalone to the north in Washington and Alaska, regions of historically healthy populations. By 1994, they had declined to the point that the sport fishery closed and they have not recovered. Their populations are so low in the waters of Washington state that they are listed as a species of concern and have been declared functionally extinct in this locality. Proposals to list the pinto abalone as endangered have been rejected because they thrive in Canada and Alaska and are already protected in Canada. Most of the range for pinto abalone lies north of where temperatures currently cause the rickettsia to be destructive (above 17 °C), so it is unknown to date if disease has had any role in their decline which is largely attributed to over-fishing. A recent study revealed extreme temperature dependence in susceptibility for the three species tested and that the white abalone have the greatest susceptibility, followed by the closely related pinto abalone as the most susceptible to the RLO (Crosson and Friedman 2018). Red, black, and pink are the next most susceptible, with green abalone being relatively resistant (Crosson and Friedman 2018). The high susceptibility of the pinto abalone and temperature dependence of the RLO shows a susceptible, at-risk species sitting right on the edge of a temperature-sensitive pathogen's expanding range in a warming ocean.

8.3 Turning the tide: marine biodiversity provides services that influence disease

Disease not only directly threatens biodiversity through reducing vulnerable species, but also indirectly can have cascading impacts through the removal of foundation or keystone species. Foundation species like seagrasses are also ecosystem engineers with potent pathogen-reducing capabilities (Lamb et al. 2017), so their loss through disease outbreaks could amplify disease risk for associated biota. Mechanisms of natural pathogen removal represent a frontier for mitigating disease in the marine environment (Raymundo et al. Chapter 9, this volume).

The oceans have a vast store of novel mechanisms that control pathogenic bacteria, viruses, and parasites. In this section, we briefly describe three scales of pathogen-fighting mechanisms that are currently untapped in today's oceans: at a whole ecosystem level in ecosystems like seagrasses, at the organismal level in the form of invertebrates acting as bio-filters, and at the microscopic scale of microbiomes.

The oceans are a microbial soup and this fact overwhelms all others in our consideration of what governs transmission and overall risk of infection for wildlife (Ben-Horin et al. Chapter 12, this volume). A vital component of marine ecosystems are the plants, invertebrates, and microbes that create powerful pathogen bio-filters. Mussels and clams can clean the water and reduce risk of infections propagating to other organisms. Their role in limiting or accelerating the transmission of disease has been largely overlooked. The tricky part is that this same filtration process doesn't always kill infections and can massively concentrate and convey pathogens (Behringer et al. Chapter 10, this volume). For example, clams and mussels can deliver a lethal dose of the pathogen *Vibrio haemolyticus* and *V. vulnificus* to humans (Froelich and Noble 2016). *Vibrio haemolyticus* causes diarrhea and is conveyed to humans by eating infected clams or oysters. *Vibrio vulnificus* is vastly worse and quite simply kills people who eat infected clams or oysters. Unfortunately, both of these are increasing in warming oceans (Baker-Austin et al. 2013). The consequences for human health are likely just the tip of the iceberg in terms of transmission—if humans are being sickened and killed by these bacteria, how are other organisms that eat mussels, clams, and oysters being affected? In our own research during the sea star epidemic, we wondered if sea stars that eat clams and mussels could be exposed to higher doses of the virus-sized pathogen. What other pathogenic micro-organisms are being similarly concentrated and conveyed to other wildlife through these bio-concentrators?

The flip side of bivalve filtration is that they can also clean the water by removing pathogens. In this way, some bivalves may actually protect wildlife from infective doses of disease. For instance, recent laboratory experiments show that oysters can remove infective zoospores of *L. zosterae* from the water and reduce pathogen risk for eelgrass (Groner et al. 2018). Both sides of this coin are illustrated in Figure 8.7, which shows that oysters are a dead-end host that can remove pathogens from the water column and kill them. In this figure, mussels are shown as amplifying infectious bacteria and increasing disease risk if consumed.

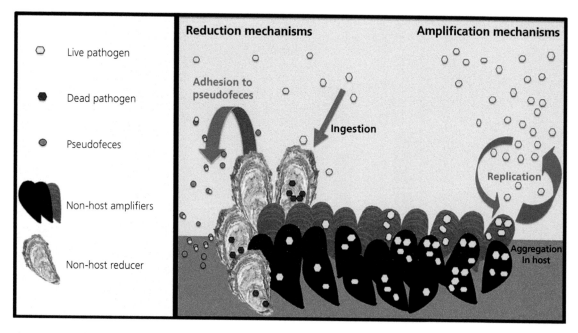

Figure 8.7 Filter feeders can influence pathogen transmission through reduction or amplification. (Figure from Burge et al. (2016).)

At a microbial scale, experimental evidence of chemical and biological pathogen regulation by seagrass and its microbiome has been shown *in vivo*. For example, phytochemicals extracted from seagrass blade tissues of multiple tropical species can kill or inhibit pathogenic bacteria that affect humans, fishes, and invertebrates (Kumar et al. 2008, Mani et al. 2012). Distinct microbial biofilms dominated by diazotrophic epiphytic cyanobacteria on seagrass blade surfaces (Hamisi et al. 2013) and antimicrobial compounds from endophytic fungi found growing within the tissues of several tropical seagrass species (Supaphon et al. 2013) have also been shown to experimentally inhibit multiple fish and human pathogens. In the field, seawater isolated from eelgrass harbors growth-inhibiting bacteria against the toxic dinoflagellate *Alexandrium tamarense* responsible for paralytic shellfish poisoning (Onishi et al. 2014).

The entire seagrass ecosystem could play a pivotal role in the removal of pathogenic bacteria, drawing parallels to literature from constructed wetland vegetation dating back to the 1950s (Wu et al. 2016). Although a seagrass monoculture alone exhibits several physical and biological characteristics of an effective pathogen removal system, an intact seagrass ecosystem comprises a diversity of bivalves, sponges, tunicates, and epiphytic organisms that influence levels of waterborne pathogenic bacteria (Burge et al. 2016). Using amplicon sequencing of the 16S ribosomal RNA gene, the presence of intact seagrass beds resulted in 50 percent reductions in the relative abundance of potential bacterial pathogens capable of causing disease in humans and marine organisms (Lamb et al. 2017). The pathogen-reducing services of seagrass beds extend to wildlife. Field surveys of more than 8,000 reef-building corals located adjacent to seagrass showed two-fold reductions in disease levels compared to paired sites without adjacent seagrass (Figure 8.8).

8.4 Summary

- This chapter highlights the disease threat to ecologically important foundation and keystone species and the resultant disruption to the balance of both tropical and temperate ecosystems.
- Four case histories of disease outbreaks that impact marine biodiversity are reviewed: eelgrass,

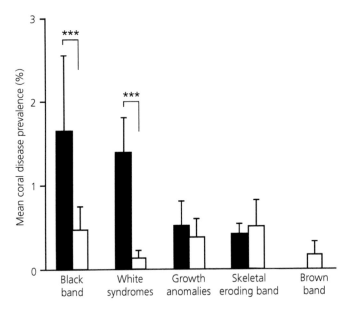

Figure 8.8 Diseases of reef-building corals are lower when adjacent to seagrass meadows. Prevalence (mean and standard error) of coral disease syndromes with adjacent seagrass meadows (white bars) compared to reefs without seagrass meadows (black bars). A total of 8,034 corals were surveyed across four different island reefs in the Spermonde Archipelago, Indonesia. Significant comparisons are indicated by asterisks. (Figure from Lamb et al. (2017).)

corals, sea stars, and abalone. Massive outbreaks in corals, abalone, and sea stars have contributed to imperilment or listing as endangered species.

- Multi-host diseases in all these cases have contributed to extreme population declines and ecosystem destruction or change.
- Vital knowledge gaps for each of the disease outbreak that have imperiled marine biodiversity are reviewed: timing and location of outbreaks, causative micro-organism, and impact of outbreak.
- Infectious diseases pose a great threat to marine biodiversity—directly through the reduction of foundation and keystone species like corals, seagrasses and seastars and indirectly through the disruption in habitat and pathogen-reducing services when foundation species or ecosystem engineers are removed.

8.5 Future directions and priorities for research

- A very poor understanding of host susceptibility to disease and the rates and pathways by which diseases are transmitted in a changing ocean impedes forecasting of outbreaks and management.
- New studies show that marine habitats like seagrass beds have powerful capability to detoxify and reduce pathogenic bacteria. There is a huge knowledge gap regarding which marine habitats provide this service and to what magnitude. Nothing is known about the mechanisms of detoxification and biofiltration. We suggest that the rising tide of marine disease may be alleviated by focused attention on natural services provided by healthy ecosystems.

References

Aeby, G. S., and D. L. Santavy. 2006. Factors affecting susceptibility of the coral *Montastraea faveolata* to black-band disease. Marine Ecology Progress Series 318:103–10.

Ainsworth, T. D., R. V. Thurber, and R. D. Gates. 2010. The future of coral reefs: a microbial perspective. Trends in Ecology & Evolution 25:233–40.

Altstatt, J. M., R. F. Ambrose, J. M. Engle, P. L. Haaker, K. D. Lafferty, and P. T. Raimondi. 1996. Recent declines of black abalone *Haliotis cracherodii* on the mainland coast of central California. Marine Ecology Progress Series 142:185–92.

Aronson, R. B., J. F. Bruno, W. F. Precht, P. W. Glynn, C. D. Harvell, L. Kaufman, C. S. Rogers, E. A. Shinn, Valentine, John F, J. M. Pandolfi, R. H. Bradbury, E. Sala, T. P. Hughes, K. A. Bjorndal, R. G. Cooke, D. McArdle, L. McClenachan, M. J. H. Newman, G. Paredes, R. R. Warner, Jackson, Jeremy B. C, T. P. Hughes, A. H. Baird, D. R. Bellwood, S. R. Connolly, C. Folke, R. Grosberg, O. Hoegh-Guldberg, J. B. C. Jackson, J. Kleypas, J. M. Lough, P. Marshall, M. Nystrom, S. R. Palumbi, J. M. Pandolfi, B. Rosen, and J. Roughgarden. 2003. Causes of coral reef degradation. Science 302:1502–4.

Baker-Austin, C., J. A. Trinanes, N. G. Taylor, R. Hartnell, A. Siitonen, and J. Martinez-Urtaza. 2013. Emerging *Vibrio* risk at high latitudes in response to ocean warming. Nature Climate Change 3:73.

Bockelmann, A.-C., V. Tams, J. Ploog, P. R. Schubert, and T. B. Reusch. 2013. Quantitative PCR reveals strong spatial and temporal variation of the wasting disease pathogen, *Labyrinthula zosterae* in northern European eelgrass (*Zostera marina*) beds. PLoS ONE 8:e62169.

Bucci C., M. Francoeur, J. McGreal, R. Smolowitz, V. Zazueta-Novoa, and G. M. Wessel. 2017. Sea Star Wasting Disease in Asterias forbesi along the Atlantic Coast of North America. PLoS ONE 12(12): e0188523.

Brandt, M. E., T. B. Smith, A. M. S. Correa, and R. Vega-Thurber. 2013. Disturbance driven colony fragmentation as a driver of a coral disease outbreak. PLoS ONE 8:e57164.

Bruno, J. F., J. J. Stachowicz, and M. D. Bertness. 2003. Inclusion of facilitation into ecological theory. Trends in Ecology & Evolution 18:119–25.

Bruno, J. F., E. R. Selig, K. S. Casey, C. A. Page, B. L. Willis, C. D. Harvell, H. Sweatman, and A. M. Melendy. 2007. Thermal stress and coral cover as drivers of coral disease outbreaks. PLoS Biology 5:e124.

Bull, J. C., E. J. Kenyon, and K. J. Cook. 2012. Wasting disease regulates long-term population dynamics in a threatened seagrass. Oecologia 169:135–42.

Burge, C. A., C. Mark Eakin, C. S. Friedman, B. Froelich, P. K. Hershberger, E. E. Hofmann, L. E. Petes, K. C. Prager, E. Weil, and B. L. Willis. 2014. Climate change influences on marine infectious diseases: implications for management and society. Annual Review of Marine Science 6:249–77.

Burge, C. A., C. J. Closek, C. S. Friedman, M. L. Groner, C. M. Jenkins, A. Shore-Maggio, and J. E. Welsh. 2016. The use of filter-feeders to manage disease in a changing world. Integrative and Comparative Biology 56:573–87.

Burke, L. M., K. Reytar, M. Spalding, and A. Perry. 2011. Reefs at Risk Revisited. World Resources Institute, Washington, DC.

Burt, J. M., M. T. Tinker, D. K. Okamoto, K. W. Demes, K. Holmes, and A. K. Salomon. 2018. Sudden collapse of a mesopredator reveals its complementary role in mediating rocky reef regime shifts. Proceedings of the Royal Society B 285:20180553.

Chandler, V. K., and J. P. Wares. 2017. RNA expression and disease tolerance are associated with a "keystone mutation" in the ochre sea star *Pisaster ochraceus*. Peer Journal 5:e3696.

Clemente, S., J. Lorenzo-Morales, J. C. Mendoza, C. López, C. Sangil, F. Alves, M. Kaufmann, and J. C. Hernández. 2014. Sea urchin *Diadema africanum* mass mortality in the subtropical eastern Atlantic: role of waterborne bacteria in a warming ocean. Marine Ecology Progress Series 506:1–14.

Crosson, L. M., and C. S. Friedman. 2018. Withering syndrome susceptibility of northeastern Pacific abalones: A complex relationship with phylogeny and thermal experience. Journal of Invertebrate Pathology 151:91–101.

Crosson, L. M., N. Wight, G. R. VanBlaricom, I. Kiryu, J. D. Moore, and C. S. Friedman. 2014. Abalone withering syndrome: distribution, impacts, current diagnostic methods and new findings. Diseases of Aquatic Organisms 108:261–70.

Daszak, P., A. A. Cunningham, and A. D. Hyatt. 2000. Emerging infectious diseases of wildlife—threats to biodiversity and human health. Science 287:443–9.

Dawkins, P. D., M. E. Eisenlord, R. M. Yoshioka, E. Fiorenza, S. Fruchter, F. Giammona, M. Winningham, and C. D. Harvell. 2018. Environment, dosage, and pathogen isolate moderate virulence in eelgrass wasting disease. Diseases of Aquatic Organisms 130:51–63.

Eisenlord, M. E., M. L. Groner, R. M. Yoshioka, J. Elliott, J. Maynard, S. Fradkin, M. Turner, K. Pyne, N. Rivlin, and R. van Hooidonk. 2016. Ochre star mortality during the 2014 wasting disease epizootic: role of population size structure and temperature. Philosophical Transactions of the Royal Society B: Biological Sciences 371:20150212.

Flint, M., J. C. Patterson-Kane, C. J. Limpus, and P. C. Mills. 2010. Health surveillance of stranded green turtles in southern Queensland, Australia (2006–2009): an epidemiological analysis of causes of disease and mortality. EcoHealth 7:135–45.

Frada, M., I. Probert, M. J. Allen, W. H. Wilson, and C. de Vargas. 2008. The "Cheshire Cat" escape strategy of the coccolithophore *Emiliania huxleyi* in response to viral infection. Proceedings of the National Academy of Sciences 105:15944–9.

Friedman, C. S., K. B. Andree, K. A. Beauchamp, J. D. Moore, T. T. Robbins, J. D. Shields, and R. P. Hedrick. 2000. "*Candidatus Xenohaliotis californiensis*", a newly described pathogen of abalone, *Haliotis* spp., along the west coast of North America. International Journal of Systematic and Evolutionary Microbiology 50:847–55.

Friedman, C. S., W. Biggs, J. D. Shields, and R. P. Hedrick. 2002. Transmission of withering syndrome in black abalone, *Haliotis cracherodii* Leach. Journal of Shellfish Research 21:817–24.

Friedman, C. S., N. Wight, L. M. Crosson, G. R. VanBlaricom, and K. D. Lafferty. 2014. Reduced disease in black abalone following mass mortality: phage therapy and natural selection. Frontiers in Microbiology 5:78.

Froelich, B. A., and R. T. Noble. 2016. Vibrio bacteria in raw oysters: managing risks to human health. Philosophical Transactions of the Royal Society B: Biological Sciences 371:20150209.

Fuess, L. E., M. E. Eisenlord, C. J. Closek, A. M. Tracy, R. Mauntz, S. Gignoux-Wolfsohn, M. M. Moritsch, R. Yoshioka, C. A. Burge, and C. D. Harvell. 2015. Up in arms: immune and nervous system response to sea star wasting disease. PLoS ONE 10:e0133053.

Godet, L., J. Fournier, M. M. van Katwijk, F. Olivier, P. Le Mao, and C. Retière. 2008. Before and after wasting disease in common eelgrass *Zostera marina* along the French Atlantic coasts: a general overview and first accurate mapping. Diseases of Aquatic Organisms 79:249–55.

Green, E. P., and A. W. Bruckner. 2000. The significance of coral disease epizootiology for coral reef conservation. Biological Conservation 96:347–61.

Groner, M. L., C. A. Burge, C. S. Couch, C. J. Kim, G.-F. Siegmund, S. Singhal, S. C. Smoot, A. Jarrell, J. K. Gaydos, and C. D. Harvell. 2014. Host demography influences the prevalence and severity of eelgrass wasting disease. Diseases of Aquatic Organisms 108:165–75.

Groner, M. L., C. A. Burge, C. J. Kim, E. Rees, K. L. Van Alstyne, S. Yang, S. Wyllie-Echeverria, and C. D. Harvell. 2016a. Plant characteristics associated with widespread variation in eelgrass wasting disease. Diseases of Aquatic Organisms 118:159–68.

Groner, M. L., J. Maynard, R. Breyta, R. B. Carnegie, A. Dobson, C. S. Friedman, B. Froelich, M. Garren, F. M. D. Gulland, S. F. Heron, R. T. Noble, C. W. Revie, J. D. Shields, R. Vanderstichel, E. Weil, S. Wyllie-Echeverria, and C. D. Harvell. 2016b. Managing marine disease emergencies in an era of rapid change. Philosophical Transactions of the Royal Society B: Biological Sciences 371(1689).

Groner, M. L., C. A. Burge, R. Cox, N. Rivlin, M. Turner, K. L. Van Alstyne, S. Wyllie-Echeverria, J. Bucci, P. Staudigel, and C. S. Friedman. 2018. Oysters and eelgrass: potential partners in a high pCO$_2$ ocean. Ecology 99(8):1802–14.

Hamisi, M., B. Díez, T. Lyimo, K. Ininbergs, and B. Bergman. 2013. Epiphytic cyanobacteria of the seagrass *Cymodocea rotundata*: diversity, diel nifH expression and nitrogenase activity. Environmental Microbiology Reports 5:367–76.

Harvell, C. D. 2019. Ocean Outbreak: Confronting the Rising Tide of Marine Disease. University of California Press, Oakland.

Harvell, C. D., C. E. Mitchell, J. R. Ward, S. Altizer, A. P. Dobson, R. S. Ostfeld, and M. D. Samuel. 2002. Climate warming and disease risks for terrestrial and marine biota. Science 296:2158–62.

Harvell, C. D., E. Jordan-Dahlgren, S. Merkel, and E. Rosenberg. 2007. Coral disease, environmental drivers, and the balance between coral and microbial associates. Oceanography 20:172–95.

Harvell, C. D., D. Montecino-Latorre, J. M. Caldwell, J. M. Burt, K. Bosley, A. Keller, S. F. Heron, A. K. Salomon, L. Lee, and O. Pontier. 2019. Disease epidemic and a marine heat wave are associated with the continental-scale collapse of a pivotal predator (*Pycnopodia helianthoides*). Science Advances 5:eaau7042.

Hewson, I., J. B. Button, B. M. Gudenkauf, B. Miner, A. L. Newton, J. K. Gaydos, J. Wynne, C. L. Groves, G. Hendler, and M. Murray. 2014. Densovirus associated with sea-star wasting disease and mass mortality. Proceedings of the National Academy of Sciences 111:17278–83.

Hewson, I., K. S. Bistolas, E. M. Quijano Cardé, J. B. Button, P. J. Foster, J. M. Flanzenbaum, J. Kocian, and C. K. Lewis. 2018. Investigating the complex association between viral ecology, environment, and northeast Pacific sea star wasting. Frontiers in Marine Science 5:77.

Jakobsson-Thor, S., G. B. Toth, J. Brakel, A.-C. Bockelmann, and H. Pavia. 2018. Seagrass wasting disease varies with salinity and depth in natural *Zostera marina* populations. Marine Ecology Progress Series 587:105–15.

Kumar, C. S., D. V. Sarada, T. P. Gideon, and R. Rengasamy. 2008. Antibacterial activity of three South Indian seagrasses, *Cymodocea serrulata*, *Halophila ovalis* and *Zostera capensis*. World Journal of Microbiology and Biotechnology 24:1989–92.

Lafferty, K. D., and A. M. Kuris. 1993. Mass mortality of abalone *Haliotis cracherodii* on the California Channel Islands: tests of epidemiological hypotheses. Marine Ecology Progress Series 96:239–48.

Lafferty, K., J. W. Porter, and S. Fort. 2004. Are diseases increasing in the ocean? Annual Review of Ecology, Evolution, and Systematics 35:31–54.

Lafferty, K. D., C. D. Harvell, J. M. Conrad, C. S. Friedman, M. L. Kent, A. M. Kuris, E. N. Powell, D. Rondeau, and S. M. Saksida. 2015. Infectious diseases affect marine fisheries and aquaculture economics. Annual Review of Marine Science 7:471–96.

Lamb, J. B., and B. L. Willis. 2011. Using coral disease prevalence to assess the effects of concentrating tourism activities on offshore reefs in a tropical marine park. Conservation Biology 25:1044–52.

Lamb, J. B., J. D. True, S. Piromvaragorn, and B. L. Willis. 2014. Scuba diving damage and intensity of tourist activities increases coral disease prevalence. Biological Conservation 178:88–96.

Lamb, J. B., D. H. Williamson, G. R. Russ, and B. L. Willis. 2015. Protected areas mitigate diseases of reef-building corals by reducing damage from fishing. Ecology 96:2555–67.

Lamb, J. B., A. S. Wenger, M. J. Devlin, D. M. Ceccarelli, D. H. Williamson, and B. L. Willis. 2016. Reserves as tools for alleviating impacts of marine disease. Philosophical Transactions of the Royal Society B: Biological Sciences 371:20150210.

Lamb, J. B., J. A. van de Water, D. G. Bourne, C. Altier, M. Y. Hein, E. A. Fiorenza, N. Abu, J. Jompa, and C. D. Harvell. 2017. Seagrass ecosystems reduce exposure to bacterial pathogens of humans, fishes, and invertebrates. Science 355:731–3.

Lamb, J. B., B. L. Willis, E. A. Fiorenza, C. S. Couch, R. Howard, D. N. Rader, J. D. True, L. A. Kelly, A. Ahmad, J. Jompa, and C. D. Harvell. 2018. Plastic waste associated with disease on coral reefs. Science 359:460–2.

Lamont, R. J., and G. Hajishengallis. 2015. Polymicrobial synergy and dysbiosis in inflammatory disease. Trends in Molecular Medicine 21:172–83.

Lips, K. R., F. Brem, R. Brenes, J. D. Reeve, R. A. Alford, J. Voyles, C. Carey, L. Livo, A. P. Pessier, and J. P. Collins. 2006. Emerging infectious disease and the loss of biodiversity in a Neotropical amphibian community. Proceedings of the National Academy of Sciences 103:3165–70.

Mani, A. E., V. Bharathi, and J. Patterson. 2012. Antibacterial activity and preliminary phytochemical analysis of seagrass Cymodocea rotundata. International Journal of Microbiological Research 3:99–103.

Martin, D. L., Y. Chiari, E. Boone, T. D. Sherman, C. Ross, S. Wyllie-Echeverria, J. K. Gaydos, and A. A. Boettcher. 2016. Functional, phylogenetic and host-geographic signatures of Labyrinthula spp. provide for putative species delimitation and a global-scale view of seagrass wasting disease. Estuaries and Coasts 39:1403–21.

McCallum, H. 2012. Disease and the dynamics of extinction. Philosophical Transactions of the Royal Society B: Biological Sciences 367:2828–39.

McCallum, H., and A. Dobson. 1995. Detecting disease and parasite threats to endangered species and ecosystems. Trends in Ecology & Evolution 10:190–4.

McCallum, H., Harvell C. D, and Dobson A. 2003. Rates of spread of marine pathogens. Ecology Letters 6:1062–7.

Menge, B. A., E. B. Cerny-Chipman, A. Johnson, J. Sullivan, S. Gravem, and F. Chan. 2016. Sea star wasting disease in the keystone predator Pisaster ochraceus in Oregon: insights into differential population impacts, recovery, predation rate, and temperature effects from long-term research. PLoS ONE 11:e0153994.

Mera, H., and D. G. Bourne. 2018. Disentangling causation: complex roles of coral-associated microorganisms in disease. Environmental Microbiology 20:431–49.

Miner, C. M., J. L. Burnaford, R. F. Ambrose, L. Antrim, H. Bohlmann, C. A. Blanchette, J. M. Engle, S. C. Fradkin, R. Gaddam, and C. D. Harley. 2018. Large-scale impacts of sea star wasting disease (SSWD) on intertidal sea stars and implications for recovery. PLoS ONE 13:e0192870.

Montecino-Latorre, D., M. E. Eisenlord, M. Turner, R. Yoshioka, C. D. Harvell, C. V. Pattengill-Semmens, J. D. Nichols, and J. K. Gaydos. 2016. Devastating transboundary impacts of sea star wasting disease on subtidal asteroids. PLoS ONE 11:e0163190.

Muehlstein, L. K., D. Porter, and F. T. Short. 1991. Labyrinthula zosterae sp. nov., the causative agent of wasting disease of eelgrass, Zostera marina. Mycologia 83(2):180–91.

Muller, E. M., and R. van Woesik. 2012. Caribbean coral diseases: primary transmission or secondary infection? Global Change Biology 18:3529–35.

Muller, E. M., and R. van Woesik. 2014. Genetic susceptibility, colony size, and water temperature drive white-pox disease on the coral Acropora palmata. PLoS ONE 9:e110759.

Mydlarz, L. D., L. E. Jones, and C. D. Harvell. 2006. Innate immunity, environmental drivers, and disease ecology of marine and freshwater invertebrates. Annual Review of Ecology, Evolution, and Systematics 37:251–88.

Nicolet, K. J., M. O. Hoogenboom, N. M. Gardiner, M. S. Pratchett, and B. L. Willis. 2013. The corallivorous invertebrate Drupella aids in transmission of brown band disease on the Great Barrier Reef. Coral Reefs 32:585–95.

Onishi, Y., Y. Mohri, A. Tuji, K. Ohgi, A. Yamaguchi, and I. Imai. 2014. The seagrass Zostera marina harbors growth-inhibiting bacteria against the toxic dinoflagellate Alexandrium tamarense. Fisheries Science 80:353–62.

Orth, R. J., T. J. Carruthers, W. C. Dennison, C. M. Duarte, J. W. Fourqurean, K. L. Heck, A. R. Hughes, G. A. Kendrick, W. J. Kenworthy, and S. Olyarnik. 2006. A global crisis for seagrass ecosystems. BioScience 56:987–96.

Paine, R. T. 1969. A note on trophic complexity and community stability. The American Naturalist 103:91–3.

Pollock, F. J., J. B. Lamb, S. N. Field, S. F. Heron, B. Schaffelke, G. Shedrawi, D. G. Bourne, and B. L. Willis. 2014. Sediment and turbidity associated with offshore dredging increase coral disease prevalence on nearby reefs. PLoS ONE 9:e102498.

Power, M. E., D. Tilman, J. A. Estes, B. A. Menge, W. J. Bond, L. S. Mills, G. Daily, J. C. Castilla, J. Lubchenco, and R. T. Paine. 1996. Challenges in the quest for keystones:

identifying keystone species is difficult—but essential to understanding how loss of species will affect ecosystems. BioScience 46:609–20.

Raimondi, P. T., C. M. Wilson, R. F. Ambrose, J. M. Engle, and T. E. Minchinton. 2002. Continued declines of black abalone along the coast of California: are mass mortalities related to El Niño events? Marine Ecology Progress Series 242:143–52.

Randall, C. J., A. G. Jordán-Garza, E. M. Muller, and R. Van Woesik. 2014. Relationships between the history of thermal stress and the relative risk of diseases of Caribbean corals. Ecology 95:1981–94.

Raymundo, L. J., W. L. Licuanan, and A. M. Kerr. 2018. Adding insult to injury: Ship groundings are associated with coral disease in a pristine reef. PLoS ONE 13:e0202939.

Richardson, L. L., K. G. Goldberg, R. B. Kuta, R. B. Aronson, G. W. Smith, K. B. Ritchie, J. C. Halas, J. S. Feingold, and S. M. Miller. 1998. Florida's mystery coral killer identified. Nature 392:557–8.

Roberts, C. M., C. J. McClean, J. E. Veron, J. P. Hawkins, G. R. Allen, D. E. McAllister, C. G. Mittermeier, F. W. Schueler, M. Spalding, and F. Wells. 2002. Marine biodiversity hotspots and conservation priorities for tropical reefs. Science 295:1280–4.

Rogers, A., J. L. Blanchard, and P. J. Mumby. 2014. Vulnerability of coral reef fisheries to a loss of structural complexity. Current Biology 24:1000–5.

Rubio-Guerri, C., M. Melero, F. Esperón, E. N. Bellière, M. Arbelo, J. L. Crespo, E. Sierra, D. García-Párraga, and J. M. Sánchez-Vizcaíno. 2013. Unusual striped dolphin mass mortality episode related to cetacean morbillivirus in the Spanish Mediterranean Sea. BMC Veterinary Research 9:106.

Schiebelhut, L. M., J. B. Puritz, and M. N. Dawson. 2018. Decimation by sea star wasting disease and rapid genetic change in a keystone species, Pisaster ochraceus. Proceedings of the National Academy of Sciences 115(27):201800285.

Schultz, J. A., R. N. Cloutier, and I. M. Côté. 2016. Evidence for a trophic cascade on rocky reefs following sea star mass mortality in British Columbia. Peer Journal 4:e1980.

Shelton, A. O., T. B. Francis, B. E. Feist, G. D. Williams, A. Lindquist, and P. S. Levin. 2017. Forty years of seagrass population stability and resilience in an urbanizing estuary. Journal of Ecology 105:458–70.

Short, F. T., L. K. Muehlstein, and D. Porter. 1987. Eelgrass wasting disease: cause and recurrence of a marine epidemic. The Biological Bulletin 173:557–62.

Short, F. T., B. W. Ibelings, and C. Den Hartog. 1988. Comparison of a current eelgrass disease to the wasting disease in the 1930s. Aquatic Botany 30:295–304.

Sullivan, B. K., T. D. Sherman, V. S. Damare, O. Lilje, and F. H. Gleason. 2013. Potential roles of Labyrinthula spp. in global seagrass population declines. Fungal Ecology 6:328–38.

Sullivan, B. K., S. M. Trevathan-Tackett, S. Neuhauser, and L. L. Govers. 2018. Host–pathogen dynamics of seagrass diseases under future global change. Marine Pollution Bulletin 134:75–88.

Supaphon, P., S. Phongpaichit, V. Rukachaisirikul, and J. Sakayaroj. 2013. Antimicrobial potential of endophytic fungi derived from three seagrass species: Cymodocea serrulata, Halophila ovalis and Thalassia hemprichii. PLoS ONE 8:e72520.

Sutherland, K. P., B. Berry, A. Park, D. W. Kemp, K. M. Kemp, E. K. Lipp, and J. W. Porter. 2016. Shifting white pox aetiologies affecting Acropora palmata in the Florida Keys, 1994–2014. Philosophical Transactions of the Royal Society B: Biological Sciences 371:20150205.

Tissot, B. N. 1995. Recruitment, growth, and survivorship of black abalone on Santa Cruz Island following mass mortality. Bulletin of the Southern California Academy of Sciences 94:179–89.

Vega Thurber, R. L., D. E. Burkepile, C. Fuchs, A. A. Shantz, R. McMinds, and J. R. Zaneveld. 2014. Chronic nutrient enrichment increases prevalence and severity of coral disease and bleaching. Global Change Biology 20:544–54.

Wares, J. P., and L. M. Schiebelhut. 2016. What doesn't kill them makes them stronger: an association between elongation factor 1-α overdominance in the sea star Pisaster ochraceus and "sea star wasting disease." Peer Journal 4:e1876.

Waycott, M., C. M. Duarte, T. J. B. Carruthers, R. J. Orth, W. C. Dennison, S. Olyarnik, A. Calladine, J. W. Fourqurean, K. L. Heck, A. R. Hughes, G. A. Kendrick, W. J. Kenworthy, F. T. Short, and S. L. Williams. 2009. Accelerating loss of seagrasses across the globe threatens coastal ecosystems. Proceedings of the National Academy of Sciences 106:12377–81.

Weil, E. 2004. Coral reef diseases in the wider Caribbean, pp. 35–68. In: E. Rosenberg and Y. Loya (eds) Coral Health and Disease. Springer, Berlin.

Wobeser, G. A. 2006. Essentials of Disease in Wild Animals. Blackwell Publishing, Ames, Iowa.

Wu, S., P. N. Carvalho, J. A. Müller, V. R. Manoj, and R. Dong. 2016. Sanitation in constructed wetlands: a review on the removal of human pathogens and fecal indicators. Science of the Total Environment 541:8–22.

Zaneveld, J. R., R. McMinds, and R. V. Thurber. 2017. Stress and stability: applying the Anna Karenina principle to animal microbiomes. Nature Microbiology 2:17121.

CHAPTER 9

Disease ecology in marine conservation and management

Laurie J. Raymundo, Colleen A. Burge, and Joleah B. Lamb

9.1 Introduction

One of the earliest efforts to describe a marine disease began with dermo, caused by the protist *Perkinsus marinus*, which affects both wild and cultured populations of the oyster *Crassostrea virginica* (described in Mackin 1951; Behringer et al. Chapter 10, this volume). The economic importance of this shellfish, whose populations suffered a precipitous decline from both overharvesting and disease, prompted a conversion from harvesting to farming the oysters, as well as one of the most comprehensive disease studies that is known within the marine disease literature (reviewed in Smolowitz 2013). The study of diseases of wild populations has often lagged behind that of aquacultured species, as culture systems are often simplified and thus easier to study, and finding disease control solutions is motived by direct economic loss associated with disease mortality (Behringer et al. Chapter 10, this volume). While we have learned a great deal about the causal agents, etiologies, and control and treatment of cultured marine organisms, culture systems frequently do not mimic open ocean, or even coastal, environments. As the impacts of emerging marine infectious diseases grow, incorporating disease ecology in management and conservation planning becomes paramount.

This chapter discusses what we currently understand regarding the role of diseases in natural marine communities, how anthropogenic stressors drive changes in the host–pathogen balance, options for disease management, and links between aquaculture and natural systems. We present two case studies examining management approaches that have been successful in reducing the impacts of disease.

9.2 Disease as an ecological component

Disease is a natural component of ecological systems, acting as a driver of population dynamics with the capacity to reshape ecosystems. In the last few decades, an increase in disease has been noted (Wilcox and Gubler 2005), including diseases affecting marine organisms (Ward and Lafferty 2004). An important aspect of the ecology of disease is the determination of whether a disease is part of the natural system, or related to a stressor (or multiple stressors). In disease ecology, the factors underlying disease are depicted with a three-part model; the "host–pathogen–environment" paradigm (Figure 9.1).

Within this model, disease outcomes operate dynamically, where a change in the environment (i.e., a temperature increase) can affect both the host (i.e., immune compromise) and pathogen (i.e., increased virulence). Human activities alter ecosystems and can thus act as drivers of disease in marine ecosystems. Anthropogenic change is often linked to emerging infectious disease (EID) (Daszak et al. 2000). EIDs are diseases that have recently and/or rapidly increased in incidence or geographic range, moved

Raymundo, L.J., Burge, C.A., and Lamb, J.B., *Disease ecology in marine conservation and management* In: *Marine Disease Ecology.* Edited by: Donald C. Behringer, Kevin D. Lafferty, and Brian. R. Silliman, Oxford University Press (2020). © Oxford University Press.
DOI: 10.1093/oso/9780198821632.003.0009

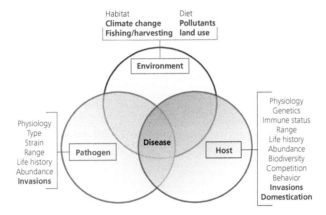

Figure 9.1 Schematic model of the host–pathogen–environment triad of ecological factors to consider in disease management. Factors in bold are those that are linked to anthropogenic activities and, thus, could be easier to control or manage.

into new host populations, or become associated with newly evolved pathogen(s) (Daszak et al. 2000, Burge et al. 2018). Examples of recent marine EIDs include ostreid herpesvirus infections in Pacific oysters (Burge et al. 2018), acroporid serratiosis in *Acropora palmata* (see Box 9.1, Figure 9.2; Burge et al. 2018), and infectious hematopoetic necrosis virus (IHNV) of finfish (Walker and Winton 2010). In the marine environment, disease is often first noted after the system is in an imbalance, characterized by large-scale mass mortalities or increased disease observations (Harvell et al. 1999, Harvell 2019). Increased general disease observations are important but cannot be linked to a specific outbreak without appropriate disease diagnosis, which highlights the need for standardized protocols for disease investigation (Frasca et al. Chapter 11, this volume; Burge et al. 2016b).

To effectively understand the cause of the imbalance in the host–pathogen interaction, the question *"What is normal?"* must first be addressed. Particularly in marine systems, a disease outbreak may occur without knowledge of what constitutes "normal" for the host and its microbiome. For organisms of direct economic importance (i.e., oysters and salmon), an understanding of the normal microbiota associated with an organism may be available, but for species with indirect economic importance (i.e., corals, sea stars), baselines are often not available. Thus, understanding marine disease must begin with determination of "normal" vs "abnormal" (i.e., diseased). If we pause to consider the diversity of

ocean habitats, the organisms inhabiting unique niches, and the potential diversity of pathogens that affect them, the lack of understanding of a normal baseline of diseases should not come as a surprise. Often, the first observation of disease will begin with the host organism and/or the population or community to which it belongs. However, first observations must be followed by a consideration of factors across scales, starting at the gene level (i.e., host or pathogen genotype), followed by the cell/tissue affected, the organism, and its population or community, and finally the environment or ecosystem (Burge et al. 2016b). Further, both host and pathogen must be considered across these scales. Though not all diseases will be caused by infectious agents (such as acute domoic acid toxicosis in sea otters; see review in Burge et al. 2016b), the ecological factors contributing to multiple diseases may be similar. A thorough investigation of host and pathogen ecology, including natural history of the host and ecological changes associated with human activities, may be useful for conservation and disease management. Anthropogenic-induced changes can directly (i.e., fishing/harvest or land/habitat use) or indirectly (i.e., climate change; Burge and Hershberger Chapter 5, this volume) impact disease (Burge et al. 2014, Lafferty 2017).

For the host, multiple ecological factors must be considered in conservation and disease management (Figure 9.1; Lafferty 2017). At a basic level, host physiology must be considered. For example, is the

Box 9.1 Acroporid serratiosis: A link between human sewage and coral disease

Acroporid serratiosis, previously known as white pox disease, was first reported in 1996 off Key West, Florida, and has since been observed throughout the Caribbean (Figure 9.2). It exclusively affects the major reef framework-building elkhorn coral *Acropora palmata*, and is characterized by irregularly shaped lesions devoid of tissue, which simultaneously appear on all colony surfaces and progress along the perimeter of lesions at a mean rate of 2.5 cm² day⁻¹ (Sutherland et al. 2004). It is most prevalent during periods of elevated water temperature. Inoculation experiments, coupled with 16s rDNA sequence analysis and microbiological characterization, identified the human fecal enteric bacterium *Serratia marcescens* as the causal agent, fulfilling Koch's postulates (Patterson et al. 2002). Outbreaks of the disease in 2002 and 2003 were associated with a bacterial strain concurrently found in human sewage. The same strain was also found in the corallivorous snail *Coralliophila abbreviata*, suggesting a possible reservoir and vector of the disease (Sutherland et al. 2011).

The disease was implicated as a principal cause of precipitous decline of *A. palmata* throughout the Florida Keys (87 percent loss between 1996 and 2002; Sutherland et al. 2004), which led to the inclusion of *A. palmata* as a candidate species on the Endangered Species List in 1999 (Diaz-Soltero 1999) and its listing as threatened in 2006 (https://ecos.fws.gov). The severity of impacts of this disease has spurred wastewater management changes in the Florida Keys. While sewage treatment plants in Key West were shown to have high concentrations of *S. marcescens* in raw sewage influent water, effluent did not contain the bacterium, demonstrating the efficacy of upgrades of the sewage treatment facility (Sutherland et al. 2010). However, much of the Florida Keys has been serviced via septic tanks, cesspools, and injection wells, which do not remove bacteria. A growing body of evidence showing the impacts of sewage on this essential reef-building coral and extent of contamination of nearshore waters pressured local authorities to devise a plan to convert to wastewater treatment plants throughout the Keys (USEPA 2013), completed in 2017.

Figure 9.2 Putative acroporid serratiosis (white pox disease) affecting *Acropora palmata* in the Florida Keys. (Photo: J. Porter.)

organism an ectotherm or endotherm? Ectotherms (i.e., organisms unable to thermoregulate), such as invertebrates and most species of fish, are more sensitive to changes in temperature, which may lead to enhanced disease expression—a topic that requires more study. Underlying genetic variability may play a role in disease susceptibility: rare, endangered, or domesticated populations may have limited genetic variation and thus less variation in susceptibility. In domesticated populations, host genetics can be leveraged to manage disease (for example, oysters: see Dègremont et al. 2015), where animals more resistant to infection may reduce infection risk. Additional factors that may be important in disease management include host range, habitat (i.e., environment), competition, behavior, and biodiversity (Morton et al. Chapter 3, this volume). Human activities can also affect the host ecology through domestication and invasions; invasions may occur through both direct human-assisted migration, such as movement of cultured organisms (see Section 9.8), and indirect mechanisms such as climate change or transport in shipping ballast water (Doney et al. 2012; Pagenkopp Lohan et al. Chapter 7, this volume).

Pathogens, and potential pathogens, are part of the normal seascape, and are linked to hosts through food webs (McLaughlin et al. Chapter 2, this volume) and the physical environment (see Section 1). A recent literature review indicates there are approximately 102 marine disease agents with notable ecological and/or economic impacts (25 viruses, 33 bacteria, 23 protists, and 21 metazoans) (Lafferty 2017; Bateman et al. Chapter 1, this volume). This is likely an underestimate of disease-causing agents, as their identification is a lengthy process, infections and disease may be cryptic or covert, and an unknown number of diseases are, as yet, undescribed. Pathogens may exist at a low level within a population, (i.e., they are "covert"), and will be undiscovered until an "overt" infection occurs. Potential pathogens may be a ubiquitous part of the environment as constituent members of communities in soils or sediments, air, or water (fresh, brackish, or sea). Each pathogen type and potential pathogen will have variations in life history and physiology (and may be dependent on host physiology) that dictate its range, abundance, strain, and potential to evolve. Important considerations in pathogen life history which may be targeted by management include: type

of transmission (horizontal or vertical); reservoirs or vectors of disease; and transmission dynamics, i.e., whether transmission is host density- or pathogen dose-dependent and/or affected by host behavior. An example of pathogen evolution is horizontal gene transfer of either antibiotic resistance or virulence that may occur in common marine bacteria (which may be virulent or benign, depending on the strain) (Little et al. Chapter 4, this volume). A recent emerging disease of shrimp, acute hepatopancreatic necrosis disease (APHD), is caused by a common bacterial species, *Vibrio parahaemolyticus*, which can also be problematic to human health. *Vibrio parahaemolyticus* contains both a plasmid coding for virulence factors (i.e., a 70-kb plasmid containing *pirAB* toxin genes; Lee et al. 2015) and antibiotic resistance (i.e., pTetB-VA1, a plasmid containing a tetracycline resistance gene; Han et al. 2015). Much like their hosts, pathogens may move or invade new populations in a process called "microbial traffic" (Morse 2004), which can occur through movement of infected hosts, vectors, intermediate hosts, or contaminated equipment. Therefore, direct management strategies may reduce pathogen introduction, including both known disease agents and those that are cryptic. Additionally, understanding pathogen ecology can help in management of a disease agent once it arrives, and in rare cases lead to successful eradication of a disease (see Boxes 9.1 and 9.2 for examples). Research focusing on testing disease management strategies is urgently needed, as few marine diseases have associated management strategies with demonstrated effectiveness.

9.3 The unique marine environment

The world's oceans are three-dimensional, open environments, linked by current patterns which can either create transmission pathways or act as barriers to transmission. Additionally, unlike air, ocean water is a microbial soup, similar in constituency to host tissues. This increases the long-term viability of pathogens in a free-living infectious state, as well as the likelihood of their long-distance transport. This capacity for transport and exchange allows for considerable contact and mixing between communities, via larvae or other mobile life stages. These mobile forms frequently serve as dispersal stages

Box 9.2 Introduction and eradication of a "parasite-like" sabellid polychaete epibiont from California

In the 1990s in California, a tiny (2-mm) introduced sabellid polychaete caused an epidemic in cultured red abalone (*Haliotis rufescens*), resulting in domed, brittle shells and deformed respiratory pores, rendering them unmarketable (Oakes and Fields 1996, Kuris and Culver 1999) (Figure 9.3). Shortly after identification within a culture facility, the sabellid was found in the intertidal zone near the outfall of the same farm (< 100 m of shoreline) within a population of susceptible gastropods (*Chlorostoma* (*Tegula*) spp; Culver and Kuris 2000).

The source of the worms was determined likely to be escaped gastropods and empty abalone shells from the culture facility; transmission from the farm was confirmed by a mark recapture "sentinel" study (Culver and Kuris 2000). Linked with sales of abalone seed, the sabellid was also identified across all commercial facilities and in some public aquariums in California (Moore et al. 2013). Previously unknown, the sabellid was named *Terebrasabella heterouncinata*, and has an unusual life history: upon settling on the host shell (typically as a mobile benthic larva) and producing a mucus sheath, the abalone secretes a nacreous layer over the sabellid, resulting in vertical shell growth (Kuris and Culver 1999). The worm then metamorphoses into its adult form: a hermaphrodite capable of self-fertilization. Transmission studies showed that severe infestations lead to the domed, brittle shells first described by farmers (Kuris and Culver 1999, Moore et al. 2007). Prior to introduction to California, *T. heterouncinata* was unrecognized even in its native habitat of South Africa; the worms likely arrived with a shipment of abalone. Though *T. heterouncinata* does not feed on abalone tissue, the sabellid was determined to be a "parasite-like" epibiont (Kuris and Culver 1999), infecting multiple gastropod species (Culver and Kuris 2004, Moore et al. 2007).

A rapid, coordinated management response among industry, academia, and regulators led to eradication of the sabellid from the infested intertidal zone and the abalone farms. First, a reduction of new infestations in the intertidal zone occurred through installation of screens in the outfall stream of the farm. In 1996, 1.6 million gastropods (primarily *Chlorotoma* spp.) were removed from the intertidal zone, a number based on the threshold host density

Figure 9.3 Sabellid polychaete disease affecting shell growth of red abalone, *Haliotis rufescens*. (Left) Normal shell; (right) infected shell. (Photo by T. Robbins, California Department of Fish and Game.)

continued

Box 9.2 *Continued*

for transmission (or the Kermack–McKendrick theorem; McKendrick 1940, Stiven 1968) (Culver and Kuris 2000). Early follow-up surveys in 1998 indicated an apparent eradication (Culver and Kuris 2000). These were followed by 9 years of negative findings, indicating eradication of *T. heterouncinata* from the intertidal zone (Moore et al.

2013). Within farms and public aquariums, hygiene practices (i.e., freshwater treatment), first recommended by Culver et al. (1997) and further refined by the California Department of Fish Wildlife (see Moore et al. 2007), were used to eradicate the pest. The last known detection occurred in 2011 (J. Moore, pers. comm.).

---

for benthic organisms, resulting in large, open populations. Thus, the potential for an infectious disease to become pan-oceanic is far greater than that for terrestrial systems of equivalent size. Lafferty (2017), for example, mentions three scleractinian coral diseases that are global in range: black band disease, caused by a cyanobacterial consortium; bacterial bleaching, associated with multiple species of the ubiquitous marine bacterium *Vibrio* spp.; and white plague, associated with a herpesvirus. Related to this potential for broad geographic range is the possibility for rapid development of epizootics (McCallum et al. 2004; reviewed in Burge et al. 2016b). The 1983–84 Caribbean epizootic of the black-spined sea urchin, *Diadema antillarum*, caused a catastrophic 95 percent reduction in the urchin population throughout its entire range in a matter of weeks (documented in Lessios et al. 1984). While the causal agent of this event was never determined, it resulted in massive and persistent consequences to Caribbean reef systems, many of which shifted from coral-dominated to algae-dominated systems, triggered, in part, by the loss of this dominant herbivore. Twenty years after the epizootic, isolated cases of recovery of urchin populations were recorded in some locations (Edmunds and Carpenter 2001, Miller et al. 2003) but not others (Chiappone et al. 2002).

The influence of marine diseases varies with spatial scale. Large-scale current patterns circulate planktonic larvae, microbes, and parasites across ocean basins, while coastal topography interacting with tidal flow can create localized eddies that can entrain planktonic organisms and restrict water exchange between adjacent communities. The capacity for long-distance dispersal of free-living infectious stages means that infection with a disease

may be decoupled from the pathogen source (Lafferty 2017), which poses a challenge for disease management if controlling the source of a pathogen is a desired strategy.

High host density, which is characteristic of many sessile benthic marine communities, is generally associated with higher disease prevalence, as spread of infectious agents between hosts is facilitated by proximity (Lafferty 2004, McCallum et al. 2005, Bruno et al. 2007). This effect may be particularly pronounced if pathogens have a wide host range and can affect multiple species, such as in coral communities, where hosts are sessile and density may be high (Figure 9.4). However, Buck et al. (2017) noted that this general paradigm is not straightforward in its impacts on benthic communities. High host density can actually reduce the risk of infection to individual hosts, providing "safety in numbers." This effect is more pronounced with water-borne pathogens which do not rely on direct transmission. Vectored and vertical transmission are considered more rare in marine systems than in terrestrial (Poulin et al. 2016), though the epidemiology of many diseases has not been fully described.

Diversity in marine organisms takes many forms. For instance, taxonomic diversity of both host and pathogenic agents is higher in marine environments than in their terrestrial counterparts (McCallum et al. 2004). Certain marine pathogens appear able to infect hosts that are taxonomically distant, which can increase their impact on a community. For example, a white syndrome outbreak in Palau in 2005 affected nine scleractinian coral species representing six families. The outbreak, putatively caused by the *Vibrio coralliilyticus* bacterium (Sussman et al. 2008), was localized to a small but diverse patch of corals and resulted in 46 percent mortality

Figure 9.4 High- vs low-density coral reef communities, illustrating high vs low potential for density-dependent pathogen transmission. (A) Tubbataha Reef, Philippines; (B) Rota Island, Micronesia. (Photos by L. Raymundo.)

over a 10-month period. Interestingly, the same bacterium was implicated in white syndrome outbreaks that exclusively targeted *Acropora cytherea* in the Marshall Islands and *Montipora aequituberculata* in the Great Barrier Reef (Sussman et al. 2008). This "plasticity'—the ability of a pathogen to infect a wide range of taxa on one reef during an outbreak, yet target a single species on another reef—is currently poorly understood, but is likely to be influenced by host immune capability, host density, vectors and reservoirs, and environmental drivers (Burge and Hershberger Chapter 5, this volume).

Life-history strategies reflect the complexity and variety of habitat and environmental choices available to marine organisms. Both coloniality and clonality are common, particularly among lower invertebrates (Shields 2017). Coloniality can create large, clonal mega-organisms lacking in genetic diversity but with the capacity to partially die, to then recover by asexually iterating their individual units (zooids, polyps), and to grow indeterminately. Older colonies are generally larger (but not always; see Hughes and Jackson 1980), which increases the probability of contact with pathogens. Thus, coloniality allows an infected host to potentially remain so for a long time, or to become re-infected seasonally, thus continuing to transmit pathogens to surrounding susceptible hosts. Likewise, colony morphology can influence contact with pathogens; highly branched, complex growth forms may entrain water and suspended particles within the colony, increas-

ing the chances of contact with host tissues (Kim and Lasker 1998). These factors require consideration when either modelling disease processes or formulating management strategies for disease outbreaks.

9.4 Environmental drivers and anthropogenic forcing

Marine disease processes are influenced by the environment, which encompasses both natural variability and anthropogenic change (reviewed by Burge et al. 2014). Natural variability in the environment may include connected abiotic (i.e., water quality: Bojko et al. Chapter 6, this volume) and biotic (i.e., food web: McLaughlin et al. Chapter 2, this volume) characteristics that a host and pathogen are exposed to. Disease is thus often seasonal, as specific environmental attributes that vary seasonally lead to disease expression. Water quality includes factors such as temperature, salinity, dissolved oxygen, carbonate chemistry, dissolved and particulate organic material, and nutrients. Temperature is the most understood driver of disease outbreaks, associated with outbreaks in plants, marine invertebrates (abalone, oysters, and corals), fishes, and marine mammals (Burge et al. 2014). Most diseases are linked with warmer summer temperatures. Rarer are the so-called cold-water diseases, such as Denman Island disease, a Pacific oyster disease caused by the microparasite *Mikrocytos mackini*, which needs winter temperatures less than 10°C for three to four consecutive months for disease outbreaks to occur (Hervio et al. 1996). Similarly, in an analysis of global coral disease prevalence, Ruiz-Moreno found that both warm and cold temperatures are linked to disease outbreaks (Ruiz-Moreno et al. 2012). Anthropogenic nutrient loading has also been associated with increases in coral disease prevalence and severity (Bruno et al. 2003, Redding et al. 2013, Vega-Thurber et al. 2014). Human-induced change often leads to shifts in both abiotic and biotic characteristics of a system, and can act on both a local (i.e., nutrient inputs or other water quality parameters, sedimentation, or overharvesting) and a global or regional level (climate change, weather pattern shifts, ocean acidification).

In an era of rapid change, attributing causation to unknown mass mortalities can be tricky, as both a pathogen and an environmental stressor can contribute to mortality. In reality, some diseases may be multistressor syndromes (i.e., summer mortality syndrome of Pacific oysters), which are typically non-infectious. Such complexity can make both management of symptoms and determination of disease causation difficult. For a presentation of these issues, Frasca et al. (Chapter 11, this volume) discusses diagnostic processes.

When we consider disease management in this era of change, questions related to environmental drivers of disease that may guide future research efforts include (but are not limited to) the following: How do both subtle and drastic changes in environmental parameters influence disease processes? What are tipping points or thresholds of these parameters, and how might they interact with each other to influence disease processes? How is climate change likely to influence marine disease? And finally: Do we currently know enough to predict answers to these questions?

9.5 Consequences for conservation and management

Diseases have emerged as a global threat to the conservation of many species (Altizer et al. 2013), at least partly because environmental conditions altered by human activities have compromised immune defenses or enhanced the virulence of pathogens (Harvell et al. 2009; Harvell and Lamb Chapter 8, this volume). Although diseases may not immediately kill their hosts, they often reduce their fitness

by deleteriously affecting fecundity, growth, behavior, and resistance to other climate-driven impacts (Harvell et al. 2009, Wobeser 2013). The need to evaluate the veracity of management practices designed to protect ecosystem health is becoming increasingly urgent, given that diseases often impact habitat-structuring species and are likely to have cascading and indirect impacts on a multitude of associated marine populations.

Disease epizootics have been linked to both anthropogenic (Bojko et al. Chapter 6, this volume) and climatic (Burge and Hershberger Chapter 5, this volume) stressors. For example, numerous studies have associated the passage of intense tropical storms with subsequent elevated levels of disease in organisms as diverse as plants (Irey et al. 2006), sea urchins (Scheibling and Lauzon-Guay 2010), and reef corals (Brandt et al. 2013). Furthermore, marine disease outbreaks are often linked to chronic exposure to pollutants such as sewage (Wear and Thurber 2015), terrestrial sediment or agricultural herbicides (Pollock et al. 2014, Renault 2015), nutrients and fertilizers (Gochfeld et al. 2012, Vega Thurber et al. 2014), and aquaculture (Lafferty et al. 2015). Relationships between marine disease and the multitude of environmental parameters that are influenced by a changing climate and increasing levels of anthropogenic activities are often complex (Altizer et al. 2013). Our understanding of the pathogens that cause most marine diseases is still unclear, particularly when compared to diseases that occur on land. Thus, developing management strategies for marine diseases has often had to occur with little or no knowledge of the causal organism, drawing heavily on experiences from terrestrial systems, which may or may not be appropriate.

9.6 Managing marine disease outbreaks

Managing local stressors more effectively may build disease resistance and resilience to global stressors. However, reducing the effects of escalating disease threats is particularly challenging in marine environments (McCallum et al. 2003, Harvell et al. 2009, Altizer et al. 2013). Managers confronted with controlling terrestrial disease outbreaks have multiple tools available, such as vaccination, quarantine, and culling, to restrict contact between healthy and infected individuals, biological and chemical controls, elimination or regulation of vectors, and genetic breeding for disease resistance or tolerance (Daszak et al. 2000). However, inherent difficulties associated with executing such disease control methods in fluid environments limit their applicability for marine species (McCallum et al. 2003). There is an increasingly critical need to apply ecological theory to practical management solutions for alleviating disease impacts on marine populations. In Table 9.1, we present examples from the available literature of tools that have been tested to incorporate disease ecology in marine conservation and management strategies. The potential role of managing food webs (McLaughlin et al. Chapter 2, this volume) to control disease outbreaks has not been investigated (hence, there is no reference to this in Table 9.1). This suggests a line of future research that could result in an additional management tool for diseases and parasitic infections for which vectors, reservoirs, or causal agents are known.

Table 9.1 Examples of tools that have been tested to incorporate disease ecology in marine conservation and management strategies.

Taxa affected	Diseases	Causative agent(s)	Management strategy examined	References
Corals	Black band disease	Cyanobacteria-based consortium	Stopping progression of band disease in corals with epoxy plugs	Aeby et al. 2015
Shellfish, salmon	Various	Bacteria, viruses, parasites	Using benthic filter feeders to "clean" water and reduce pathogen load	Reviewed in Burge et al. 2016a, Lamb et al. 2017
Corals	White syndromes Skeletal eroding band Brown band disease	*Vibrio* spp. and others Halofolliculinid ciliate *Porpostoma* ciliates	Establishment of marine protected areas and reserves	Raymundo et al. 2009, Page et al. 2009, Lamb et al. 2015, 2016
Corals, sea turtles	White syndromes Skeletal eroding band Brown band disease (corals)	*Vibrio* spp. and others Halofolliculinid ciliate *Porpostoma* ciliates	Reducing physical injury from human activities—including fishing gear, concentrating tourism, ship groundings	Lamb and Willis 2011, Lamb et al. 2014, 2015, 2016, Work et al. 2015, Raymundo et al. 2018
Shrimp, fish	AHPND	*Vibrio parahaemolyticus*	Biological control	Reviewed in Shields 2017
Shrimp	TSV MSGS IHHNV	Virus Unknown Virus	Selective breeding for resistance; specific pathogen-free stocks	Thitamadee et al. 2016, reviewed in Shields 2017
Shellfish	Various	Various	Culling	Elston and Ford 2011, reviewed in Shields 2017
Shrimp	WSSV, TSV, YHV	Viruses	Vaccination	Reviewed in Shields 2017
Bivalves, crustaceans	Various	Various	Quarantine	Reviewed in Shields 2017
Shrimp	Viral diseases	Viral pathogens	Elimination of vector species in culture	Thitamadee et al. 2016
Shrimp	WSSV, YHV, TSV	Viral pathogens	Increasing biosecurity monitoring and surveillance	Flegel 2012, reviewed in Shields 2017
Corals	White syndromes	*Vibrio* spp. and others	Increasing forecasting and predictive power using satellite imagery—temperature, sediment exposure, plastic waste from terrestrial sources	Maynard et al. 2011, 2015a, 2016, Pollock et al. 2014, Lamb et al. 2016, 2018
Corals	Serratiosis	*Serratia marcescens* bacterium	Improving sewage treatment and waste water discharge	Sutherland et al. 2010, Sutherland et al. 2011

(continued)

Table 9.1 Continued

Taxa affected	Diseases	Causative agent(s)	Management strategy examined	References
Corals	Bacterial diseases	Bacterial pathogens	Probiotics	Teplitsky and Ritchie 2009
Corals, lobsters	White syndromes	Bacterial pathogens	Disease surveillance, monitoring	Maynard et al. 2011, Shields 2017
Corals	Black band disease	Cyanobacteria-based consortium	Rapid pathogen tests	Discussed in Pollock et al. 2011
Abalone, oysters, shrimp	Abalone herpesvirus, oyster dermo, shrimp AHPND	*Perkinsus marinus; Vibrio parahaemo-lyticus*	Improving health certification protocols for transported animals	Thitamadee et al. 2016
Corals	Bacterial bleaching	*Vibrio coralliilyticus*	Phage therapy	Efrony et al. 2007, Teplitsky and Ritchie 2009
Corals	Florida Reef Tract white disease	Unknown, possible bacterium	*In-situ* chiseled trench in skeleton behind disease progression front, amoxicillin, amoxicillin + trenching	Neely and Lewis 2018
Corals	Florida Reef Tract white disease	Unknown, possible bacterium	UV radiation applied to lesions	Enochs and Kolodziej 2018
Corals	Florida Reef Tract white disease	Unknown, possible bacterium	*Ex-situ* testing of antibiotics, chlorine with various barriers (epoxy, clay), with and without trenching	Neely 2018

9.7 Managing human impacts on natural systems

Terrestrial disease management strategies are not necessarily directly transferrable to the marine environment. Marine disease control is challenging due to the fluidity of open ocean systems; options such as quarantine, culling, and vaccination are not viable in most situations. Thus, diseases affecting marine communities require the development of innovative approaches and novel tools that integrate marine and terrestrial management areas. In this section, we discuss three promising strategies for the management of marine diseases: the use of marine protected areas, the development of early warning systems for disease outbreaks, and the use of natural systems as pathogen "filters."

9.7.1 Marine protected areas and spatial management

Tissue damage has been shown to promote disease development by providing a site for pathogen invasion in many taxonomic groups, including humans and megafauna (Wobeser 2013), insects (Ferrandon et al. 2007), plants and trees (Underwood 2012), fishes (Austin and Austin 2007), and marine invertebrates such as sponges and corals (Mydlarz et al. 2006). Moreover, invertebrate immune responses are depleted during wound regeneration, resulting in reduced capacity to develop an immune response following exposure to a foreign substance, further increasing the likelihood of disease development (Mydlarz et al. 2006). Protecting flora and fauna from physical disturbances associated with human use has prompted spatial management solutions, such as restricting site access or activities allowed within designated areas (De'ath et al. 2012, Newsome and Moore 2012).

Protected areas have been instrumental in reducing damage in terrestrial and aquatic environments (Leung and Marion 2000, Sobel and Dahlgren 2004), yet links between damage reduction and disease mitigation have only been assessed in a handful of studies. The efficacy of marine reserves as management tools for preventing disease outbreaks has been assessed numerous times in coral populations, though results vary. For example, no-take marine reserves have been shown to mitigate coral disease by maintaining functionally diverse fish assemblages (Raymundo et al. 2009) and by reducing physical damage associated with fishing activities and derelict gear (Lamb et al. 2015, Lamb et al. 2016). It is postulated that high densities of herbivorous fish within marine reserves could limit the growth of algae (Bellwood et al. 2003), which can act as reservoirs of pathogens (Nugues et al. 2004, Smith et al. 2006). Moreover, exclusion of activities that injure corals inside marine reserves, such as destructive fishing methods and gears (e.g., Asoh et al. 2004, Yoshikawa and Asoh 2004), and high-intensity tourism (Lamb and Willis 2011, Lamb et al. 2014), is likely to mitigate disease by reducing entry points for opportunistic coral pathogens (Page and Willis 2008, Nicolet et al. 2013, Katz et al. 2014, Lamb et al. 2014). Other studies have found little evidence that protected areas mitigate coral disease (Coelho and Manfrino 2007, McClanahan et al. 2009, Page et al. 2009), although authors cautioned that either poor compliance with fishing restrictions or the presence of environmental influences that permeate reserve borders could have negated reserve effectiveness in their studies. It is also plausible that protected areas might facilitate the spread of disease by increasing densities or cover of susceptible coral hosts (McCallum et al. 2005, Bruno et al. 2007, Myers and Raymundo 2009), or by increasing densities of fishes that are either vectors for coral pathogens or cause feeding injuries that increase coral susceptibility to opportunistic pathogens (Aeby and Santavy 2006, Raymundo et al. 2009).

Well-managed marine reserves may assist adaptation to impacts of climate change by increasing community resilience (Figure 9.5a; Roberts et al. 2017); however, there is mounting evidence that climate-related stressors may undermine resistance to disease afforded by reserve protection. For example, although marine reserves mitigated coral disease following a severe cyclone, they were ineffective in moderating disease when sites were exposed to higher than average terrestrial runoff from a degraded river catchment (Lamb et al. 2016). This is further supported by Hughes et al. (2017), who reported that water quality and marine reserves

had no influence on the unprecedented bleaching on the Great Barrier Reef in 2016, suggesting that local protection may provide little or no protection from coral diseases associated with temperature.

The capacity of spatial closures and marine reserves to ameliorate disease will also depend upon the mechanism of disease pathogenesis. Reserves are likely inadequate for mitigating marine disease under several conditions. They may not, for example, protect highly mobile mega-fauna and fishes that traverse boundaries. Frequently, impacts are displaced outside protected area boundaries (Agardy et al. 2011) and could thus further degrade adjacent ecosystems, overriding the management benefits. Furthermore, range shifts due to anthropogenic and climate-driven processes may cause both pathogens and hosts to move out of protected areas (Hannah et al. 2007), potentially reducing the relevance of fixed spatial locations as conservation strategies for moderating disease. The benefits and limitations presented here represent only some of the considerations needed to inform the development of spatial management strategies for moderating marine disease.

Placement of protected areas is a key consideration for maximizing their effectiveness (Halpern 2003). Exposure to disturbances is an essential determinant of the vulnerability of marine ecosystems (Marshall and Johnson 2007) and such disturbances are not often spatially uniform (Devlin et al. 2013). Therefore, it is imperative to consider local patterns of exposure to disturbances when establishing protected areas for disease management. However, exposure to disturbance is rarely considered in marine protected area planning (Game et al. 2008) (but see Maynard et al. 2015b), and considerable debate is generated regarding the efficacy of protecting high- versus low-risk areas (Game et al. 2008).

9.7.2 Early warning systems and forecasting marine disease outbreaks

Early warning systems can potentially form an important component of management options of disease outbreaks, particularly in locations that are very difficult to access (Lamb et al. 2016). For example, a forecasting system linking global ocean and atmospheric climate models to malaria risk in

Botswana enabled the prediction of anomalously high probability areas so that strategies for mitigation could be initiated (Thomson et al. 2006). Forecasting is well established in crop disease management and leads to improved timing of pesticide application and deployment of planting strategies to lower disease risk (Schaafsma and Hooker 2007). On coral reefs, forecasting programs for coral bleaching are core to marine resilience programs (Eakin et al. 2010) and are guiding the development of climate-driven disease forecasting algorithms for white syndromes in reef-building corals (Maynard et al. 2011, Maynard et al. 2015a) and shell disease of American lobsters (Maynard et al. 2016). However, temporal and spatial variability in thermal anomalies could complicate efforts to predict disease outbreaks, due to uncertainty pertaining to future patterns and trajectories of stress, as well as a general lack of knowledge regarding the role of temperature stress in disease etiologies.

Satellite-derived water quality data have been critical in assessing the drivers of coral disease following acute sediment exposure from seafloor dredging (Pollock et al. 2014) and chronic exposure to terrestrial runoff (Lamb et al. 2016). Such data offer the potential for identifying and forecasting locations at increased risk of outbreaks triggered by poor water quality. Likewise, algal blooms cause mass mortality via anoxia or toxic exposure, which has an obvious immediate impact on marine populations; chronic hypoxia or exposure to algal toxins could be equally detrimental in the development of disease. For instance, nutrient-enriched primary productivity is linked with increases in the severity of amoebic gill disease of fish (Nowak 2001) and the promotion of the debilitating tumor-forming disease fibropapillomatosis in sea turtles (Van Houtan et al. 2010). Nutrient enrichment increases the prevalence and severity of multiple coral diseases in controlled laboratory and field settings (Bruno et al. 2003, Redding et al. 2013, Vega Thurber et al. 2014). Recent research found that plastic waste increased disease likelihood on coral reefs and used estimates of plastic waste entering the oceans to predict and forecast disease levels across Asia Pacific (Lamb et al. 2018). These studies suggest that established links between environmental drivers and disease outbreaks may be useful in developing forecasting tools across larger spatial scales, thus improving the use of targeted surveillance and management action.

9.7.3 Natural ecosystem "filters"

Marine and terrestrial environments are often regarded as two separate ecosystems, and managed as independent entities (Alvarez-Romero et al. 2011). Significant benefits to the health of marine

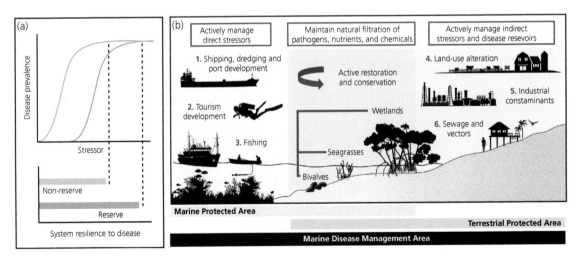

Figure 9.5 Conceptual diagram illustrating active strategies for mitigating marine disease outbreaks by increasing ecosystem resilience to disease (see also Table 9.1). (a) Effect of marine reserves on resilience to disease; (b) linkages between terrestrial and marine systems and management options for each. (Diagram from Lamb et al. (2016).)

Figure 9.6 A seagrass bed adjacent to a coastal fishing village in Indonesia. The seagrass bed acts as a filter, mitigating the delivery of human pathogens and nutrients to nearshore coral reefs. (Photo by J. Lamb.)

organisms may arise from investing conservation efforts in connecting marine and terrestrial systems (Figure 9.5b; Lamb et al. 2016). For example, ecosystem sequestration of pathogens by healthy marine habitats is a novel tool to mitigate disease outbreaks in the ocean. Lamb et al. (2017) found that seagrass ecosystems reduced sewage-sourced bacterial pathogens of humans, fishes, and invertebrates by 50 percent and coral diseases were reduced two-fold where reefs were adjacent to seagrasses (Figure 9.6). Similarly, filter feeders, such as bivalves, sponges, and polychaetes, have huge potential for reducing disease transmission to the marine environment from terrestrial sources, by filtering out pathogenic micro-organisms from the water column (Faust et al. 2009, Burge et al. 2016a). Mangroves and constructed wetlands have also been used as bio-filters for natural sewage control and are effective

filtration systems for excess sedimentation, nutrients, and organic matter (Yang et al. 2008). Therefore, conserving marine littoral zones may reduce levels of disease-causing pollutants entering coastal environments. An important area for future research would be to assess the mechanisms by which these habitats sequester pollutants and alleviate marine disease outbreaks.

9.8 Managing aquaculture systems and their interactions with natural systems

The human global population continues to expand as fish stocks dwindle, many to the point of collapse (Worm et al. 2009). Many countries are turning to aquaculture to address food security issues (Goldberg and Naylor 2005). While this can alleviate overharvesting, increase protein availability, and boost income, ecological and biosecurity issues associated with aquaculture are of concern, as many practices are poorly regulated and can have disastrous consequences for local coastal ecosystems (Naylor et al. 2000). Behringer et al. (Chapter 2, this volume) explores the topic of aquacultured fisheries in detail; here, we focus on interactions between aquaculture and natural systems.

Cultured organisms are often grown in artificial, high-density monocultures with greatly altered water quality, which stresses organisms and reduces their immune function. Regular prophylaxis, such as treatment with antibiotics, may be necessary to prevent infectious disease outbreaks. Food provisioning is often unregulated, as are the means by which prophylactics and growth hormones are delivered. Excess food and waste are released in effluent, impacting coastal water quality. Water exchange between culture systems and adjacent nearshore marine communities provides an avenue for pathogens, species invasions, and degraded water (Murray 2013). Further, many aquaculture facilities are built on converted or reclaimed wetlands, mangroves, seagrass, or coral communities, which can alter local hydrodynamics and water quality, and remove the innate microbial filtering capacity of these ecosystems (Lamb et al. 2017). Lastly, cultured organisms are frequently traded between facilities and may not be native to the area where they are cultured. This facilitates interactions between cultured and native organisms that can have deleterious consequences for either or both, but certainly constitutes an invasive species risk, referred to as "biological pollution" by Naylor et al. (2001).

Nonetheless, though unregulated aquaculture can wreak havoc on nearshore marine communities, much of what we know about marine diseases developed via experiences with cultured species (reviewed in Shields 2017). Economic losses from disease are estimated in billions of dollars, providing a strong incentive to understand disease epizootiology and develop effective control measures (Lafferty et al. 2015). Most diseases of cultured finfish originated from wild stock, due to high connectivity between aquatic environments and multiple transmission pathways (Kurath and Winton 2011). Pathogens may be latent in a natural population,

but become infectious when an unusually stressful event occurs or when introduced to the artificial and dense stocking conditions of aquaculture. This appears to be the case with aquarium corals, which are susceptible to heavy infestations of microcrustaceans and worms, yet these organisms are not reported to infest wild corals in significant densities (Sweet et al. 2011). Complex transmission dynamics may develop over time between farmed and wild populations. For instance, sea lice are a common ectoparasite of adult salmon, but rarely infect wild juveniles due to a life-history stage that prevents contact with infected adults. However, the presence of salmon farms altered the transmission dynamics of sea lice by increasing exposure of wild juveniles. This increased their mortality from predation, as predators selectively consumed infected prey (Krkosek et al. 2005). More rarely, aquaculture may introduce pathogens to wild naïve populations, with devastating effects. Non-native cultured red drum, for example, were the source of the bacterial pathogen *Streptococcus iniae* which infected two species of wild fish in the Red Sea (Corloni et al. 2002). Changing climate patterns are predicted to influence disease dynamics as well. *Perkinsus marinus*, the protist parasite causing dermo in the oyster *Crassostrea virginica*, is favored under a warming environment; parasitic infection intensity increases, and host survival decreases, with higher temperatures, suggesting a long-term increase in the impact of this disease (Malek and Byers 2018).

While managing diseases within aquaculture facilities involves well-established best-management practices for controlling water quality, inoculation, quarantine, and culling (Sindermann 1984, Tucker and Hargreaves 2008), the challenge for reducing spillover impacts on wild communities remains, as ineffective or non-existing regulations plague many facilities. In Lingayen Gulf, the Philippines, unregulated culture of the milkfish *Chanos chanos* has resulted in construction of over 1,170 fish pens within an 8-ha area of shallow estuarine water (Figure 9.7) (Travaglia et al. 2004). Water quality has degraded, with resultant increases in eutrophication in both coastal and river waters (Aban et al. 2008; Garren et al. 2008). Controlling or treating effluent should be a priority, as wastewater transports microbes, excess nutrients, and organic matter,

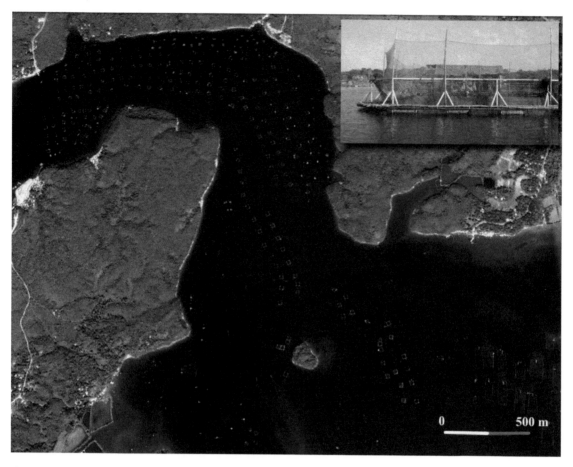

Figure 9.7 Aerial view of unregulated proliferation of fish pens of cultured milkfish, *Chanos chanos*, in Bolinao, Lingayen Gulf, the Philippines. Inset shows a single fish pen. (Aerial view taken from Google Earth; inset photo by L. Raymundo.)

antibiotics, sediment, and fecal waste (LaPatra 2003). Aside from the direct impact of introduction of pathogenic microbes into wild systems, degrading water quality can further stress organisms, increasing their susceptibility to opportunistic pathogens.

Finally, unregulated transport and trading of farmed animals provides another avenue for introduction of exotics and micro-organisms transported with them. Exotic mollusc aquaculture, in particular, has existed for hundreds of years, constituting what may be the greatest mode of introduction of any taxon worldwide (McKindsey et al. 2007). Failure to manage exotic introductions can be disastrous for both aquaculture ventures and natural populations; transport of South African abalone into California in the 1980s unwittingly introduced a parasitic polychaete that deforms shells. This introduction

impacted market prices and caused the failure of several farms. An infection of a population of gastropods was detected soon after but was successfully eradicated (Kuris and Culver 1999; see Box 9.2, Figure 9.3). Impacts of the abalone herpeslike virus (AbHV), which causes a highly lethal disease, abalone viral ganglioneuritis (AVG), was devastating to both cultured and wild populations in Australia between 2006 and 2012. The disease first emerged in farms and was later found in nearby wild populations, resulting in devastating population reductions and economic loss. Strict regulation of the transport of animals between farms is the current most effective management option (Corbeil et al. 2010).

Health certification programs exist for the movement of cultured animals, though these programs are predicated by knowledge of specific diseases/pathogens. Additionally, it is prudent to develop protocols for transport of aquatic organisms, such as the health certification procedure outlined by Humphrey et al. (1988) for cultured giant clams. These clams were transported throughout the western Pacific as part of a reintroduction program in the 1980s, sponsored by the Australian Centre for International Agriculture Research. Such protocols exist for many cultured species and/or industries. For some infectious diseases (of both terrestrial and marine vertebrates and invertebrates), the World Organisation for Animal Health (OIE; acronym based on a previous name: Office International des Epizooties) requires reporting of 117 OIE-Listed Diseases (OIE 2019). To date, the OIE has focused on diseases of economically important organisms, though a portal for reporting non-listed wildlife diseases also exists. Additionally, the OIE provides diagnostic manuals, treatment options (where they exist), and reference laboratories for listed diseases (e.g., "White spot disease in shrimp") and other specific pathogens of interest (e.g., "Infection with ostreid herpesvirus 1 microvariants"). In the USA, the USDA-APHIS is the governing body for movement of cultured animals for international import and export; for non-cultured animals, agencies such as NOAA and the US Fish & Wildlife Service are the equivalent. Movement between US states is based on the laws and regulations of the individual state; states may, in fact, share bodies of water, with laws regulating transfer of animals across state lines via the body of water. Before a culturist or researcher moves animals (either within the state or across state lines), it is important to contact the appropriate authorities (e.g., the California Department of Fish & Wildlife, Maryland Department of Natural Resources) to determine the necessary health examinations and health history required prior to the movement of animals.

9.9 Managing disease for restoration

Ecological restoration has rapidly become an essential management option to reverse the decline of marine systems, as it has been for terrestrial systems for decades. At present, disease outbreaks in restored marine populations, particularly corals, pose a threat to the success of many efforts, as restoration may involve introducing naïve organisms to new environments. Thus, disease management should be a necessary consideration in the development of restoration science. Managing and treating disease, as well as building resistance, have historically been necessary foci of aquaculture ventures, to ensure continued financial benefits (discussed in Section 9.8). Developing strategies that apply to restoration could naturally progress from this, drawing on lessons learned from culturing organisms for food and trade. For instance, Carnegie and Burreson (2011) reported increased natural resistance in wild *Crassostrea virginica* populations to *Haplosporidium*, which causes MSX and has hampered restoration efforts for the oyster. The disease is enzootic but decreasing in wild oysters growing in polyhaline conditions, while still very lethal in naïve sentinel populations growing in low-salinity environments. The authors recommended using this naturally resistant population as a source of restoration efforts; Lipcius et al. (2015) further recommended selecting polyhaline sites for restoration. Thus, lessons learned from disease management of the economically important cultured oyster is informing a re-establishment strategy in wild populations.

Disease management in marine restoration is likely to require a diverse toolkit. This may include managing environmental quality, reintroducing organisms, tracking the disease history of source populations and reintroduced organisms, and selectively breeding for disease resistance. *Ex-situ* facilities used to culture organisms for reintroduction are well equipped to monitor and control disease using traditional practices such as culling, quarantine, and active intervention. Care should be taken to use only clinically healthy organisms in restoration in order to reduce the risk of introducing pathogens to naïve populations and reduce post-introduction stress, as transplantation itself is a stressful event which may increase disease susceptibility.

Corals, which are currently a focus of restoration, may benefit from the concept of assisted evolution (van Oppen et al. 2015). This technique uses genetic tools to culture colonies that are resilient to a changing climate, with the goal of reintroducing them to natural reef habitat where it is hypothesized they will survive better than their predecessors. While the main thrust of such efforts has been producing corals that are resistant to increasing temperatures, disease resistance must be considered as well, as temperature stress and subsequent disease outbreaks may be linked (Bruno et al. 2007, Brandt and McManus 2009). Further, coral taxa may vary in susceptibility to temperature stress and disease; certain taxa show low-temperature resilience but high subsequent disease susceptibility, or vice versa (Smith et al. 2013). This suggests evolutionary trade-offs in coping with these two major stressors. Differences in susceptibility may even vary between genotypes, highlighting the importance of genetic analysis of cultured populations and applying this information to identify potentially resistant strains for restoration (Miller et al. 2019). Thus, a focus for restoration research should encompass population genetics tools to understand how these stress-coping strategies manifest in species selected for culture. Protocols could then be incorporated that reduce stress, such as timing outplanting to avoid heat stress during the recovery phase, and selective breeding for disease resistance.

9.10 Summary

- Disease is a natural part of ecosystems and communities. However, marine diseases are on the rise, so it is essential to establish what is considered a normal level of disease for a given species or community in order to assess disease risk. Anthropogenic stress is associated with many emerging infectious diseases (EIDs) and current marine disease ecology research efforts involve understanding the influence of such stressors on the three interacting components of disease: the host, the pathogen, and their environment.

- Marine systems differ fundamentally from terrestrial systems in that they are more open and connected; sessile, clonal colonial forms with larval dispersal stages are ubiquitous; host density and diversity may be very high; and microbes are diverse and abundant in ocean water.

- Marine disease processes are influenced by the environment, which includes both natural variability and anthropogenic change, as well as both biotic (species and community interactions) and abiotic factors (temperature, water quality).

- Managing marine diseases requires novel approaches, as terrestrial strategies may not be applicable. Three novel approaches—establishment of marine protected areas, development of early warning or forecasting tools, and the maintenance and use of natural ecosystem filters—are promising strategies.

- Aquaculture can have significant negative impacts on wild ecosystems via introduction of pathogens and exotic species, degraded water quality, and conversion of coastal habitats. The reverse—impacts of wild species on farmed organisms—can also have deleterious consequences for cultured animals, though more rarely.

- Disease management and regulatory procedures have been developed, but implementation is still an issue. This is particularly challenging when international transport and exchange occur with cultured organisms.

- Ecological restoration is a growing field in conservation which would benefit from a consideration of disease susceptibility and resistance. Genotyping to identify distinct cultured populations should become standard protocol in restoration, to examine natural differences in resistance to disease and other stressors and to provide guidance in selective breeding for resistance.

References

Aban, S., De Vera, R., Garcia, A. 2008. Environmental effect of fishpen culture operation of milkfish (*Chanos chanos*) to the eutrophication of brackish water rivers in sector 2 of Lingayen Gulf, Philippines. *AGRIS* 7(1):219–26.

Aeby, G.S., Santavy, D.L. 2006. Factors affecting susceptibility of the coral *Montastraea faveolata* to black-band disease. *Marine Ecology Progress Series* 318:103–10.

Aeby, G.S., Work, T.M., Runyon, C.M., Shore-Maggio, A., Ushijima, B., Videau, P., Beurmann, S., Callahan, S.M. 2015. First record of black band disease in the Hawaiian archipelago: response, outbreak status, virulence, and a method of treatment. *PLoS One* 10:1–17.

Agardy, T., Di Sciara, D.L., Christie, P. 2011. Mind the gap: addressing the shortcomings of marine protected areas through large scale marine spatial planning. *Marine Policy* 35:226–32.

Altizer, S., Ostfeld, R.S., Johnson, P.T.J., Kutz, S., Harvell, C.D. 2013. Climate change and infectious diseases: from evidence to a predictive framework. *Science* 341:514–19.

Alvarez-Romero, J.G., Pressey, R.L., Ban, N.C., Vance-Borland, K., Willer, C., Klein, J., Gaines, S.D. 2011. Integrated land–sea conservation planning: the missing links. *Annual Review of Ecology, Evolution, and Systematics* 42:381–409.

Asoh, K., Yoshikawa, T., Kosaki, R.M.R., Marschall, E.A. 2004. Damage to cauliflower coral by monofilament fishing lines in Hawaii. *Conservation Biology* 18:1645–50.

Austin, B., Austin, D.A. 2007. *Bacterial Fish Pathogens: Disease of Farmed and Wild Fish*. Basel: Springer Nature.

Bellwood, D.R., Hoey, A.S., Choat, J.H. 2003. Limited functional redundancy in high diversity systems: resilience and ecosystem function on coral reefs. *Ecology Letters* 6:281–5.

Brandt, M.E., McManus, J.W. 2009. Disease incidence is related to bleaching extent in reef-building corals. *Ecology* 90:2859–67.

Brandt, M.E., Smith, T.B., Correa, A.M.S., Vega-Thurber, R. 2013. Disturbance driven colony fragmentation as a driver of a coral disease outbreak. *PLoS ONE* 8:e57164.

Bruno, J.F., Petes, L.E., Harvell, C.D., Hettinger, A., Harvell, C.D., Hettinger, A. 2003. Nutrient enrichment can increase the severity of coral diseases. *Ecology Letters* 6:1056–61.

Bruno, J.F., Selig, E.R., Casey, K.S., Page, C.A., Willis, B.L., Harvell, C.D., Sweatman, H., Melendy, A.M. 2007. Thermal stress and coral cover as drivers of coral disease outbreaks. *PLoS Biology* 5(6):e124.

Buck, J.C., Hechinger, R.F., Wood, A.C., Stewart, T.E., Kuris, A.M., Lafferty, K.D. 2017. Host density increases parasite recruitment but decreases host risk in a snail–trematode system. *Ecology* 98:2029–38.

Burge, C.A., Eakin, C.M., Friedman, C.S., Froelich, B., Hershberger, P.K., Hoffman, E.E., Petes, L.E., Prager, K.C., Weil, E., Willis, B.L., Ford, S.E., Harvell, C.D. 2014. Climate change influences on marine infectious disease: implications for management and society. *Annual Review in Marine Science* 6:249–77.

Burge, C.A., Closek, C.J., Friedman C.S., Groner, M.L., Jenkins, C.M., Shore-Maggio, A., Welsh, J.E. 2016a. The use of filter-feeders to manage disease in a changing world. *Integrative and Comparative Biology* 56(4):573–87.

Burge, C.A., Friedman, C.S., Getchell, R., House, M., Lafferty, K.D., Mydlarz, L.D., Prager, K.C., Sutherland, K.P., Renault, T., Kiryu, I., Vega-Thurber, R. 2016b. Complementary approaches to diagnosing marine diseases: a union of the modern and the classic. *Philosophical Transactions of the Royal Society B: Biology* 371:20150207.

Burge, C.A., Shore-Maggio, A., Rivlin, N.D. 2018. Ecology of emerging infectious diseases of invertebrates. In:

Hajek, A.E. (ed.), *Ecology of Invertebrate Diseases*, pp. 587–625. Oxford: John Wiley & Sons.

Carnegie, R.B., Burreson, E.M. 2011. Declining impact of an introduced pathogen: *Haplosporidium nelsoni* in the oyster *Crassostrea virginica* in Chesapeake Bay. *Marine Ecology Progress Series* 432:1–15.

Chiappone, M., Swanson, D.W., Miller, S.L., Smith, S.G. 2002. Large-scale surveys on the Florida Reef Tract indicate poor recovery of the long-spined sea urchin *Diadema antillarum*. *Coral Reefs* 21:155–78.

Coelho, V.R., Manfrino, C. 2007. Coral community decline at a remote Caribbean island: marine no-take reserves are not enough. *Aquatic Conservation: Marine and Freshwater Ecosystems* 17:666–85.

Corbeil, S., Colling, A., Williams, L.M., Wong, F.Y.K., Savin, K., Warner, S., Murdoch, B., Cogan, N.O.I., Sawbridge, T.I., Fegan, M., Mohammad, I., Sunarto, A., Handlinger, J., Pyecroft, S., Douglas, M., Chang, P.H., Crane, M.S.J. 2010. Development and validation of a TaqMan® PCR assay for the Australian abalone herpes-like virus. *Diseases of Aquatic Organisms* 92:1–10.

Corloni, A., Diamant, A., Eldar, A., Kvitt, H., Zlotkin, A. 2002. *Streptococcus iniae* infections in Red Sea cage-cultured and wild fishes. *Diseases of Aquatic Organisms* 49:165–70.

Culver, C.S., Kuris, A.M. 2000. The apparent eradication of a locally established introduced marine pest. *Biological Invasions* 2(3):245–53.

Daszak, P., Cunningham, A.A., Hyatt, A.D. 2000. Emerging infectious diseases of wildlife—threats to biodiversity and human health. *Science* 287(5452):443–9.

De'ath, G., Fabricius, K.E., Sweatman, H., Puotinen, M. 2012. The 27-year decline of coral cover on the Great Barrier Reef and its causes. *Proceedings of the National Academy of Sciences* 109:17995–9.

Dégremont, L., Garcia, C., Allen, S.K. 2015. Genetic improvement for disease resistance in oysters: a review. *Journal of Invertebrate Pathology* 131:226–41.

Devlin, M.J., Da Silva, E.T. , Petus, C., Wenger, A., Zeh, D., Tracey, D., Álvarez-Romero, J.G., Brodie, J. 2013. Combining *in-situ* water quality and remotely sensed data across spatial and temporal scales to measure variability in wet season chlorophyll-a: Great Barrier Reef lagoon (Queensland, Australia). *Ecological Processes* 2:1–22.

Diaz-Soltero, H. 1999. Endangered and threatened species: a revision of candidate species list under the Endangered Species Act. *Federal Register* 64(210):33466–8.

Doney, S.C., Ruckelshaus, M., Duffy, J.E., Barry, J.P., Chan, F., English, C.A., Galindo, H.M., Grebmeier, J.M., Hollowed, A.B., Knowlton, N., Polovina, J., Rabalais, N.N., Sydeman, W.J., Talley, L.D. 2012. Climate change impacts on marine ecosystems. *Annual Review of Marine Science* 4:11–37.

Eakin, C.M., Nim, C.J., Brainard, R.E., Aubrect, C., Elvidge, C., Gledhill, D.K., Muller-Karger, F., Mumby, P.J., Skirving, W.J., Strong, A.E., Wang, M., Weeks, S., Wentz, F., Ziskin, D. 2010. Monitoring coral reefs from space. *Oceanography* 23(4):118–33.

Edmunds, R.C., Carpenter, P.J. 2001. Recovery of *Diadema antillarum* reduces macroalgal cover and increases abundance of juvenile corals on a Caribbean reef. *Proceedings of the National Academy of Sciences* 98:5067–71.

Efrony, R., Loya, Y., Bacharach, E., Rosenberg, E. 2007. Phage therapy of coral disease. *Coral Reefs* 26:7–16.

Elston, R.A., Ford, S.E. 2011. Shellfish diseases and health management. In: Shumway, S.E. (ed.), *Shellfish Aquaculture and the Environment*, pp. 359–94. Chichester: John Wiley & Sons.

Enochs, I.C., Kolodziej, G. 2018. Ultraviolet deactivation of coral disease lesions. Final Report. Tallahassee: Florida Department of Environmental Protection.

Faust, C., Stallknecht, D., Swayne, D., Brown, J. 2009. Filter-feeding bivalves can remove avian influenza viruses from water and reduce infectivity. *Proceedings. Biological Sciences* 276:3727–35.

Ferrandon, D., Imler, J.-L. , Hetru, C., Hoffmann, J.A. 2007. The *Drosophila* systemic immune response: sensing and signalling during bacterial and fungal infections. *Nature Reviews Immunology* 7:862–74.

Flegel, T.W. 2012. Historic emergence, impact and current status of shrimp pathogens. *Asian Journal of Invertebrate Pathology* 110:166–73.

Game, E.T., McDonald-Madden, E., Puotinen, M.L., Possingham, H.P. 2008. Should we protect the strong or the weak? Risk, resilience, and the selection of marine protected areas. *Conservation Biology* 22:1619–29.

Garren, M., Smriga, S., Azam, F. 2008. Gradients of coastal fish farm effluents and their effect on coral reef microbes. *Environmental Microbiology* 10:2299–312.

Gochfeld, D., Easson, C., Freeman, C., Thacker, R., Olson, J. 2012. Disease and nutrient enrichment as potential stressors on the Caribbean sponge *Aplysina cauliformis* and its bacterial symbionts. *Marine Ecology Progress Series* 456:101–11.

Goldburg, R. Naylor, R. 2005. Future seascapes, fishing, and fish farming. *Frontiers in Ecology and the Environment* 3:21–8.

Halpern, B.S. 2003. The impact of marine reserves: do reserves work and does reserve size matter? *Ecological Applications* 13:117–37.

Han, J.E., Mohney, L.L., Tang, K.F., Pantoja, C.R., Lightner, D.V. 2015. Plasmid mediated tetracycline resistance of *Vibrio parahaemolyticus* associated with acute hepatopancreatic necrosis disease (AHPND) in shrimps. *Aquaculture Reports* 2:17–21.

Hannah, L., Midgley, G., Andelman, S., Araújo, M., Hughes, M., Martinez-Meyer, E., Pearson, R., Williams, P. 2007. Protected area needs in a changing climate. *Frontiers in Ecology and the Environment* 5:131–8.

Harvell, C.D., Kim, K., Burkholder, J.M, Colwell, R.R., Epstein, P.R., Grimes, D.J., Hofmann, E.E., Lipp, E.K., Osterhaus, A.D.M.E., Overstreet, R.M., Porter, J.W., Smith, G.W., Vasta, G.R. 1999. Emerging marine diseases—climate links and anthropogenic factors. *Science* 285(5433):1505–10.

Harvell, D., Altizer, S., Cattadori, I.M., Harrington, L., Weil, E. 2009. Climate change and wildlife diseases: when does the host matter the most? *Ecology* 90:912–20.

Harvell, C.D. 2019. *Ocean Outbreak: Confronting the Rising Tide of Marine Disease*. Berkeley: University of California Press.

Hervio, D., Bower, S.M., Meyer, G.R. 1996. Detection, isolation and experimental transmission of *Mikrocytos mackini*, a microcell parasite of Pacific oysters *Crassostrea gigas* (Thunberg). *Journal of Invertebrate Pathology* 67:72–9.

Hughes, T.P., Jackson, J.B.C. 1980. Do corals lie about their age? Some demographic consequences of partial mortality, fission, and fusion. *Science* 209:713–15.

Hughes, T.P., Kerry, J.T., Álvarez-Noriega, M., Álvarez-Romero, J.G., Anderson, K.D., Baird, A.H., Babcock, R.C., Beger, M., Bellwood, D.R., Berkelmans, R. 2017. Global warming and recurrent mass bleaching of corals. *Nature* 543:373–7.

Humphrey, J.D., Copland, J., Lucas, J. 1988. Disease risks associated with translocation of shellfish, with special reference to the giant clam *Tridacna gigas*. In: Copland, J.W., Lucas, J.S. (eds), *Giant Clams in Asia and the Pacific*, pp. 241–4. Canberra: Australian Centre for International Agricultural Research.

Irey, M., Gottwald, T.R., Graham, J.H., Riley, T.D., Carlton, G. 2006. Post-hurricane analysis of citrus canker spread and progress towards the development of a predictive model to estimate disease spread due to catastrophic weather events. *Plant Health Progress* 7(1):10.

Katz, S.M., Pollock, F.J., Bourne, D.G., Willis, B.L. 2014. Crown-of-thorns starfish predation and physical injuries promote brown band disease on corals. *Coral Reefs* 33(3):705–16.

Kim, K., Lasker, H.R. 1998. Allometry of resource capture in colonial cnidarians and constraints on modular growth. *Functional Ecology* 12:646–54.

Krkošek, M., Lewis, M.A., Volpe, J.P. 2005. Transmission dynamics of parasitic sea lice from farm to wild salmon. *Proceedings of the Royal Society B: Biological Sciences* 272:689–96.

Kurath, G., Winton, J. 2011. Complex dynamics at the interface between wild and domestic viruses of finfish. *Current Opinion in Virology* 1:73–80.

Kuris, A., Culver, C. 1999. An introduced sabellid polychaete pest infesting cultured abalones and its potential spread to other California gastropods. *Invertebrate Biology* 118:391–403.

Lafferty, K.D. 2004. Fishing for lobsters indirectly increases epidemics in sea urchins. *Ecological Applications* 14:1566–73.

Lafferty, K.D. 2017. Marine infectious disease ecology. *Annual Review of Ecology and Systematics* 48:473–96.

Lafferty, K.D., Harvell, C.D., Conrad, J.M., Friedman, C.S., Kent, M.L., Kuris, A.M., Powell, E.N., Rondeau, D., Saksida, S.M. 2015. Infectious diseases affect marine fisheries and aquaculture economics. *Annual Review of Marine Science* 7:471–96.

Lamb, J.B., Willis, B.L. 2011. Using coral disease prevalence to assess the effects of concentrating tourism activities on offshore reefs in a tropical marine park. *Conservation Biology* 25:1044–52.

Lamb, J.B., True, J.D., Piromvaragorn, S., Willis, B.L. 2014. Scuba diving damage and intensity of tourist activities increases coral disease prevalence. *Biological Conservation* 178:88–96.

Lamb, J.B., Williamson, D.H., Russ, G.R., Willis, B.L. 2015. Protected areas mitigate diseases of reef-building corals by reducing damage from fishing. *Ecology* 96(9):2555–67.

Lamb, J.B., Wenger, A.S., Devlin, M.J., Ceccarelli, D.M., Williamson, D.H., Willis, B.L. 2016. Reserves as tools for alleviating impacts of marine disease. *Philosophical Transactions of the Royal Society B* 371(1689):1–11.

Lamb, J.B., van de Water, J.A.J.M., Bourne, D.G., Altier, C., Hein, M.Y., Fiorenza, E.A., Abu, N., Jompa, J., Harvell, C.D. 2017. Seagrass ecosystems reduce exposure to bacterial pathogens of humans, fishes, and invertebrates. *Science* 355:731–3.

Lamb, J.B., Willis, B.L., Fiorenza, E.A., Couch, C.S., Howard, R., Rader, D.N., True, J.D., Kelly, L.A., Ahmad, A., Jompa, J., Harvell, C.D. 2018. Plastic waste associated with disease on coral reefs. *Science* 359:460–2.

LaPatra, S.E. 2003. The lack of scientific evidence to support the development of effluent limitations guidelines for aquatic animal pathogens. *Aquaculture* 226:191–9.

Lee, C.T., Chen, I.T., Yang, Y.T., Ko, T.P., Huang, Y.T., Huang, J.Y., Huang, M.F., Lin, S.J., Chen, C.Y., Lin, S.S., Lightner, D.V., Wang, H.C., Wang, A.H., Wang, H.C., Hor, L.I., Lo, C.F. 2015. The opportunistic marine pathogen *Vibrio parahaemolyticus* becomes virulent by acquiring a plasmid that expresses a deadly toxin. *Proceedings of the National Academy of Sciences* 112:10798–803.

Lessios, H.A., Cubit, J.D., Robertson, D.R., Shulman, M.J., Parker, M.R., Garrity, S.D., Levings, S.C. 1984. Mass mortality of *Diadema antillarum* on the Caribbean coast of Panama. *Coral Reefs* 3:173–83.

Leung, Y.F., Marion, J.L. 2000. Recreation impacts and management in wilderness: a state-of-knowledge review. In: Cole, D.N., McCool, S.F., Borrie, W.T., O'Loughlin, J. (eds), *Wilderness Science in a Time of Change Conference Proceedings. Volume 5: Wilderness Ecosystems, Threats, and Management*, pp. 23–48. May 23–27, Missoula, MT. Proceedings RMRS-P-15-VOL-5. Ogden, UT: US Department of Agriculture, Forest Service, Rocky Mountain Research Station.

Lipcius, R.N., Burke, R., McCulloch, D., Schreiber, S., Schulte, D., Seitz, R., Shen, J. 2015. Overcoming restoration paradigms: value of the historical record and metapopulation dynamics in native oyster restoration. *Frontiers in Marine Science* 2:1–15.

Mackin, J.G. 1951. Histopathology of infection of *Crassostrea virginica* (Gmelin) by *Dermocystidium marinum*. *Bulletin of Marine Science Gulf Caribbean* 1:72–87.

Malek, J.C., Byers, J.E. 2018. Responses of an oyster host (*Crassostrea virginica*) and its protozoan parasite (*Perkinsus marinus*) to increasing air temperature. *Peer Journal* 6:e5046.

Marshall, P.A., Johnson, J.E. 2007. The Great Barrier Reef and climate change: vulnerability and management implications. In: Johnson, J.E., Marshall, P.A. (eds), *Climate Change and the Great Barrier Reef*, pp. 774–801. Townsville, Queensland: Great Barrier Reef Marine Park Authority.

Maynard, J.A., Anthony, K.R.N., Harvell, C.D., Burgman, M.A., Beeden, R., Sweatman, H., Heron, S.F., Lamb, J.B., Willis, B.L. 2011. Predicting outbreaks of a climate-driven coral disease in the Great Barrier Reef. *Coral Reefs* 30:485–95.

Maynard, J., van Hooidonk, R., Eakin, C.M., Puotinen, M., Garren, M., Williams, G., Heron, S.F., Lamb, J., Weil, E., Willis, B. 2015a. Projections of climate conditions that increase coral disease susceptibility and pathogen abundance and virulence. *Nature Climate Change* 5:688–94.

Maynard, J.A., Beeden, R., Puotinen, M., Johnson, J.E., Marshall, P., Hooidonk, R., Heron, S.F., Devlin, M., Lawrey, E., Dryden, J. 2015b. Great Barrier Reef no-take areas include a range of disturbance regimes. *Conservation Letters* 9(3):191–9.

Maynard, J.A., Van Hooidonk, R., Harvell, C.D., Eakin, C.M., Liu, G., Willis, B.L., Williams, G.J., Groner, M.L., Dobson, A., Heron, S.F., Glenn, R., Reardon, K., Shields, J.D. 2016. Improving marine disease surveillance through sea temperature monitoring, outlooks and projections. *Philosophical Transactions of the Royal Society B: Biological Sciences* 371(1689). pii: 20150208.

McCallum, H., Harvell, C.D., Dobson, A. 2003. Rates of spread of marine pathogens. *Ecology Letters* 6:1062–7.

McCallum, H., Kuris, A., Harvell, C.D., Lafferty, K.D., Smith, G.W., Porter, J. 2004. Does terrestrial epidemiology apply to marine systems? *Trends in Ecology and Evolution* 19:585–91.

McCallum, H., Gerber, L., Jani, A. 2005. Does infectious disease influence the efficacy of marine protected areas? A theoretical framework. *Journal of Applied Ecology* 42:688–98.

McClanahan, T., Weil, E., Maina, J. 2009. Strong relationship between coral bleaching and growth anomalies in massive *Porites*. *Global Change Biology* 15:1804–16.

McKendrick, A.G. 1940. The dynamics of crowd infection. *Edinburgh Medical Journal* 47:117–36.

McKindsey, C.W., Landry, T., O'Beirn, F.X., Davies, I.M. 2007. Bivalve aquaculture and exotic species: a review of ecological considerations and management issues. *Journal of Shellfish Research* 26:281–94.

Miller, M.W., Colburn, P.J., Pontes, E., Williams, D.E., Bright, A.J., Serrano, X.M., Peters, E.C. 2019. Genotypic variation in disease susceptibility among cultured stocks of elkhorn and staghorn corals. *Peer Journal* 7:e6751.

Miller, R.J., Adams, A.J., Ogden, N.B., Ogden, J.C., Ebersole, J.P. 2003. *Diadema antillarum* 17 years after mass mortality: is recovery beginning on St. Croix? *Coral Reefs* 22:181–7.

Moore, J.D., Juhasz, C.I., Robbins, T.T., Grosholz, E.D. 2007. The introduced sabellid polychaete *Terebrasabella heterouncinata* in California: transmission, methods of control and survey for presence in native gastropod populations. *Journal of Shellfish Research* 26(3):869–76.

Moore, J.D., Marshman, B.C., Robbins, T.T., Juhasz, C I. 2013. Continued absence of sabellid fan worms (*Terebrasabella heterouncinata*) among intertidal gastropods at a site of eradication in California, USA. *California Fish and Game* 99(33):115–21.

Morse, S.S. 2004. Factors and determinants of disease emergence. *Revue Scientifique et Technique. Office International des Épizooties* 23:443–52.

Murray, A.G. 2013. Epidemiology of the spread of viral diseases under aquaculture. *Current Opinion in Virology* 3:74–8.

Mydlarz, L.D., Jones, L.E., Harvell, C.D. 2006. Innate immunity, environmental drivers, and disease ecology of marine and freshwater invertebrates. *Annual Review of Ecology, Evolution, and Systematics* 37:251–88.

Myers, R.L., Raymundo, L.J. 2009. Coral disease in Micronesian reefs: a link between disease prevalence and host abundance. *Diseases of Aquatic Organisms* 87:97–104.

Naylor, R.L., Goldburg, R.J., Primavera, J.H., Kautsky, N., Beveridge, M.C., Clay, J., Folke, C., Lubchenco, J., Mooney, H., Troell, M. 2000. Effect of aquaculture on world fish supplies. *Nature* 405:1017–24.

Naylor, R., Williams, S.L., Strong, D.R. 2001. Aquaculture—a gateway for exotic species. *Science* 294:1655–6.

Neely, K. 2018. *Ex-situ* disease treatment trials. Final Report. Tallahassee: Florida Department of Environmental Protection Coral Reef Conservation Program.

Neely, K., Lewis, C. 2018. *In-situ* disease intervention. Final Report. Tallahassee: Florida Department of Environmental Protection Coral Reef Conservation Program.

Newsome, D., Moore, S.A. 2012. *Natural Area Tourism: Ecology, Impacts and Management*, p. 58. Bristol, UK: Channel View Publications.

Nicolet, K., Hoogenboom, M., Gardiner, N., Pratchett, M., Willis, B. 2013. The corallivorous invertebrate *Drupella* aids in transmission of brown band disease on the Great Barrier Reef. *Coral Reefs* 32:585–95.

Nowak, BF. 2001. Qualitative evaluation of risk factors for amoebic gill disease in cultured Atlantic salmon. *Proceeding of an International Conference*, February 8–10, 2000, Paris, pp. 148–55.

Nugues, M.M., Smith, G.W., Hooidonk, R.J., Seabra, M.I., Bak, R.P. 2004. Algal contact as a trigger for coral disease. *Ecology Letters* 7:919–23.

Oakes, F.R., Fields, R.C. 1996. Infestation of *Haliotis rufescens* shells by a sabellid polychaete. *Aquaculture* 140:139–43.

OIE. 2019. OIE-listed diseases, infections and infestations in force in 2019. https://www.oie.int/animal-health-in-the-world/oie-listed-diseases-2019/.

Page, C.A., Willis, B.L. 2008. Epidemiology of skeletal eroding band on the Great Barrier Reef and the role of injury in the initiation of this widespread coral disease. *Coral Reefs* 27:257–72.

Page, C.A., Baker, D.M., Harvell, C., Golbuu, Y., Raymundo, L., Neale, S.J., Rosell, K.B., Rypien, K.L., Andras, J.P., Willis, B.L. 2009. Influence of marine reserves on coral disease prevalence. *Diseases of Aquatic Organisms* 87:135–50.

Patterson, K.L., Porter, E.E., Ritchie, K.B., Polson, S.W., Mueller, E., Peterson, E., Santavy, D.L., Smith, G.W. 2002. The etiology of white pox, a lethal disease of the Caribbean elkhorn coral, *Acropora palmata*. *Proceedings of the National Academy of Sciences* 99:8725–30.

Pollock, F.J., Morris, P.J., Willis, B.L., Bourne, D.G. 2011. The urgent need for robust coral disease diagnostics. *PLoS Pathogens* 7(10):e1002183.

Pollock, F.J., Lamb, J.B., Field, S.N., Heron, S.F., Schaffelke, B., Shedrawi, G., Bourne, D.G., Willis, B.. 2014. Sediment and turbidity associated with offshore dredging increase coral disease prevalence on nearby reefs. *PLoS ONE* 9:e102498.

Poulin, R., Blasco-Costa, I., Randhawa, H.S. 2016. Integrating parasitology and marine ecology: seven challenges towards greater synergy. *Journal of Sea Research* 113:3–10.

Raymundo, L.J., Halford, A.R., Maypa, A.P., Kerr, A.M. 2009. Functionally diverse reef-fish communities ameliorate coral disease. *Proceedings of the National Academy of Sciences* 106:17067–70.

Raymundo, L.J., Licuanan, W.Y., Kerr, A.M. 2018. Adding insult to injury: ship groundings are associated with coral disease in a pristine reef. *PLoS ONE* 13(9):e0202939.

Redding, J.E., Myers-Miller, R.L., Baker, D.M., Fogel, M., Raymundo, L.J., Kim, K. 2013. Link between sewage-derived nitrogen pollution and coral disease severity in Guam. *Marine Pollution Bulletin* 73:57–63.

Renault, T. 2015. Immunotoxicological effects of environmental contaminants on marine bivalves. *Fish and Shellfish Immunology* 46(1):88–93.

Roberts, C.M., O'Leary, B.C., McCauley, D.J., Cury, P.M., Duarte, C.M., Lubchenco, J., Pauly, D., Sáenz-Arroyo, A., Sumaila, U.R., Wilson, R.W., Worm, B., Castilla, J.C. 2017. Marine reserves can mitigate and promote adaptation to climate change. *Proceedings of the National Academy of Sciences* 114(24):6167–75.

Ruiz-Moreno, D., Willis, B.L., Page, A.C., Weil, E., Cróquer, A., Vargas-Angel, B., Jordan-Garza, A.G., Jordán-Dahlgren, E., Raymundo, L., Harvell, C.D., 2012. Global coral disease prevalence associated with sea temperature anomalies and local factors. *Diseases of Aquatic Organisms* 100(3):249–61.

Schaafsma, A.W., Hooker, D.C. 2007. Climatic models to predict occurrence of *Fusarium* toxins in wheat and maize. *International Journal of Food Microbiology* 11(1–2):116–25.

Scheibling, R.E., Lauzon-Guay, J.-S. 2010. Killer storms: North Atlantic hurricanes and disease outbreaks in sea urchins. *Limnology and Oceanography* 55:2331–8.

Shields, J. 2017. Prevention and management of infectious diseases in aquatic invertebrates. In: Hajek, A., Shapiro-Ilan, D. (eds), *Ecology of Invertebrate Diseases*, Chapter 15. Hoboken, NJ: John Wiley & Sons.

Sinderman, C.J. 1984. Disease in marine aquaculture. *Helgoländer Meeresuntercuchungen* 37:505–32.

Smith, J.E., Shaw, M., Edwards, R.A., Obura, D., Pantos, O., Sala, E., Sandin, S.A., Smriga, S., Hatay, M., Rohwer, F.L. 2006. Indirect effects of algae on coral: algae-mediated, microbe-induced coral mortality. *Ecology Letters* 9:835–45.

Smith, T.B., Brandt, M.E., Calnan, J.M., Nemeth, R.S., Blondeau, J., Kadison, E., Taylor, M., Rothenberger, P. 2013. Convergent mortality response of Caribbean coral species to seawater warming. *Ecosphere* 4(7):87.

Smolowitz, R. 2013. A review of current state of knowledge concerning *Perkinsus marinus* effects on *Crassostrea virginica* (Gmelin) (the eastern oyster). *Veterinary Pathology* 50:404–11.

Sobel, J. Dahlgren, C. 2004. *Marine Reserves: A Guide to Science, Design, and Use*. Washington, DC: Island Press.

Stiven, A.E. 1968. The components of a threshold in experimental epizootics of *Hydramoeba hydroxena* in populations of *Chlorhydra viridissima*. *Journal of Invertebrate Research* 11:348–57.

Sussman, M., Willis, B.L., Victor, S., Bourne, D.G., Ahmed, N. 2008. Coral pathogens identified for white syndrome (WS) epizootics in the Indo-Pacific. *PLoS ONE* 3:e2393.

Sutherland, K., Porter, J., Torres, C. 2004. Disease and immunity in Caribbean and Indo-Pacific zooxanthellate corals. *Marine Ecology Progress Series* 266:273–302.

Sutherland, K.P., Porter, J.W., Turner, J.W., Thomas, B.J., Looney, E.E., Luna, T.P., Meyers, M.K., Futch, J.C., Lipp, E.K. 2010. Human sewage identified as likely source of white pox disease of the threatened Caribbean elkhorn coral, *Acropora palmata*. *Environmental Microbiology* 12:1122–31.

Sutherland, K.P., Shaban, S., Joyner, J.L., Porter, J.W., Lipp, E.K. 2011. Human pathogen shown to cause disease in the threatened elkhorn coral *Acropora palmata*. *PLoS ONE* 6(8):e23468.

Sweet, M., Jones, R., Bythell, J. 2011. Coral diseases in aquaria and in nature. *Journal of the Marine Biology Association of the United Kingdom* 92(4):791–801.

Teplitski, M., Ritchie, K. 2009. How feasible is the biological control of coral diseases? *Trends in Ecology and Evolution* 24:378–85.

Thitamadee, S., Prachumwat, A., Srisala, J., Jaroenlak, P., Salachan, P.V., Sritunyalucksana, K., Flegel, T.W., Itsathitphaisarn, O. 2016. Review of current disease threats for cultivated penaeid shrimp in Asia. *Aquaculture* 452:69–87.

Thomson, M.C., Doblas-Reyes, F.J., Mason, S.J., Hagedorn, R., Connor, S.J., Phindela, T., Morse, A.P, Palmer, T.N. 2006. Malaria early warnings based on seasonal climate forecasts from multi-model ensembles. *Nature* 439(7076):576–9.

Travaglia, C., Profeti, G., Aguilar-Manjarrez, J., Lopez, N. 2004. Mapping coastal aquaculture and fisheries structures by satellite imaging radar. Case study of the Lingayen Gulf, Philippines. *FAO Fisheries Technical Paper No. 459*. Rome: FAO.

Tucker, C.S., Hargreaves, J.A. (eds). 2008. *Environmental Best Management Practices for Aquaculture*. Oxford: Blackwell Publishing.

Underwood, W. 2012. The plant cell wall: a dynamic barrier against pathogen invasion. *Frontiers in Plant Science* 3:85.

US EPA. 2013. Florida Keys National Marine Sanctuary Water Quality Protection Program. https://floridakeys.noaa.gov/wqpp/.

Van Houtan, K.S., Hargrove, S.K., Balazs, G.H. 2010. Land use, macroalgae, and a tumor-forming disease in marine turtles. *PLoS ONE* 5:e12900.

van Oppen, M.J. H., Oliver, J.K., Putnam, H.M., Gates, R.D. 2015. Building coral reef resilience through assisted evolution. *Proceedings of the National Academy of Sciences* 112(8):2307–13.

Vega-Thurber, R.L., Burkepile, D.E., Fuchs, C., Shantz, A.A., McMinds, R., Zaneveld, J.R. 2014. Chronic nutrient enrichment increases prevalence and severity of coral disease and bleaching. *Global Change Biology* 20(2):544–54.

Walker, P.J., Winton, J.R. 2010. Emerging viral diseases of fish and shrimp. *Veterinary Research* 41(6):51.

Ward, J.R., Lafferty, K.D. 2004. The elusive baseline of marine disease: are diseases in ocean ecosystems increasing? *PLoS Biology* 2(4):e120.

Wear, S.L., Thurber, R.V. 2015. Sewage pollution: mitigation is key for coral reef stewardship. *Annals of the New York Academy of Sciences* 1355(1):15–30.

Wilcox, B.A., Gubler, D.J. 2005. Disease ecology and the global emergence of zoonotic pathogens. *Environmental Health and Preventive Medicine* 10(5):263–72.

Wobeser, G.A. 2013. *Essentials of Disease in Wild Animals.* Chichester: John Wiley & Sons.

Work, T.M., Balazs, G.H., Summers, T.M., Hapdei, J.R., Tagarino, A.P. 2015. Causes of mortality in green turtles from Hawaii and the insular Pacific exclusive of fibropapillomatosis. *Diseases of Aquatic Organisms* 115:103–10.

Worm, B., Hilborn, R., Baum, J.K., Branch, T.A., Collie, J.S., Costello, C., Fogarty, M.J., Fulton, E.A., Hutchings, J.A., Jennings, S., Jensen, O.P., Lotze, H.K., Mace, P.M., McClanahan, T.R., Minto, C., Palumbi, S.R., Parma, A.M., Ricard, D., Rosenberg, A.A., Watson, R., Zeller, D. 2009. Rebuilding global fisheries. *Science* 325:578–85.

Yang, Q., Tam, N.F., Wong, Y.S., Luan, T.G., Su, WS, Lan, C.Y., Shin, P.K., Cheung, S.G. 2008. Potential use of mangroves as constructed wetland for municipal sewage treatment in Futian, Shenzhen, China. *Marine Pollution Bulletin* 57(6–12):735–43.

Yoshikawa, T., Asoh, K. 2004. Entanglement of monofilament fishing lines and coral death. *Biological Conservation* 117:557–60.

CHAPTER 10

Disease in fisheries and aquaculture

Donald C. Behringer, Chelsea L. Wood, Martin Krkošek, and David Bushek

10.1 Introduction

Economic or human health impacts often drive scientific research priorities that lead to fundamental knowledge. This is especially true for infectious marine diseases caused by macro- (e.g., trematodes, crustaceans) or micro-parasites (e.g., bacteria, protists, fungi, and viruses), which we will henceforth refer to collectively as "parasites". For example, most information about *Hematodinium*, a genus of parasitic dinoflagellates that infect many decapod crustacean species, has been obtained from studies on crab and lobster fisheries (Small, 2012). Similarly, the billions of dollars in disease-driven losses suffered by the shrimp aquaculture industry have inspired research on the many parasites that infect penaeid shrimps (Stentiford et al. 2012). Furthermore, it is common to justify research on diseases of foundational species and ecosystem engineers, such as coral reefs (Graham, 2014), seagrass beds (Waycott et al. 2009), and oyster reefs (Beck et al. 2011), by pointing to how foundational species support fisheries. Of course, conservation, ecosystem health concerns, and pure curiosity also drive research, but the economic and human health implications of disease for fisheries and aquaculture are potent motivators.

Wild, fished, and cultivated populations differ in ways that affect disease transmission (Figure 10.1). For example, fishing often removes the largest individuals, which may be the oldest and least susceptible to parasites. On the other hand, aquaculture often relies on disease-resistant stocks which could act to dilute transmissible stages, thereby reducing parasite transmission to wild stocks; aquaculture also uses disease-tolerant stocks which can be reservoirs of parasites that spill over to wild species (Ben-Horin et al. 2018). The high stocking densities required by aquaculture increase the contact rates among individuals, greatly increasing the potential rates of transmission. Closed aquaculture systems are designed to prevent inputs of infectious agents through strict quarantine and reliance on a supply of specific-pathogen-free (SPF) stocks, although elimination of disease is a goal not always easy to achieve. For these reasons, we often see more diseases in aquaculture than in unfished stocks, and even fewer diseases in fished stocks (Wood et al. 2010).

In this chapter, we review the effects of disease on fisheries and aquaculture, and the reciprocal effects that fisheries and aquaculture have on disease emergence and dynamics. As a case study, we consider the unique disease ecology of diadromous (migrate between fresh and saltwater environments) host organisms. The remainder of our review focuses on disease management and the human health implications of disease in fisheries and aquaculture. The final section presents key points and recommends future directions for research and management of marine diseases.

Behringer, D.C., Wood, C.L., Krkošek, M., and Bushek, D., *Disease in fisheries and aquaculture* In: *Marine Disease Ecology*. Edited by: Donald C. Behringer, Kevin D. Lafferty, and Brian. R. Silliman, Oxford University Press (2020). © Oxford University Press. DOI: 10.1093/oso/9780198821632.003.0010

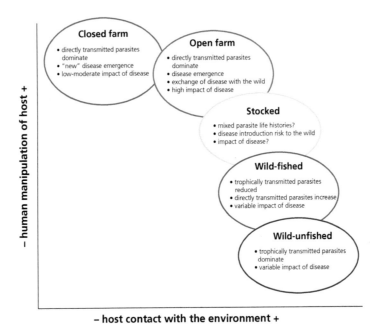

Figure 10.1 Marine aquaculture and fishery populations exist along the intersection of two axes—host contact with the environment and human manipulation of hosts—which drive fundamental population dynamics and interactions with disease. For some of the aquaculture and fishery systems along this "continuum" we have a solid understanding of the relationships between these axes, but for others there is much to learn.

10.2 Effects of disease on fisheries and aquaculture

Infectious diseases cost fishery and aquaculture industries billions of dollars annually through reduced growth and abundance, mass mortality, and reduced product quality (Lafferty et al. 2015). These effects are better understood in aquaculture environments than they are for wild fisheries, but it is likely that the two are interconnected via parasite spill-over/spill-back dynamics, or the introduction of parasites or naïve hosts into new regions. For example, Lafferty and Ben-Horin (2013) documented the potential for spill-over of withering syndrome from abalone farms to wild populations, and Burreson et al. (2004) found that a previously cryptic local parasite of native oysters infected a non-native oyster being cultivated in North Carolina, USA. In the following subsections, we investigate the biological and economic effects of infectious diseases on fisheries and aquaculture.

10.2.1 Decreased growth

Growth reductions in response to parasitic infection are well known among domesticated fishes and shellfish. For example, infection by the ectoparasitic copepods *Lepeophtheirus salmonis* and *Caligus* spp., known commonly as sea lice (Box 10.1), is a major cost to Atlantic salmon (*Salmo salar*) aquaculture industries (Johnson et al. 2004; Costello, 2009), and perkinsosis caused by infection with alveolate protozoans from the genus *Perkinsus* has impacted molluscan aquaculture globally (Villalba et al. 2004). For wild Atlantic salmon, infection with sea lice can affect growth, delay age at maturity (Vollset et al. 2014), and impair the fitness of infected individuals competing for resources in food-limited environments (Godwin et al. 2015; Godwin et al. 2017). In juvenile abalone of the species *Haliotis rufescens* and *Haliotis discus hannai*, withering syndrome caused by the rickettsia-like bacterium *Candidatus Xenohaliotis californiensis* reduced energy available

Box 10.1 Case study: Sea lice and salmon

The dynamics of parasite spill-over and spill-back between wild and farmed fish are well exemplified by salmon and their ectoparasites; "sea lice" are marine copepods (*Lepeoptheirus salmonis* and *Caligus* spp.) that feed on the epidermis, musculature, and blood of their hosts (Boxaspen, 2006; Costello, 2006). These common parasites of wild Atlantic and Pacific salmon can spread to farmed salmon (spill-over) via larval dispersal in the shared marine environment (Figures 10.2 and 10.3). Larval nauplii hatch into the water column from the eggstrings of adult females attached to a host. The

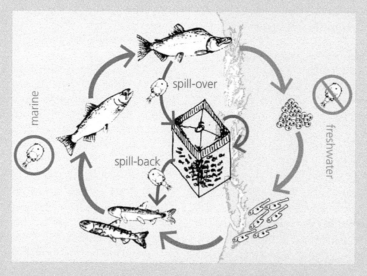

Figure 10.2 Life cycle of sea lice and salmon. Sea lice inhabit the marine phase of the salmon life cycle (blue) but not the freshwater phase (green), resulting in a separation of infected wild adult fish and uninfected wild juvenile fish (a characteristic called migratory allopatry) (Krkošek et al. 2007b). Sea lice can also spread from returning wild adult salmon to farmed salmon (spill-over), from which they can then spread back to wild juvenile salmon (spill-back).

Figure 10.3 Adult female salmon lice (*Lepeophtheirus salmonis*) infesting a juvenile pink salmon (*Oncorhynchus gorbuscha*).

nauplii are non-feeding and then develop into copepodites that must find a host fish or die. The parasitic phase involves newly attached larval lice that feed around an attachment point, then a pre-adult and adult phase where lice can move on and among their hosts, in search of mates and to evade predation (Connors et al. 2008b; Connors et al. 2011).

Sea lice impose substantial costs on the salmon farming industry in treatment and lost growth (Costello 2009), and can also spread back to wild salmon (spill-back) (Krkošek et al. 2006), which has created concerns for fisheries and conservation (Krkošek et al. 2007a; Vollset et al. 2018). In Europe, replicated experimental studies following the survival of large cohorts of tagged treated and untreated juvenile Atlantic salmon indicate that parasitism by sea lice during the first month at sea suppresses the recruitment of adult salmon by an average of 18 percent in Norway and 39 percent

continued

Box 10.1 *Continued*

across Europe (Gargan et al. 2012; Krkošek et al. 2013; Vollset et al. 2015), with substantial variation in the magnitude of treatment effects depending on the baseline survival rate of wild salmon (Vollset et al. 2015). Control of sea lice has been accomplished via in-feed pesticides for farmed fish, which can control epidemics in both wild and farmed fish populations (Peacock et al. 2013), although lice have also evolved resistance in areas of North America, Europe, and Chile.

The ecological pathways by which sea lice affect overall survival of wild salmon probably bring about a variety of indirect effects on foraging success (Godwin et al. 2015; Godwin et al. 2018), growth (Godwin et al. 2017), and predator evasion (Krkošek et al. 2011b), which can create complex dynamics in food webs of juvenile salmon by shifting

predation rates among host species (Peacock et al. 2014). Direct mortality may also occur under high parasite abundances (Morton and Routledge, 2005). Maintaining abundant populations of wild salmon, the source of lice for farmed salmon, is ironically beneficial for parasite control because wild fish provide a refuge from selection for increased resistance and virulence, as well as providing connectivity to the broader genetic pool of parasites maintained among wild populations in the offshore feeding grounds of wild salmon (Kreitzman et al. 2016). Coordinated area-based management of sea lice is also considered to be important because the broad dispersal of larvae connects populations of parasites among farms at regional scales (Stucchi et al. 2011; Adams et al. 2012).

---

for growth by up to 49 percent in laboratory experiments (González et al. 2012). Aquaculturists are commonly advised to look for signs of disease when reductions in growth rate occur under otherwise normal conditions (e.g., Getchis, 2014), but reductions in growth are more challenging to detect in wild populations, which are not monitored as closely. Many seafood products have higher value at larger sizes, so lost growth, in addition to reduced yield, also means reduced value.

10.2.2 Decreased fecundity

Infectious diseases can also affect the fecundity of fishes and shellfish in aquaculture or natural environments. Specifically, parasites may reduce fecundity by decreasing the development of gonadal tissue, the maturation of gametes, or the viability of gametes. For example, myxosporeans reduce the reproductive rate of cultured sea bass by either destroying germinative tissue or feeding directly on spermatozoa (Alvarezpellitero and Sitjabobadilla, 1993). The rhizocephalan barnacle *Loxothylacus panopaei*, a castrating parasite of crabs, has expanded its range into the western North Atlantic, raising concerns regarding its impact on fisheries (Kruse and Hare, 2007). Parasites may also contribute to reduced *per capita* fecundity of sockeye salmon (*Oncorhynchus nerka*)

in British Columbia due to pre-spawn mortality, where thermal stress may interact with disease to cause mortality of females before they reach spawning habitats (Martins et al. 2012; Miller et al. 2014). Several species of bucephalid trematodes are known to castrate molluscan shellfish (Tranter, 1958; Cheng and Burton, 1965; Sakaguchi, 1966; Seed, 1969; Boyden, 1971), but the impacts on population dynamics are unknown. Interestingly, the production of reproductively sterile triploid oysters is used, in part, as a mechanism to bring them to market size before disease can limit growth or increase mortality (Nell, 2002). Most aquaculture stocks do not depend on natural reproduction, so parasite impacts on fecundity are more likely to impact wild fisheries (or capture-based aquaculture, like net pens).

10.2.3 Increased mortality

Infectious disease is known to cause incremental and mass mortality events in wild and farmed fishes (Lafferty et al. 2015). A prominent example among farmed fishes is the 2005–2010 outbreak of infectious salmon anemia virus (IHNV) among Atlantic salmon in Chile, which caused a 75 percent decrease in production (Asche et al. 2009). For wild fish, the pilchard herpes epidemics of Australia may have resulted from the introduction of a new herpes strain from feed imported for tuna ranching operations. This resulted in mass mortality of Australian pilchard *Sardinops sagax* stocks, which caused a range expansion of their competitors (primarily the Australian anchovy *Engraulis australis* (Ward et al. 2001)) and decreased the survival and reproduction of penguins, which prey on pilchards (Jones et al. 1997; Dann et al. 2000; Gaughan et al. 2000). Nevertheless, most fish kills are usually related to oxygen depletion associated with algal blooms.

Mass mortalities of molluscan shellfish in wild and farmed populations have also been reported throughout the world. Mortalities of the eastern oyster *Crassostrea virginica* have been particularly well documented to be caused by parasites, including the protozoans *Haplosoridium nelsoni* (which causes MSX disease) and *Perkinsus marinus* (which causes dermo disease), and the α-proteobacterium *Roseovarius crassostreae* (which causes ROD) (Ford and Tripp, 1996). In addition to eastern oysters, perkinsosis affects manila clams globally, as well as a diverse array of gastropods and other bivalve molluscs (Villalba et al. 2004). European flat oysters, *Ostrea edulis*, and Chilean oysters, *Ostrea chilensis*, were nearly wiped out by bonamiasis caused by protozoans from the genus *Bonamia* (Hudson and Hill, 1991; Cranfield et al. 2005). Global production of Pacific oysters is currently threatened by microvariant strains of the oyster herpesvirus OsHV-1 (Pernet et al. 2016). This virus has caused mass mortalities of larvae and juvenile oysters in Europe, Australia, New Zealand, China, and Japan. Along the Pacific Coast of North America, withering syndrome, combined with overfishing and habitat degradation, has facilitated the decline of several species of abalone (Crosson et al. 2014) to the point that some are now listed as endangered. This problem may have been exacerbated by conservation efforts to restore wild populations with aquaculture-produced animals (Friedman and Finley, 2003) because abalone farm operations have been identified as the sources of infectious parasites for wild populations (Lafferty and Ben-Horin, 2013; Crosson et al. 2014). Mass mortalities get public attention, particularly when they take items off the menu or cause economic harm.

10.2.4 Decreased marketability

While dead fish on the shore can bring rapid public attention and response, most impacts from parasites on fishery and aquaculture industries are felt through decreased production or decreased product value. The latter can occur when a parasite reduces marketability. For example, consumers may avoid cod, pollock, or herring visibly infected with seal worm (*Pseudoterranova decipiens*) larvae (nematodes with complex life cycles that include fish and marine mammal hosts) (McClelland, 2002), or avoid fish filets because of postmortem degradation associated with the myxosporean *Kudoa thyrsites* (Moran et al. 1999) or protozoans from the genus *Ichthyophonus* (White et al. 2013). Visible infections can also reduce the dockside value of molluscan shellfish, causing growers to avoid areas known to harbor the parasite. Examples include lumpy gonads in Pacific oysters, *Crassostrea gigas*, caused by the paramyxean *Martelloides chungmeuensis* (Itoh et al. 2004), or Denman Island disease that forms green pustules on the surfaces of oyster tissues (Elston et al. 2012). When infected with epizootic shell disease (ESD), the American lobster *Homarus americanus* can become so distasteful in appearance (Box 10.2) that it is rejected by the live lobster market, but readily sold for canning at a lower profit (Barris et al. 2018). Similarly, fouling organisms such as polychaete worms that create mud blisters on oyster shells can reduce marketability by affecting the appearance of raw oysters served on the half shell, even though there may be little impact on taste (Nell, 2007). Consumers will not buy products that disgust them.

Clearly, parasites can affect the bottom-line for both wild capture and aquaculture industries, but the most spectacular examples of their impacts come from aquaculture settings, where outbreaks can be catastrophic and losses readily quantified (e.g., counting carcasses and capital invested). For example, losses from parasites have exceeded US $3 billion annually in the penaeid shrimp aquaculture industry (Stentiford et al. 2012). Economic losses to fisheries from disease might also be massive, but assigning a cost is difficult due to the many other factors that affect fishery populations, such as larval supply, population connectivity, recruitment,

Box 10.2 Case study: American lobster and ESD

Epizootic shell disease (ESD) was first observed in the American lobster *Homarus americanus* (Figure 10.4) in 1997 (Castro and Angell, 2000). The etiological agent(s) remain unknown, but evidence suggests that ESD may not be caused by a single pathogen; instead, the disease is considered a "syndrome" or dysbiosis resulting from environmental stress (temperature and contamination) combined with an imbalance of the exoskeleton microbiome (Castro et al. 2012; Meres et al. 2012; Shields, 2013). Prevalence of ESD has been reported at nearly 85 percent among ovigerous female lobsters, which appear to suffer higher mortality from ESD than males or non-ovigerous females (Hoenig et al. 2017), presumably because they molt less and cannot rid themselves of the disease (Stevens, 2009). The resulting impact on reproductive output aligns with the recruitment failure reported for the lobster fishery in southern New England (Figure 10.5)

(Wahle et al. 2009) and a call by managers for closure of the lobster fishery in that region.

While the fishery in southern New England remains open, landings are only a fraction of that in the years prior to ESD's emergence (Howell, 2012). In contrast, the fishery to the north in the Gulf of Maine is still at peak productivity and remains highly lucrative relative to other local fisheries that are depleted or in decline, causing local fishing communities to focus almost entirely on the American lobster. This has led to fears of a "gilded trap" scenario whereby fishermen are enticed to focus on this single species, increasing their own financial vulnerability should a disease or other catastrophe impact the fishery (Steneck et al. 2011). ESD has recently been increasing in prevalence in the Gulf of Maine, compounding fears of its continued spread north with increasing seawater temperatures.

Figure 10.4 American lobster *Homarus americanus* from the Gulf of Maine (USA) infected with ESD. (Photo credit Jason Goldstein.)

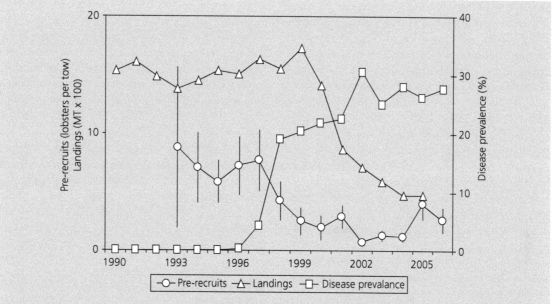

Figure 10.5 Time series of inshore Rhode Island commercial lobster landings, pre-recruit abundance (mean ± 1 SE), and shell disease prevalence from autumn near-shore trawl surveys. (Figure from Wahle et al. (2009) used with permission.)

overfishing, and natural mortality. Moreover, Lafferty et al. (2015) point out that the economic effect of disease is compounded in the indirect or secondary seafood economy, which includes processing, product shipping, fishing gear, vessel maintenance and manufacturing, and seafood retail. All of this can be worth three or more times the value of the landed product (Dyck and Sumaila, 2010). As fisheries decline and aquaculture increases, infectious diseases and their impacts will likely become even more apparent.

10.2.5 Disease and catch variability

Infectious disease outbreaks, or epizootics, can cause fishery collapse or aquaculture crash and may drive oscillations in both wild and captive populations. Molluscan fisheries and aquaculture systems have experienced dramatic shifts and even closures attributed to epizootics. One well-documented example is the collapse of the "traditional" eastern oyster fishery along the mid-Atlantic coast of North America

following the emergence of MSX. The traditional fishery was an extensive aquaculture system that transplanted submarket "seed" oysters from natural beds to privately held leases for subsequent harvest. In 1957, 75–95 percent of the oysters on leases in Delaware Bay perished from MSX and landings fell dramatically (Ford and Haskin, 1982). In 1990, as the population appeared to be recovering, perkinsosis emerged and doubled the background natural rate of mortality across Delaware Bay (Powell et al. 2008). In response, fishery management shifted from transplanting oysters for extensive aquaculture to a direct market fishery that is sustained under a strict total allowable catch (Powell et al. 2008). Few collapsed fisheries have been so successful at reducing variability.

Two intriguing examples of population oscillations come from wild Pacific salmon that spawn every other year (pink salmon, *Oncorhynchus gorbuscha*) or every 4 or 5 years (e.g., sockeye salmon, *O. nerka*) and have marked numerical dominance of some lineages relative to others. The primary working hypothesis to explain the sustained dominance of year classes is delayed density dependence, which may be caused by parasite transmission among year-classes (Ricker, 1997; Krkošek et al. 2011a). Another intriguing example arises in populations of the green sea urchin *Strongylocentrotus droebachiensis* in the rocky intertidal zone around Nova Scotia, Canada. Epizootics of paramoebiasis caused by the amoeba *Paramoeba invadens* occur with regularity and are associated with increased water temperature, the intensity of North Atlantic hurricanes, and the proximity of the storms to the coast of Nova Scotia, which combine to facilitate the introduction and spread of the parasite through the urchin population (Scheibling and Hennigar, 1997; Scheibling and Lauzon-Guay, 2010). Marine parasites with delayed responses to host density can drive oscillations in their host populations just like predators can drive oscillations in prey populations.

10.2.6 Positive effects of disease on fisheries and aquaculture

Although counter-intuitive, infectious diseases could improve fisheries. This could occur as indirect effects within food webs, where a parasite affecting one species may release another commercially valuable species from predation mortality or competition. For example, coho salmon (*Oncorhynchus kisutch*) feed on mixed species schools of juvenile pink (*O. gorbuscha*) and chum (*O. keta*) salmon but have a preference for pink salmon (Hargreaves and Lebrasseur, 1985). However, parasitism by sea lice (Box 10.1) increases predation risk for both juvenile pink and chum salmon. In this case, theoretical models have revealed that elevated levels of parasitism may allow predation to become increasingly focused on pink salmon, thereby releasing chum salmon from predation. If the release from predation is larger than the direct effects of parasitism on mortality of chum salmon, then parasitism may not add to overall mortality or may even improve survival, explaining why chum salmon have not suffered declines in response to the same epidemics that decimated pink salmon populations (Peacock et al. 2018). Such an effect exemplifies the pivotal role of parasites in driving community structure (McLaughlin et al. Chapter 2, this volume).

Aquaculture is often proffered as a solution to revitalize fishery production after a collapse, but examples documenting actual fishery recoveries are lacking. Nevertheless, when disease leads to fishery collapse, disease-resistant lines are frequently developed. In this way, disease can stimulate the development and proliferation of aquaculture even though the ability of aquaculture to restore the fishery is unclear. For example, the disease-induced collapse of the eastern oyster fishery stimulated the development of disease-resistant oyster stocks. Those stocks have facilitated and stimulated the development of eastern oyster aquaculture. Hence, aquaculture has the potential to compensate for fishery loses, but whether it can lead to recovery of a fishery is less certain.

10.3 Effects of fisheries and aquaculture on disease

As discussed in section 10.2, infectious diseases can impact fishery and aquaculture production, but the reciprocal is also true. Fisheries targeting wild populations can remake the parasite assemblages not only of the targeted fish species, but of the entire community in which the target species is embedded. The effects of fisheries on parasite assemblages are complex and include both direct and indirect effects, so our investigation is organized into effects on parasite populations, metapopulations, communities, and entire ecosystems (Figure 10.6).

10.3.1 Population level

Fisheries affect disease by influencing the target population's density, average body size, age structure, and average susceptibility to infection. Density often drives transmission rates, and if fishing reduces host density, we might expect reduced transmission (McCallum et al. 2001; McCallum et al. 2005). Exceptions might occur for social fish species that maintain high contact rates even at low host densities (Lafferty and Gerber, 2002; Johnson et al. 2011), for parasites with strong dispersal capabilities (Kuris and Lafferty, 1992), and for parasites that induce long-term immunity and therefore thrive when fishing increases recruitment of susceptible individuals into the population (Choisy and Rohani, 2006;

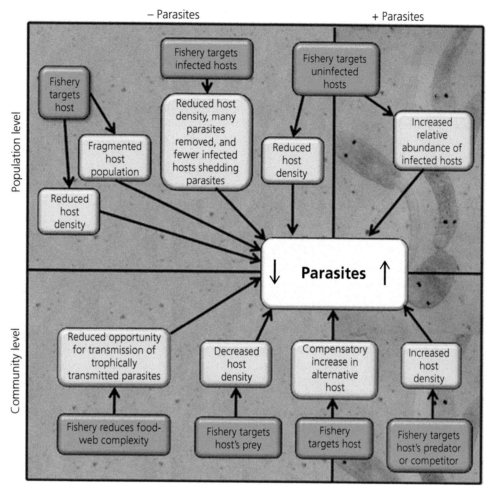

Figure 10.6 Impacts of fishing on rates of marine parasitism. Impacts described in the left region of the diagram tend to decrease rates of parasitism (– parasites), while impacts described in the right region tend to increase rates of parasitism (+ parasites). Impacts described in the top region of the diagram tend to act on host populations, while impacts described in the bottom region tend to act on communities in which hosts are embedded. Actions taken by fisheries are denoted in dark blue boxes and epidemiologically relevant impacts of each action are denoted in light blue boxes. (Reproduced with permission from Wood et al. (2010).)

Holt and Roy, 2007). For example, oyster densities in Delaware Bay (USA) declined by orders of magnitude, yet naïve oysters introduced into the bay are rapidly infected and quickly die from MSX or dermo due to a lack of immunity (Ford and Bushek, 2012). McCollough et al. (2007) deployed SPF oysters adjacent to infected populations and 5 km away from any known populations and found that the distance, and therefore the local density, had no impact on the rate of transmission and development of disease. This lack of density-dependence may occur because *Perkinsus marinus*, the agent of dermo disease, is directly transmitted through the water column as oysters filter water and host dilution reduces the per capita filtration of infective stages, or alternatively, by increasing contact rate if populations are reduced from fishing or habitat degradation (Ben-Horin et al. 2018). Other examples also support the importance of host density in the fishing–parasitism relationship. A monogenean parasite of an exploited rocky reef fish was more abundant in protected areas of central Chile, where the host was more abundant, than in matched open-access areas (Wood et al. 2013).

These fishing-driven reductions in host abundance can lead to local parasite extirpation and even global extinction, if host populations are driven below the parasite's transmission threshold (Dobson and May, 1987; Dunn et al. 2009; Colwell et al. 2012). Indeed, the first empirical evidence for fishing-induced extinction of a marine parasite was recently presented for *Stichocotyle nephropis*, an aspidogastrean parasite of near-threatened and endangered skates and rays that may now be extinct (MacKenzie and Pert, 2018). Conversely, a recent set of modelling papers found that densities of benthic filter feeders may never be low enough to suppress transmission from abundant, long-lived, highly dispersive parasites with direct-transmission life cycles (Bidegain et al. 2016; Bidegain et al. 2017). Therefore, fisheries often reduce parasite transmission by reducing host density, but exceptions to this rule commonly occur.

The effect of fishing on parasites may be obscured in complex food webs. When a fishery targets the predator (Sala, 1998; Pinnegar et al. 2000; Dulvy et al. 2004; Baskett et al. 2007) or competitor (McClanahan, 1992; Baskett et al. 2007) of a focal host, compensatory increases in the abundance of that host may facilitate increases in parasite abundance (Packer et al. 2003). For example, Wood et al. (2014; 2015) found that the directly transmitted parasites of seven small coral reef fish species increased where fishing predators released hosts from predation. Loss of natural predators (e.g., otters, lobsters, fishes) can increase host density and preceded deadly epizootics in both urchins (Lafferty, 2004) and black abalone (Lafferty and Kuris, 1993) in California's Channel Islands. In the Galapagos Islands of Ecuador, fishing removes predators of both urchins and the commensal crabs that prey on parasitic snails, leading to a net decrease in parasitic snails on urchins (Sonnenholzner et al. 2011). Trophic cascades therefore influence the abundance not only of lower trophic species, but also of the species that parasitize them (Morton et al. Chapter 3, this volume).

Wild capture fisheries that target large individuals can reduce parasite transmission rates. Large body size is a trait prized in many fisheries, so much so that fishing can shift size distributions (Bianchi, 2000;

Friedlander and DeMartini, 2002; Jennings and Blanchard, 2004; Pauly et al. 2005; Sala and Knowlton, 2006; McClenachan, 2008) and community composition (Pauly et al. 2005; Pauly and Watson, 2005; Essington et al. 2006). Large-bodied individuals often carry high parasite burdens because these individuals have had more time to accumulate long-lived parasites (Pacala and Dobson, 1988; Dobson et al. 1995; Arai and Zelmer, 1998; Lo et al. 1998; Poulin, 2000; Poulin, 2007) and they consume more food, filter more water, move greater distances, and present a greater surface area to the environment, which increases the likelihood of encountering infectious stages (Guegan and Hugueny, 1994; Arneberg, 2002; Coile et al. 2014). Targeting large-bodied individuals often leads to intensified exploitation of high-trophic-level fishes (Pauly et al. 2005; Pauly and Watson, 2005; Essington et al. 2006); this may also disproportionately reduce parasite transmission, since high-trophic-level hosts tend to support more parasite species than do low-trophic-level hosts (Rasmussen and Randhawa, 2018). Not all fisheries exclusively remove large-bodied, highly parasitized hosts, and other fishing strategies can result in alternative outcomes for parasite transmission. A counter argument is that removing the larger and generally older survivors may reduce the development of disease resistance or tolerance in the population (Groner et al. 2016), but, either way, the size selectivity of a fishery is likely to mediate the effects of that fishery on rates of parasitism.

Where parasitic infection renders captured fish less marketable, removal of uninfected individuals and release of infected individuals can result in intensified transmission. Few examples exist where such discrimination is documented, but it may also occur fortuitously where other traits are correlated with infection status (Kuris and Lafferty, 1992). For example, in some parts of Alaska, female tanner crabs (*Chionoecetes bairdi*) are infected with *Hematodinium* at higher rates than are male crabs (Urban and Byersdorfer, 2002); given that females are released and males retained in this fishery, this could lead to a higher prevalence of infection and intensified transmission (Kuris and Lafferty, 1992). Similarly, the Atlantic sea scallop *Placopecten magellanicus* fishery is suffering from an epizootic of the nematode *Sulcascaris sulcata* that infects the adductor muscle.

Although there is no evidence that they pose a risk to consumers, shellfish dealers avoid landings with a high prevalence of nematode infections. This causes fisherman to avoid infested areas and potentially increases the prevalence of infection population-wide (NFSC, 2018). The preferential removal of infected or uninfected individuals has rarely been documented, but could be a powerful mechanism mediating the impact of a fishery on rates of parasitism.

Fishing might increase host susceptibility, thereby increasing parasite transmission. For instance, one grouper species demonstrated a heightened immune response to nematode infection in the absence of fishing (Wood et al. 2014). Dead, melanized nematode "husks" (i.e., evidence of an effective immune response) were more abundant among grouper from unfished islands, whereas living nematodes were more abundant among grouper from fished islands. This pattern might arise if fishing exerts stress sufficient to erode fish body condition or immune capacity (Bly et al. 1997; Hoole, 1997). Indeed, among lobsters, confinement within traps or handling by fishers (e.g., during catch and release of undersized individuals) can increase transmission of a virus between Caribbean spiny lobsters *Panulirus argus* (Behringer et al. 2012) or susceptibility to the bacterial pathogens that cause shell disease in the southern rock lobster *Jasus edwardsii* (Freeman and MacDiarmid, 2009; Quinn et al. 2012). This mechanism would be especially important in fisheries where a large proportion of the individuals caught are eventually released (e.g., fisheries where females are not retained).

The effects of fishing on disease dynamics discussed above assume that exploited populations remain genetically unchanged by fishing, but ample evidence exists to suggest that exploited populations evolve in response to fishing pressure (Law, 2007). Fishing can lead to increased survivorship for those individuals that mature earlier and at smaller sizes (Kuparinen and Merila, 2007), so given the disproportionate number of parasite individuals and species carried by large-bodied hosts (see above), this could reduce parasite transmission and abundance. Alternatively, for a parasite that preferentially infects juveniles, if a fishery targets only mature individuals, this would skew the remaining population toward those most susceptible to infection (Behringer, 2012). Similarly, if a host population is evolving a "faster" life-history strategy in response to elevated fishing mortality, reduced reproductive investment might accompany reduced time-to-maturity and size-at-maturity, resulting in greater susceptibility to infection and greater parasite transmission (as in Tieleman et al. 2005; Martin et al. 2006; Johnson et al. 2012). Whether these mechanisms operate in fished populations remains to be tested.

10.3.2 Metapopulation level

Dispersal processes operate differently in marine relative to terrestrial ecosystems, given their contrasting opportunities for and constraints on movement. Parasite propagules survive longer in the sea and can be dispersed by currents, which is probably why there are few vector-mediated parasites in aquatic environments (McCallum et al. 2004). A metapopulation perspective gives some insight into the unique disease dynamics of exploited marine organisms. Even sessile organisms can exchange parasite propagules over great distances, resulting in epizootics that may spread quickly and persist (McCallum et al. 2003; Sokolow et al. 2009). This metapopulation dynamic can "rescue" parasite populations that might otherwise be negatively affected by fishing pressure on their host (see section 10.3.1), but amplify the negative effects for isolated populations (Kuris and Lafferty, 1992). For example, in a study of several parasite species affecting exploited fishes and invertebrates of central Chile, parasite species with limited dispersal were more impacted by fishing, presumably because they were not "rescued" by outside propagules (Wood et al. 2013). The distance among metapopulations is therefore especially consequential for transmission rates among marine parasites.

10.3.3 Community level

Fishing can even reduce parasite diversity. For instance, parasite taxon richness was 50 percent higher for coral reef fishes of unfished islands than for those of fished islands (Wood et al. 2014; Wood et al. 2015). In the same coral reef system, parasite diversity was sensitive to host diversity on unfished islands, but insensitive to host diversity on fished islands, suggesting that fishing removes the vulnerable parasite species that depend on rare hosts, leaving behind those that are resilient to anthropogenic stressors (Wood et al. 2018); those species lost tended to be complex life-cycle parasites (i.e., species that require multiple hosts to complete their life cycle), whereas the resilient species tended to be directly transmitted (Wood et al. 2014; Wood et al. 2015; Wood et al. 2018). These results were echoed in a meta-analysis, where the abundance of directly transmitted parasites increased in response to fishing whether their hosts were fished or not, whereas complex life-cycle parasites were more abundant if their hosts were unfished and less abundant if their hosts were fished (Wood and Lafferty, 2015). These

compositional shifts have implications for host health that remain unexplored. For example, does a shift from infection by complex life-cycle parasites to infection by directly transmitted parasites entail a loss or gain in fitness for hosts? Work conducted at the community level (Morton et al. Chapter 3, this volume) suggests that—although there may be a few "winners" among parasites in fished ecosystems, and these might present a threat to host health—many parasite species will decline alongside their hosts.

10.3.4 Ecosystem level

What do the population- and community-level changes in parasite abundance, diversity, and species composition explored in previous sections portend

Figure 10.7 In many second intermediate hosts (hosts of the second larval stage) of trematode parasites—like the California killifish (*Fundulus parvipinnis*)—the parasite induces behavioral changes to facilitate transmission to the final host (in this example, bird predators). (a) In the absence of parasites, fish display evasive and camouflaging behavior that minimizes the likelihood of bird predation. (b) When trematode metacercariae (larval stage) infect killifish, the fish perform behaviors that make them conspicuous to bird predators, effectively increasing the availability of fish resources to birds (Lafferty and Morris, 1996). In this way, parasites may provide a "subsidy" to predators. Such behavioral manipulations are common across the diversity of parasite life. (T. Saxby/UMCES/IAN Image Library; inset: C. Clark. Reproduced with permission from Wood and Johnson (2015).)

for fished ecosystems (Wood and Johnson, 2015)? Despite the complex effects of fishing on host populations and in turn on parasite transmission, one outcome is strongly supported from the empirical results summarized above: fishing tends to reduce the abundance of complex life-cycle parasites. Some of these parasites manipulate host behavior to increase predation risk by a final host (Poulin, 2010; Hughes et al. 2012). For example, California killifish infected with trematode (*Euhaplorchis californiensis*) metacercariae are 10–30 times more likely to be consumed by a bird (the definitive host) than are uninfected killifish (Lafferty and Morris, 1996) (Figure 10.7). On a broader scale, it might be that we have complex life-cycle parasites to thank for the number of apex predators currently in existence (Lefevre et al. 2009; Weinersmith and Faulkes, 2014; Wood and Johnson, 2015) because parasites shunt energy to higher trophic levels. Because few data exist to test how the effects of fishing on parasitism will influence marine ecosystems, this and other interesting hypotheses deserve research attention.

10.4 Unique disease ecology of diadromous fisheries

The disease dynamics of diadromous fishes differ significantly from those of non-diadromous species. Host migration between fresh and marine systems creates challenges for parasite transmission and persistence. It is difficult for ectoparasites to adapt to different salinities, and migration is well known to create challenges for parasites (Altizer et al. 2011). On the other hand, migration between freshwater and marine systems extends infection risk over a broader range of parasite species and host stressors, and so may increase infectious disease risk for diadromous fishes relative to purely marine or purely freshwater species. Thus, the complexity of a diadromous life history is likely to increase the complexity of host–parasite interactions in ways that are not necessarily intuitive.

Transmission of most infectious diseases of fishes occurs through the water column, and so the characteristics of the aquatic environment can either facilitate or constrain transmission. For example, the endangered catadromous European eel *Anguilla anguilla* contends with infection risk from the parasitic

nematode *Anguillicoloides crassus*, which decreases as salinity increases (Lefebvre and Crivelli, 2012), whereas disease caused by the bacterium *Vibrio vulnificus* is low in freshwater (Amaro et al. 1995). For anadromous salmonids, the transmission of whirling disease caused by the myxozoan *Myxobolus cerebralis* is constrained to freshwater environments due to an obligate intermediate oligochaete host that inhabits the sediments of lakes and rivers (Bartholomew et al. 2005). In contrast, salmonids experience transmission of sea lice—the ectoparasitic copepods *L. salmonis* and *Caligus* spp.—primarily in marine environments (Box 10.1), because the infective and adult stages of the parasite rapidly die in salinities less than 25 psu (Bricknell et al. 2006; Connors et al. 2008a). As diadromous species migrate between freshwater and saltwater environments, escape from one parasite is often met with entrapment by another.

The migration between freshwater and marine environments entails major changes to the thermal and osmotic environments of diadromous fishes, which can cause stress and divert host resources away from immune function, elevating the risk of infectious disease. For Atlantic salmon, the process of smoltification and subsequent transition from freshwater to seawater suppresses immune function via broad-spectrum downregulation of immune genes and is commonly followed by increased occurrences of viral (e.g., infectious pancreatic necrosis virus) and bacterial (e.g., *Moritella viscosa*) infections that can lead to increased mortality (Johansson et al. 2016). For Pacific salmon, such as Chinook (*Oncorhynchus tshawytscha*), the spawning migration from the ocean to freshwater rivers involves major changes to thermal environments, especially with rising river temperatures, which can result in reduced immune function, increased infection intensity and parasite richness, and increased pre-spawn mortality (Dolan et al. 2016; Teffer et al. 2018). Transitioning between freshwater and saltwater is physiologically demanding and can also carry the added challenge of increased susceptibility to parasites.

Beyond stress-induced susceptibility to disease, migration of hosts can have other ecological consequences for infectious disease (Altizer et al. 2011). For diadromous fishes, the amphidromous galaxids of New Zealand may experience less parasitism from trematodes than their freshwater relatives (Poulin et al. 2012)—an example of "migratory escape," where migration releases hosts from localized infection pressure. Diadromy can also result in allopatric age classes that live in sympatry only during brief periods, which allow the transmission of parasites from older to younger cohorts. For example, adult and juvenile pink salmon use both freshwater and marine environments, but are only sympatric during brief periods in coastal marine environments when out-migrating juveniles intersect with return-migrating adults—an example of "migratory allopatry" that limits opportunities for disease transmission to juveniles at particular stages of ontogeny (Krkošek et al. 2007b). Mechanisms of "migratory culling," where infected individuals are lost from migratory groups (Bradley and Altizer, 2005), and "migratory stalling" (Peacock et al. 2018), where infection slows migration and thereby escalates disease spread, may also be influential for diadromous fishes, whose migrations may be more arduous and span more heterogeneous habitat than other fishes. The ecological consequences of diadromy on parasitism can be positive or negative for the host, and context dependent.

10.5 Managing disease in fisheries and aquaculture

10.5.1 Managing disease in fisheries

Despite the economic toll that parasites take from fisheries of all types (Lafferty et al. 2015), they often remain unaccounted for in stock assessment models (Hoenig et al. 2017). Mortality due to parasitic infection is typically lumped in with other forms of natural mortality, and the indirect impacts of parasites on growth, reproduction, longevity, or marketability (Kuris and Lafferty, 1992) are seldom considered outside of academic exercises. For fisheries where disease has been explicitly considered in the fishery assessment, it has revealed striking outcomes; for example, in the American lobster fishery, the emergence of ESD corresponded closely with recruitment failure in southern New England (Wahle et al. 2009) (Box 10.2). Incorporating parasite effects explicitly into fisheries management can

provide valuable insights to improve management strategies.

Omitting disease from fishery models stems from the erroneous assumption that disease is not sufficiently influential to warrant consideration, or from a lack of adequate epidemiological data on the disease. Data deficiencies can result from a lack of support for parasite monitoring studies, inability to detect infected individuals, or the inaccessibility of infected stages or segments of the population. As noted in section 10.3.1, large-bodied fishes may carry a greater parasite load, but some parasites preferentially infect juveniles (Behringer, 2012). This is particularly problematic for managers who want to consider disease in their management plans, because juvenile stages of many marine fishes and invertebrates are excluded from fishery-dependent gear (e.g., via escape gaps) or the juvenile stages are difficult to sample because they are cryptic or spatially segregated from the fished population (Behringer, 2012; Shields, 2012). Where the effects of disease have been included in models, it has typically been motivated by a desire to test disease management strategies, such as culling or dilution.

Managing for disease in fisheries is undoubtedly challenging, considering that methods applied in terrestrial environments—such as vaccinating and quarantining—are not viable strategies for the marine environment (Burge et al. 2014; Raymundo et al. Chapter 9, this volume). Active measures (e.g., culling) have been suggested but rarely put into practice. However, theory suggests that simply fishing a host population to a point below the threshold for density-dependent transmission could reduce or eliminate the prevalence of a parasite (Dobson and May, 1987; Kuris and Lafferty, 1992; Ben-Horin et al. 2016). Similarly, the act of harvesting from an infected host population should in itself remove parasites from the system and reduce their abundance and diversity (Dobson and May, 1987; Wood et al. 2010). However, a meta-analysis testing the above theories found the situation to be more context-dependent (Wood et al. 2010). Fishing does indeed appear to reduce parasite diversity and the typical targeting of large individuals does remove more parasites, but after controlling for size, the relationship depends on the life cycle of the parasite (Wood and Lafferty, 2015). For example, clearing oyster beds of oysters infected with dermo and leaving the area fallow before repopulating it with uninfected juveniles was a recommended control strategy for decades (Andrews and Ray, 1988) but failed because long-distance dispersal of transmissible stages countered any local reductions in parasite abundance (McCollough et al. 2007), as predicted by Kuris and Lafferty (1992). Thus, there is no one-size-fits-all strategy and the host–parasite population dynamics must be understood within the ecological context of the fishery to identify an appropriate strategy.

Some fisheries manage disease through regulation. Regulations focused on fishery diseases are aimed at: (1) maintaining or improving the condition and productivity of a fishery stock, (2) limiting or reducing the proliferation or distribution of diseases outside of their established geographic range or host specificity, and (3) protecting human health from current or potential zoonotic diseases. Disease in fisheries is primarily managed through stock-specific regulations, limitations on the trade/transport of fishery products, and mandatory disease reporting systems. But such regulations are not common and are difficult to enforce.

Stock-specific fishery regulations aimed at managing disease are rarely put into practice, although they are commonly considered in the scientific literature. The concept of culling infected individuals is often suggested, considering its intuitive potential benefits, but culling is contentious for several reasons. First, many parasites show no gross signs until infection is advanced, if at all, and technologies that offer rapid and cost-effective diagnostics (Ben-Horin et al. 2016; Shields, 2017) are of little use to a fisherman culling his catch. Second, for parasites that target juveniles (Behringer, 2012), culling infected hosts likely violates fishery regulations. Although it would seem logical to add regulations to permit the culling of infected individuals, the subjective nature of infection assessments and the risk that the regulation would be abused might outweigh the benefits. Moreover, culling the most heavily infected individuals, and those likely to have the most virulent strain of a given parasite, could select for persistence of less virulent strains of the parasite that are able to expand their prevalence in less dense host populations (Bolzoni and De

Leo, 2013). Identifying appropriate, effective, and enforceable fishery management regulations that mitigate disease remains a difficult challenge in need of theoretical, experimental, and empirical research.

On the contrary, regulations for the transport of fishery products are in place at national and international levels. At a national level, for example, Fisheries and Oceans Canada requires inspection and regulates the movement of fishes and fish products from culture or wild fishery industries through the Fish Health Protection Regulations under the Fisheries Act of Canada. At the international level, the World Organisation for Animal Health (originally the "Office International des Epizooties" (OIE)) is an intergovernmental organization with 182 member countries, created to combat terrestrial and aquatic animal diseases around the world. Among other objectives, OIE publishes health standards for international trade in animals and animal products, some of which are specific to aquatic animals, including the Aquatic Animal Health Code (http://www.oie.int/standard-setting/aquatic-code/) and the Manual of Diagnostic Tests for Aquatic Animals (http://www.oie.int/standard-setting/aquatic-manual/). The former includes a list of diseases that member countries must report if discovered in their sovereign waters for the first time. The report outlines procedures to deal with the disease and puts in place limitations on the sale or trade of the infected species.

10.5.2 Fishery restoration and its implications for disease

Can we use what we know about how fishing influences parasite transmission to pre-empt or prepare for unwanted disease outcomes as we plan fishery restoration projects (see also Raymundo et al. Chapter 9, this volume)? The first step is to appreciate that parasitism is a natural part of healthy, functioning ecosystems, and that ecosystem "health" may be positively correlated with the richness of its parasite fauna (Hudson et al. 2006). With reduced fishing pressure, we would expect a more diverse assemblage of parasite taxa to return (Wood and Lafferty, 2015). Most of these taxa will not have an appreciable impact on the sustainability or profitability of fisheries, but a small subset could (Lafferty et al. 2015).

Managers might focus on those parasite taxa that are (1) economically important to the fishery (e.g., those that are pathogenic or that render fisheries products unmarketable) and (2) likely to change with changing fishing pressure (e.g., specific to the fished host). With an understanding of the life cycle of this parasite or subset of parasites, managers might anticipate potential changes in transmission and attempt to suppress the parasite in alternative ways (e.g., not discarding infected individuals, conserving predators of another required host). In short, understanding a parasite's ecology might suggest ecological levers for its control.

10.5.3 Managing disease in aquaculture

Regulations on disease management in aquaculture vary across countries and jurisdictions. The OIE global standards that apply to fisheries also apply to aquaculture and are recognized by the World Trade Organization. As noted in section 10.5.2, the OIE lists notifiable pathogens of concern to help with surveillance, respond to outbreaks, and prevent spread. Many parasites of national, regional, and local concern are, however, not listed. Regulations for those not listed are set by entities within each jurisdiction at each level, but are commonly not harmonious, due to variations in detectability and uncertainty in distributions. Several solutions include more surveillance, better technical training, eliminating fundamental knowledge gaps, and applying risk analysis (Carnegie et al. 2016). Because more of the host's life cycle is under control in aquaculture compared to fisheries, it seems intuitive that managing disease should be easier, but this is frequently not the case because host–parasite relationships can change under culture conditions (e.g., high density and reduced genetic diversity). Cultivation changes host behavior, but most importantly for disease transmission is an increase in host contact rate due to overcrowding. This includes contact not only with living infected individuals and the infectious elements they may be releasing, but also with infected carcasses that remain among the cultured population. Disease management in agriculture provides a terrestrial analog for aquaculture, but the aquatic environment poses a unique set of challenges.

Many marine organisms are naturally aggregated in their distributions, such as gregarious spiny lobsters that shelter in groups, fishes that form schools, or the larvae of sessile organisms such as oysters that preferentially settle near conspecifics. The increased density resulting from aggregation can facilitate transmission of infectious diseases, unless these organisms have developed behaviors to reduce infection risk (Behringer et al. 2006; Behringer et al. 2018). Despite behavioral adaptations, aquaculture can increase densities far beyond any observed in nature and also induce stress (Murray and Peeler, 2005). As a result, parasites not known to be mortality drivers in the wild can have major impacts on aquaculture. Examples include white spot syndrome virus (WSSV) of penaeid shrimp (Murray and Peeler, 2005), infectious salmon anemia virus (ISAV) of salmonids (Nylund et al. 2003), and ROD in juvenile oysters. The primary management strategy for ROD highlights the stress generated by high-density aquaculture; the disease can be prevented or eliminated by a combination of reducing the density of oysters in nursery systems, increasing flow rate through the system, or diluting source water (Boettcher Miller et al. 2006). Understanding the ecological processes that may limit disease in wild populations may provide disease management solutions for aquaculture.

Contradicting the paradigm that crowding leads to increased transmission, Ben-Horin et al. (2018) modeled dermo (*P. marinus*) disease in eastern oysters (*C. virginica*) and showed that high densities of cultivated oysters may actually reduce disease transmission if infected animals that have accumulated infections are harvested before they become sources of infectious propagules. While Bidegain et al. (2017) found that there is no oyster population density low enough to reduce the basic reproduction number (R_0) for *P. marinus* below 1 and cause extinction of the disease, their model also demonstrated that at historically high population densities (> 300 oysters m^{-2}) R_0 falls below 1, which indicates that high abundances of filter-feeding oysters can control pathogen densities and therefore the impact of infection on the population. Although crowding typically facilitates transmission of disease, mitigating factors such as harvest or predation (e.g., filter-feeding) can override its effect.

10.5.4 Microbial ecology and disease management in aquaculture—the microbiome

Since *circa* 2010, the microbiome has been increasingly recognized as important to disease management in aquatic organisms (Akhter et al. 2015). There are clear differences in the microbiomes of cultured organisms compared to their wild counterparts (Llewellyn et al. 2014) that can affect their susceptibility to parasites. Such differences probably arise from variations in the physical environment, diet, genetics, seasonality, and timing of exposure to microbes. This microbial diversity can be overwhelming and obscure relationships with disease, but progress in understanding the development and application of probiotics is increasing (Martinez Cruz et al. 2012). For example, Wanka et al. (2018) developed a protocol to rapidly isolate and screen naturally occurring microbes for probiotic potential. They screened 248 isolates from three commonly cultivated flatfishes and identified a dozen microbes that inhibit *Tenacibaculum maritimum*, a filamentous Gram-negative bacterium that causes an ulcerative disease (Avendaño-Herrera et al. 2006). Studies such as these indicate there is great promise in understanding how the microbiome influences disease and in developing methods to manipulate the microbiome in aquaculture settings.

Efforts to develop probiotics that generate a more disease-resistant microbiome for shellfish are less common and tend to focus on larval or juvenile stages (Karim et al. 2013). Similar to finfish, the application of antibiotic treatments to populations cultivated in open waters is virtually impossible because there is no feasible or safe way to apply antibiotics in field grow-out conditions. However, some benefit can be achieved by manipulating aerial exposure during intertidal periods or moving shellfish to higher or lower salinity environments (Ford and Tripp, 1996). More recently, because many diseases of fishes and shellfish are caused by bacteria or other microbes, phage therapy has been investigated as a prophylactic treatment (Richards, 2014). As the application of probiotics increases, there

may be unanticipated interactions with wild stocks that should be closely monitored.

10.5.5 The role of hatcheries in disease management for fisheries and aquaculture

Fish and shellfish hatcheries are widely used for stocking aquaculture operations and restoring or enhancing fisheries (Kitada, 2018). Hatcheries risk spreading disease with larvae, fry, fingerlings, seed, or other stages. One way to reduce disease spread is to develop SPF stocks along with an array of biosecurity protocols to be incorporated into regulations and best management practices (Oidtmann et al. 2011). Biosecurity helps isolate stocks from sources of infection, which may or may not require the production and maintenance of pathogen-free brood stock (especially when vertically transmitted pathogens are involved), filtering or sterilizing source water, and sterilizing food. For instance, the shrimp aquaculture industry began developing SPF protocols, modeled after the poultry industry (Alday-Sanz et al. 2018). This process spread to other aquaculture species. Today, hatcheries produce several SPF fishes and shellfish, many of which are regulated nationally and internationally (e.g., by the Fisheries Act of Canada or the OIE, as noted in section 10.5.1). Hatcheries provide additional opportunities for managing disease via the production of disease-resistant strains (Frank-Lawale et al. 2014; Yáñez et al. 2014). However, the potential impact of releasing selectively bred lines on natural populations has led to concern, most prominently for Pacific salmon (Waples et al. 2016), but for other species as well (Liu et al. 2010; Varney et al. 2018). Risk analysis studies should compare not only the risk of such actions versus inaction, but also the societal and ethical values involved.

10.6 Human health and other human impacts on the feedbacks among marine disease, fisheries, and aquaculture

Infectious-disease risk for human consumers of fisheries and aquaculture products is a concern in the developing world. Globally, the food-borne trematodiases rank as one of the most burdensome

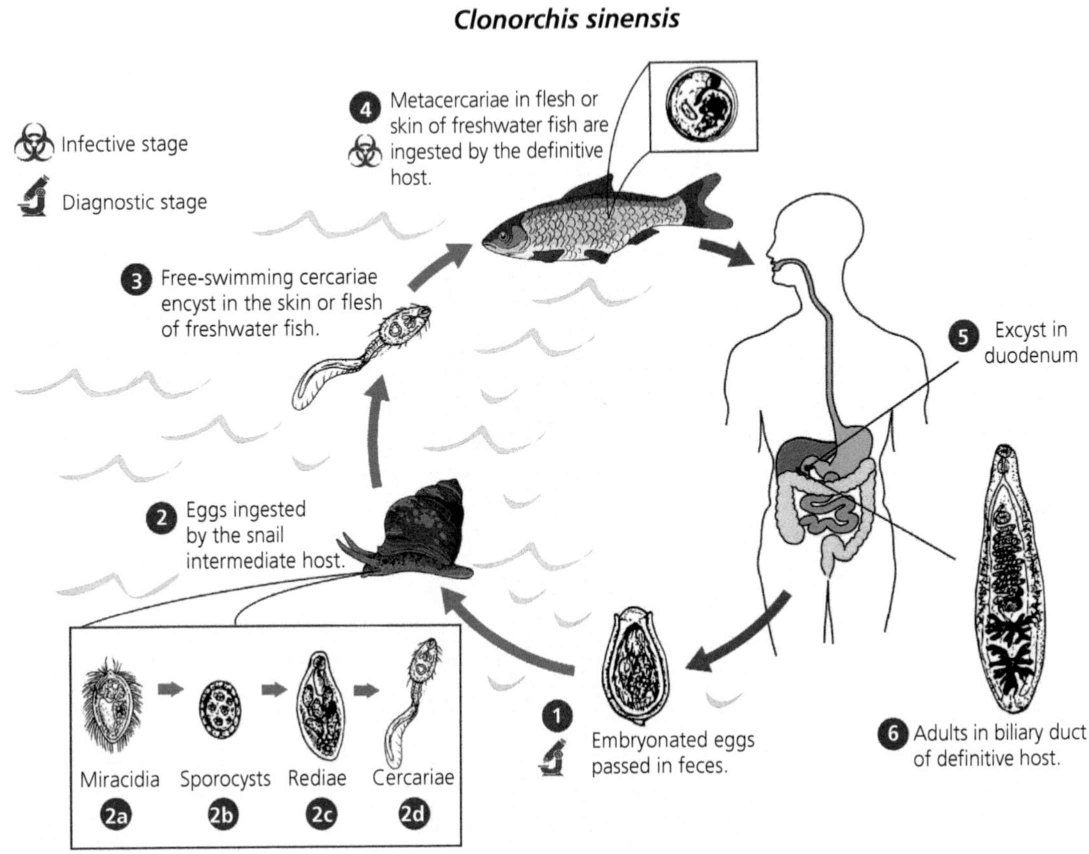

Clonorchis sinensis

Figure 10.8 The Chinese liver fluke, *Clonorchis sinensis*, is transmitted in undercooked freshwater fish and shellfish and causes foodborne trematodiasis in human patients. (Life cycle diagram courtesy of the US Centers for Disease Control and Prevention—Division of Parasitic Diseases and Malaria.)

food-borne illnesses of humans (Figure 10.8), affecting ~ 56 million people and causing ~ 665,000 years of healthy life to be lost annually (Furst et al. 2012a). These infections can affect the liver, lungs, and intestine, causing a range of health impacts, including bile duct cancer, tuberculosis-like pulmonary symptoms, and potentially fatal ectopic infections, including infections of the central nervous system (Furst et al. 2012b). Recent increases in the food-borne trematodiasis burden have been blamed on increasing production of cultured, freshwater fishes (WHO, 1995; Keiser and Utzinger, 2005; Furst et al. 2012a; Furst et al. 2012b). Given this, we expect the ongoing global transition from wild- to farmed-fish production (Naylor et al. 2009; World Bank, 2013; OECD/FAO, 2015), and particularly the increasing

contribution of cultured freshwater fishes from Asia (World Bank, 2013), to drive continued increases in the global food-borne trematodiasis burden (Keiser and Utzinger, 2005). Less is known about food-borne trematodiasis from marine and estuarine systems.

The developed world does not escape the reach of food-borne illness from fisheries and aquaculture products. Around 4 million annual cases of hepatitis A or E arise from eating raw or undercooked shellfish, resulting in 40,000 deaths (Shuval, 2003). In the USA, seafood-borne *Vibrio* spp., *Salmonella enterica*, norovirus, and *Listeria monocytogenes* infections cause over 55,000 cases of gastrointestinal illness annually, 5 percent of which require hospitalization; this results in US $882 million in economic losses per year (Batz et al. 2012). The rising popularity of sushi in the USA and Europe (Panel on Biological Hazards, 2010; Llarena-Reino et al. 2015; Isle, 2017) may be driving elevated rates of anisakiasis, a gastrointestinal illness caused by nematode worms that encyst in the flesh of marine fishes (Acha and Szyfres, 2003; Painter et al. 2013; Bao et al. 2017). Infection risk for these pathogens could shift with climate change (Harvell et al. 1999; Harvell et al. 2002), increasing human population density in coastal cities (Glasoe and Christy, 2004), and globalization of markets for fish and shellfish products (Keiser and Utzinger, 2005).

In the developed world, we have interrupted the transmission cycles of human-infecting parasites by disentangling human activities from nature: our water is purified, our sewers direct waste away from water and food sources, and our food is processed. But some activities keep humans intimately connected with ecosystems, and therefore risk ensnaring people in parasite life cycles; consumption of raw and undercooked fish and shellfish is one of those activities. Significant shifts in disease pressure could result from accelerating global change, with implications for the influence of fisheries and aquaculture products on human health. An ecological understanding of the life cycles, seasonality, and mechanisms of transmission of zoonotic diseases can help to minimize impacts on human health.

10.7 Summary

Below we provide a series of conclusions and recommendations for future research and management approaches. We believe that following these recommendations will lead to a greater understanding of the mutual feedbacks between infectious diseases and the function and productivity of the fishery and aquaculture sectors.

10.7.1 Conclusions

- Disease can play a pivotal role in the sustainability of fishery and aquaculture operations.
- Economics often drive disease research as disease often drives aquaculture and fishery economics

through reductions in target species growth, fecundity, survival, and catchability.

- Parasites are natural, important members of communities and ecosystems, but fishery and aquaculture activities can alter the natural relationship between host and parasite, resulting in changes to parasite or host abundance, diversity, distribution, or ecology that can reverberate from population to ecosystem. These changes can be negative or positive for the host or the parasite depending on the specifics of the host–parasite relationship (e.g., parasite life cycle) and the context of the fishery or aquaculture effect (e.g., history of fishing or type of aquaculture operation).
- Diadromous and estuarine fisheries provide unique case studies of the effects of environmental variability on disease ecology. Parasites that move or are moved between fresh, salt, or estuarine habitats must successfully survive or transmit themselves under variable environmental conditions. On the other hand, their hosts face a broader array of parasite taxa than their freshwater or marine counterparts, but alternatively, could also experience "migratory escape" from parasites by moving between fresh and saltwater.
- Regulations exist to limit the spread of disease among fisheries, but it is particularly difficult to manage an established parasite. Despite challenges to management, it is increasingly recognized as important to explicitly include disease or its effects in fishery assessments.
- Some of the same regulations that exist to limit the spread of disease in fisheries apply to aquaculture products (e.g., O.I.E. regulations). However, the element of control afforded the aquaculture industry, such as isolated culture facilities and use of SPF or disease-resistant stocks, make managing disease more tractable and terrestrial agriculture provides an analog for disease management in aquaculture.
- Zoonotic diseases arising from fisheries or aquaculture are rare in the developed world, but not so in the developing world. As the global population increasingly turns to aquaculture to supply its protein requirements, this might change.

10.7.2 Recommendations

• Increase inclusion of the direct and indirect effects of disease in fishery models to allow for better forecasting and prediction.

• Management efforts should focus on surveillance and monitoring for disease emergence, spread, or change in prevalence to inform forecasting models and management actions.

• Future research should aim to assess how environmental change affects disease dynamics and host–parasite ecology.

• Aquaculture disease research should increase focus on the host microbiome but also include the whole-system microbiome, which includes the pathobiome.

• Aquaculture health research should broaden the focus from eradication of parasites and isolation of target organisms to include an ecological approach to parasite management.

References

Acha, P. N., and Szyfres, B. 2003. Anisakiasis. *In* Zoonoses and communicable diseases common to man and animals, pp. 231–236. Ed. by M. R. Periago. Pan American Health Organization, Washington, DC.

Adams, T., Black, K., MacIntyre, C., MacIntyre, I., and Dean, R. 2012. Connectivity modelling and network analysis of sea lice infection in Loch Fyne, west coast of Scotland. Aquaculture Environment Interactions, 3: 51–63.

Akhter, N., Wu, B., Memon, A. M., and Mohsin, M. 2015. Probiotics and prebiotics associated with aquaculture: a review. Fish & Shellfish Immunology, 45: 733–741.

Alday-Sanz, V., Brock, J., Flegel, T. W., McIntosh, R., Bondad-Reantaso, M. G., Salazar, M., and Subasinghe, R. 2018. Facts, truths and myths about SPF shrimp in aquaculture. Reviews in Aquaculture, November 10, 1–9.

Altizer, S., Bartel, R., and Han, B. A. 2011. Animal migration and infectious disease risk. Science, 331: 296–302.

Alvarezpellitero, P., and Sitjabobadilla, A. 1993. Pathology of myxosporea in marine fish culture. Diseases of Aquatic Organisms, 17: 229–238.

Amaro, C., Biosca, E. G., Fouz, B., Alcaide, E., and Esteve, C. 1995. Evidence that water transmits *Vibrio vulnificus* biotype-2 infections to eels. Applied and Environmental Microbiology, 61: 1133–1137.

Andrews, J. D., and Ray, S. M. 1988. Management strategies to control the disease caused by *Perkinsus marinus*. American Fisheries Society Special Publication, 18: 257–264.

Arai, H. P., and Zelmer, D. A. 1998. The contributions of host age and size to the aggregated distribution of parasites in yellow perch, *Perca flavescens*, from Garner Lake, Alberta, Canada. Journal of Parasitology, 84: 24–28.

Arneberg, P. 2002. Host population density and body mass as determinants of species richness in parasite communities: comparative analyses of directly transmitted nematodes of mammals. Ecography, 25: 88–94.

Asche, F., Hansen, H., Tveteras, R., and Tveteras, S. 2009. The salmon disease crisis in Chile. Marine Resource Economics, 24: 405–411.

Avendaño-Herrera, R. T., Toranzo, A. E., and Magariños, B. 2006. Tenacibaculosis infection in marine fish caused by *Tenacibaculum maritimum*: a review. Diseases of Aquatic Organisms, 71: 255–266.

Bao, M., Pierce, G. J., Pascual, S., Gonzalez-Munoz, M., Mattiucci, S., Mladineo, I., Cipriani, P., et al. 2017. Assessing the risk of an emerging zoonosis of worldwide concern: anisakiasis. Nature Scientific Reports, 7: 43699.

Barris, B. N., Shields, J. D., Small, H. J., Huchin-Mian, J. P., O'Leary, P., Shawver, J. V., Glenn, R. P., et al. 2018. Laboratory studies on the effect of temperature on epizootic shell disease in the American lobster, *Homarus americanus*. Bulletin of Marine Science, 94: 887–902.

Bartholomew, J. L., Kerans, B. L., Hedrick, R. P., Macdiarmid, S. C., and Winton, J. R. 2005. A risk assessment based approach for the management of whirling disease. Reviews in Fisheries Science, 13: 205–230.

Baskett, M., Micheli, F., and Levin, S. 2007. Designing marine reserves for interacting species: insights from theory. Biological Conservation, 137: 163–179.

Batz, M., Hoffman, S., and Morris, J. 2012. Ranking the disease burden of 14 pathogens in food sources in the United States using attribution data from outbreak investigations and expert elicitation. Journal of Food Protection, 75: 1278–1291.

Beck, M. W., Brumbaugh, R. D., Airoldi, L., Carranza, A., Coen, L. D., Crawford, C., Defeo, O., et al. 2011. Oyster reefs at risk and recommendations for conservation, Restoration, and Management, 61: 107–116.

Behringer, D. C. 2012. Diseases of wild and cultured juvenile crustaceans: insights from below the minimum landing size. Journal of Invertebrate Pathology, 110: 225–233.

Behringer, D., Butler IV, M., and Shields, J. 2006. Ecology: avoidance of disease by social lobsters. Nature, 441: 421.

Behringer, D. C., Butler, M. J., Moss, J., and Shields, J. D. 2012. PaV1 infection in the Florida spiny lobster (*Panulirus argus*) fishery and its effects on trap function and disease transmission. Canadian Journal of Fisheries and Aquatic Sciences, 69: 136–144.

Behringer, D. C., Karvonen, A., and Bojko, J. 2018. Parasite avoidance behaviours in aquatic environments. Philosophical Transactions of the Royal Society B: Biological Sciences, 373: 20170202.

Ben-Horin, T., Lafferty, K. D., Bidegain, G., and Lenihan, H. S. 2016. Fishing diseased abalone to promote yield and conservation. Philosophical Transactions of the Royal Society B: Biological Sciences, 371: 20150211.

Ben-Horin, T., Burge, C. A., Bushek, D., Groner, M. L., Proestou, D. A., Huey, L. I., Bidegain, G., et al. 2018. Intensive oyster aquaculture can reduce disease impacts on sympatric wild oysters. Aquaculture Environment Interactions, 10: 557–567.

Bianchi, G. 2000. Impact of fishing on size composition and diversity of demersal fish communities. ICES Journal of Marine Science, 57: 558–571.

Bidegain, G., Powell, E., Klinck, J., Ben-Horin, T., and Hofmann, E. 2016. Microparasitic disease dynamics in benthic suspension feeders: infective dose, non-focal hosts, and particle diffusion. Ecological Modelling, 328: 44–61.

Bidegain, G., Powell, E. N., Klinck, J. M., Hofmann, E. E., Ben-Horin, T., Bushek, D., Ford, S. E., et al. 2017. Modelling the transmission of *Perkinsus marinus* in the eastern oyster *Crassostrea virginica*. Fisheries Research, 186: 82–93.

Bly, J. E., Quiniou, S. M. A., and Clem, L. W. 1997. Environmental effects on fish immune mechanisms. Developments in Biological Standardization, 90: 33–43.

Boettcher Miller, K., Smolowitz, R., Lewis, E. J., Allam, B., Dickerson, H. W., Ford, S., et al. 2006. Juvenile oyster disease (JOD) in *Crassostrea virginica*: synthesis of knowledge and recommendations. Journal of Shellfish Research, 25: 683–686.

Bolzoni, L., and De Leo, G. A. 2013. Unexpected consequences of culling on the eradication of wildlife diseases: the role of virulence evolution. The American Naturalist, 181: 301–313.

Boxaspen, K. 2006. A review of the biology and genetics of sea lice. ICES Journal of Marine Science, 63: 1304–1316.

Boyden, C. R. 1971. A comparative study of the reproductive cycles of the cockles *Cerastoderma edule* and *C. glaucum*. Journal of the Marine Biological Association of the United Kingdom, 51: 605–622.

Bradley, C. A., and Altizer, S. 2005. Parasites hinder monarch butterfly flight: implications for disease spread in migratory hosts. Ecology Letters, 8: 290–300.

Bricknell, I. R., Dalesman, S. J., O'Shea, B., Pert, C. C., and Luntz, A. J. M. 2006. Effect of environmental salinity on sea lice *Lepeophtheirus salmonis* settlement success. Diseases of Aquatic Organisms, 71: 201–212.

Burge, C. A., Mark Eakin, C., Friedman, C. S., Froelich, B., Hershberger, P. K., Hofmann, E. E., Petes, L. E., et al.

2014. Climate change influences on marine infectious diseases: implications for management and society. Annual Review of Marine Science, 6: 249–277.

Burreson, E. M., Stokes, N. A., Carnegie, R. B., and Bishop, M. J. 2004. *Bonamia* sp. (Haplosporidia) found in nonnative oysters *Crassostrea ariakensis* in Bogue Sound, North Carolina. Journal of Aquatic Animal Health, 16: 1–9.

Carnegie, R. B., Arzul, I., and Bushek, D. 2016. Managing marine mollusc diseases in the context of regional and international commerce: policy issues and emerging concerns. Philosophical Transactions of the Royal Society B: Biological Sciences, 371: 20150215.

Castro, K., and Angell, T. E. 2000. Prevalence and progression of shell disease in Rhode Island waters and the offshore canyons. Journal of Shellfish Research, 19: 691–700.

Castro, K. M., Cobb, J. S., Gomez-Chiarri, M., and Tlusty, M. 2012. Epizootic shell disease in American lobsters *Homarus americanus* in southern New England: past, present and future. Diseases of Aquatic Organisms, 100: 149–158.

Cheng, T. C., and Burton, R. W. 1965. Relationships between *Bucephalus* sp. and *Crassostrea virginica*: histopathology and sites of infection. Chesapeake Science, 6: 3–16.

Choisy, M., and Rohani, P. 2006. Harvesting can increase severity of wildlife disease epidemics. Proceedings of the Royal Society B, 273: 2025–2034.

Coile, A., Welicky, R., and Sikkel, P. 2014. Female *Gnathia marleyi* (Isopoda: Gnathiidae) feeding on more susceptible fish hosts produce larger but not more offspring. Parasitology Research, 113: 3875–3880.

Colwell, R. K., Dunn, R. R., and Harris, N. C. 2012. Coextinction and persistence of dependent species in a changing world. Annual Review of Ecology and Systematics, 43: 183–203.

Connors, B. M., Juarez-Colunga, E., and Dill, L. M. 2008a. Effects of varying salinities on *Lepeophtheirus salmonis* survival on juvenile pink and chum salmon. Journal of Fish Biology, 72: 1825–1830.

Connors, B. M., Krkošek, M., and Dill, L. M. 2008b. Sea lice escape predation on their host. Biology Letters, 4: 455–457.

Connors, B. M., Lagasse, C., and Dill, L. M. 2011. What's love got to do with it? Ontogenetic drivers of dispersal in a marine ectoparasite. Behavioral Ecology, 22: 588–593.

Costello, M. J. 2006. Ecology of sea lice parasitic on farmed and wild fish. Trends in Parasitology, 22: 475–483.

Costello, M. J. 2009. The global economic cost of sea lice to the salmonid farming industry. Journal of Fish Diseases, 32: 115–118.

Cranfield, H. J., Dunn, A., Doonan, I. J., and Michael, K. P. 2005. *Bonamia exitiosa* epizootic in *Ostrea chilensis* from

Foveaux Strait, southern New Zealand between 1986 and 1992. ICES Journal of Marine Science, 62: 3–13.

Crosson, L. M., Wight, N., Vanblaricom, G. R., Kiryu, I., Moore, J. D., and Friedman, C. S. 2014. Abalone withering syndrome: distribution, impacts, current diagnostic methods and new findings. Diseases of Aquatic Organisms, 108: 261–270.

Dann, P., Norman, F. I., Cullen, J. M., Neira, F. J., and Chiaradia, A. 2000. Mortality and breeding failure of little penguins, *Eudyptula minor*, in Victoria, 1995–96, following a widespread mortality of pilchard, *Sardinops sagax*. Marine and Freshwater Research, 51: 355–362.

Dobson, A. P., and May, R. M. 1987. The effects of parasites on fish populations—theoretical aspects. International Journal of Parasitology, 17: 363–370.

Dobson, A. P., Hudson, P. J., and Grenfell, B. T. 1995. Macroparasites: observed patterns in naturally fluctuating animal populations. Cambridge University Press, Cambridge, UK.

Dolan, B. P., Fisher, K. M., Colvin, M. E., Benda, S. E., Peterson, J. T., Kent, M. L., and Schreck, C. B. 2016. Innate and adaptive immune responses in migrating spring-run adult chinook salmon, *Oncorhynchus tshawytscha*. Fish & Shellfish Immunology, 48: 136–144.

Dulvy, N. K., Freckleton, R. P., and Polunin, N. V. C. 2004. Coral reef cascades and the indirect effects of predator removal by exploitation. Ecology Letters, 7: 410–416.

Dunn, R. R., Harris, N. C., Colwell, R. K., Koh, L. P., and Sodhi, N. S. 2009. The sixth mass coextinction: are most endangered species parasites and mutualists? Proceedings of the Royal Society B, 276: 3037–3045.

Dyck, A. J., and Sumaila, U. R. 2010. Economic impact of ocean fish populations in the global fishery. Journal of Bioeconomics, 12: 227–243.

Elston, R. A., Moore, J., and Abbott, C. L. 2012. Denman Island disease (causative agent *Mikrocytos mackini*) in a new host, Kumamoto oysters *Crassostrea sikamea*. Diseases of Aquatic Organisms, 102: 65–71.

Essington, T. E., Beaudreau, A. H., and Wiedenmann, J. 2006. Fishing through marine food webs. Proceedings of the National Academy of Sciences of the United States of America, 103: 3171–3175.

Ford, S., and Bushek, D. 2012. Development of resistance to an introduced marine pathogen by a native host. Journal of Marine Research, 70: 205–223.

Ford, S. E., and Haskin, H. H. 1982. History and epizootiology of *Haplosporidium nelsoni* (MSX), an oyster pathogen in Delaware Bay, 1957–1980. Journal of Invertebrate Pathology, 40: 118–141.

Ford, S. E., and Tripp, M. R. 1996. Diseases and defense mechanisms. *In* The eastern oyster: *Crassostrea virginica*, pp. 581–660. Ed. by V. S. Kennedy, R. I. E. Newell, and A. F. Eble. Maryland Sea Grant, College Park.

Frank-Lawale, A., Allen, S. K., and Dégremont, L. 2014. Breeding and domestication of eastern oyster (*Crassostrea virginica*) lines for culture in the Mid-Atlantic, USA: line development and mass selection for disease resistance. Journal of Shellfish Research, 33, 153–165.

Freeman, D. J., and MacDiarmid, A. B. 2009. Healthier lobsters in a marine reserve: effects of fishing on disease incidence in the spiny lobster, *Jasus edwardsii*. Marine and Freshwater Research, 60: 140–145.

Friedlander, A., and DeMartini, E. 2002. Contrasts in density, size, and biomass of reef fishes between the northwestern and the main Hawaiian islands: the effects of fishing down apex predators. Marine Ecology Progress Series, 230: 253–264.

Friedman, C. S., and Finley, C. A. 2003. Anthropogenic introduction of the etiological agent of withering syndrome into northern California abalone populations via conservation efforts. Canadian Journal of Fisheries and Aquatic Sciences, 60: 1424–1431.

Furst, T., Keiser, J., and Utzinger, J. 2012a. Global burden of human food-borne trematodiasis: a systematic review and meta-analysis. The Lancet. Infectious Diseases, 12: 210–221.

Furst, T., Sayasone, S., Odermatt, P., Keiser, J., and Utzinger, J. 2012b. Manifestation, diagnosis, and management of foodborne trematodiasis. BMJ, 344: e4093.

Gargan, P., Forde, G., Hazon, N., Russell, D. J. F., and Todd, C. D.. 2012. Evidence for sea lice-induced marine mortality of Atlantic salmon (*Salmo salar*) in western Ireland from experimental releases of ranched smolts treated with emamectin benzoate. Canadian Journal of Fisheries and Aquatic Sciences 69: 343–353.

Gaughan, D. J., Mitchell, R. W., and Blight, S. J. 2000. Impact of mortality, possibly due to herpesvirus, on pilchard *Sardinops sagax* stocks along the south coast of Western Australia in 1998–99. Marine and Freshwater Research, 51: 601–612.

Getchis, T. L., (ed.). 2014. Northeastern U.S. Aquaculture Management Guide: A Manual for the Identification and Management of Aquaculture Production Hazards. Northeastern Regional Aquaculture Center, University of Maryland.

Glasoe, S., and Christy, A. 2004. Coastal urbanization and microbial contamination of shellfish growing areas. ICES Document PSAT04-09.

KrkošekGodwin, S. C., Dill, L. M., Reynolds, J. D., and Krkošek, M. 2015. Sea lice, sockeye salmon, and foraging competition: lousy fish are lousy competitors. Canadian Journal of Fisheries and Aquatic Sciences, 72: 1113–1120.

Godwin, S. C., Dill, L. M., Krkošek, M., Price, M. H. H., and Reynolds, J. D. 2017. Reduced growth in wild juvenile sockeye salmon *Oncorhynchus nerka* infected with sea lice. Journal of Fish Biology, 91: 41–57.

Godwin, S., Krkošek, M., Reynolds, J. D., Rogers, L., and Dill, L. M. 2018. Heavy sea louse infection is associated with decreased stomach fullness in wild juvenile sockeye salmon. Canadian Journal of Fisheries and Aquatic Sciences, 75: 1587–1595.

González, R. C., Brokordt, K., and Lohrmann, K. B. 2012. Physiological performance of juvenile *Haliotis rufescens* and *Haliotis discus hannai* abalone exposed to the withering syndrome agent. Journal of Invertebrate Pathology, 111: 20–26.

Graham, N. A. 2014. Habitat complexity: coral structural loss leads to fisheries declines. Current Biology, 24: R359–361.

Groner, M. L., Maynard, J., Breyta, R., Carnegie, R. B., Dobson, A., Friedman, C. S., Froelich, B., et al. 2016. Managing marine disease emergencies in an era of rapid change. Philosophical Transactions of the Royal Society B: Biological Sciences, 371: 20150364.

Guegan, J. F., and Hugueny, B. 1994. A nested parasite species subset pattern in tropical fish: host as major determinant of parasite infracommunity structure. Oecologia, 100: 184–189.

Hargreaves, N. B., and Lebrasseur, R. J. 1985. Species selective predation on juvenile pink (*Oncorhynchus gorbuscha*) and chum salmon (*O. keta*) by coho salmon (*O. kisutch*). Canadian Journal of Fisheries and Aquatic Sciences, 42: 659–668.

Harvell, C. D., Epstein, P. R., Colwell, R. R., Vasta, G. R., Osterhaus, A. D. M. E., Overstreet, R. M., Smith, G. W., et al. 1999. Emerging marine diseases: climate links and anthropogenic factors. Science, 285: 1505–1510.

Harvell, C. D., Mitchell, C. E., Ward, J. R., Altizer, S., Dobson, A. P., Ostfeld, R. S., and Samuel, M. D. 2002. Climate warming and disease risks for terrestrial and marine biota. Science, 296: 2158–2162.

Hoenig, J. M., Groner, M. L., Smith, M. W., Vogelbein, W. K., Taylor, D. M., Landers, D. F., Swenarton, J. T., et al. 2017. Impact of disease on the survival of three commercially fished species. Ecological Applications, 27: 2116–2127.

Holt, R. D., and Roy, M. 2007. Predation can increase the prevalence of infectious disease. The American Naturalist 169: 690–699.

Hoole, D. 1997. The effects of pollutants on the immune response of fish: implications for helminth parasites. Parassitologia, 39: 219–225.

Howell, P. 2012. The status of the southern New England lobster stock. Journal of Shellfish Research, 31: 573–579.

Hudson, E. B., and Hill, B. J. 1991. Impact and spread of bonamiasis in the UK. Aquaculture, 93: 279–285.

Hudson, P., Dobson, A., and Lafferty, K. 2006. Is a healthy ecosystem one that is rich in parasites? Trends in Ecology and Evolution, 21: 381–385.

Hughes, D. P., Brodeur, J., and Thomas, F. 2012. Host manipulation by parasites. Oxford University Press, Oxford, UK.

Isle, R. 2017. Sushi in America. Food & Wine. https://www.foodandwine.com/articles/sushi-in-america.

Itoh, N., Komiyama, H., Ueki, N., and Ogawa, K. 2004. Early developmental stages of a protozoan parasite, *Marteilioides chungmuensis* (Paramyxea), the causative agent of the ovary enlargement disease in the Pacific oyster, *Crassostrea gigas*. International Journal for Parasitology, 34: 1129–1135.

Jennings, S., and Blanchard, J. 2004. Fish abundance with no fishing: predictions based on macroecological theory. Journal of Animal Ecology, 73: 632–642.

Johansson, L. H., Timmerhaus, G., Afanasyev, S., Jorgensen, S. M., and Krasnov, A. 2016. Smoltification and seawater transfer of Atlantic salmon (*Salmo salar* L) is associated with systemic repression of the immune transcriptome. Fish & Shellfish Immunology, 58: 33–41.

Johnson, M. B., Lafferty, K. D., van Oosterhout, C., and Cable, J. 2011. Parasite transmission in social interacting hosts: monogenean epidemics in guppies. PLOS ONE, 6: e22634.

Johnson, P. T. J., Rohr, J. R., Hoverman, J. T., Kellermanns, E., Bowerman, J., and Lunde, K. B. 2012. Living fast and dying of infection: host life history drives interspecific variation in infection and disease risk. Ecology Letters, 15: 235–242.

Johnson, S. C., Treasurer, J. W., Bravo, S., Nagasawa, K., and Kabata, Z. 2004. A review of the impact of parasitic copepods on marine aquaculture. Zoological Studies, 43: 229–243.

Jones, J. B., Hyatt, A. D., Hine, P. M., Whittington, R. J., Griffin, D. A., and Bax, N. J. 1997. Australasian pilchard mortalities. World Journal of Microbiology & Biotechnology, 13: 383–392.

Karim, M., Zhao, W., Rowley, D., Nelson, D., and Gomez-Chiarri, M. 2013. Probiotic strains for shellfish aquaculture: protection of eastern oyster, *Crassostrea virginica*, larvae and juveniles against bacterial challenge. Journal of Shellfish Research, 32: 401–408.

Keiser, J., and Utzinger, J. 2005. Emerging foodborne trematodiasis. Emerging Infectious Diseases, 11: 1507–1514.

Kitada, S. 2018. Economic, ecological and genetic impacts of marine stock enhancement and sea ranching: a systematic review. Fish and Fisheries, 19: 511–532.

Kreitzman, M., Ashander, J., Driscoll, J., Bateman, A., Chan, K., Lewis, M. A., and Krkošek, M. 2016. An evolutionary ecosystem service: wild salmon sustain the effectiveness of parasite control on salmon farms. Conservation Letters, 11: e12395.

Krkošek, M., Lewis, M. A., Morton, A., Frazer, L. N., and Volpe, J. P. 2006. Epizootics of wild fish induced by farm fish. Proceedings of the National Academy of Sciences of the USA, 103: 15506–15510.

Krkošek, M., Ford, J. S., Morton, A., Lele, S., Myers, R. A., and Lewis, M. A. 2007a. Declining wild salmon populations in relation to parasites from farm salmon. Science, 318: 1772–1775.

Krkošek, M., Gottesfeld, A., Proctor, B., Rolston, D., Carr-Harris, C., and Lewis, M. A. 2007b. Effects of host migration, diversity, and aquaculture on sea lice threats to Pacific salmon populations. Proceedings of the Royal Society B, 274: 3141–3149.

Krkošek, M., Hilborn, R., Peterman, R. M., and Quinn, T. P. 2011a. Cycles, stochasticity and density dependence in pink salmon population dynamics. Proceedings of the Royal Society B, 278: 2060–2068.

Krkošek, M., Connors, B., Mages, P., Peacock, S., Ford, H., Ford, J. S., Morton, A., Volpe, J. P., Hilborn, R., Dill, L. M., and Lewis, M. A. 2011b. Fish farms, parasites, and predators: implications for salmon population dynamics. Ecological Applications, 21: 897–914.

Krkošek, M., Revie, C., Gargan, P., Skilbrei, O. T., Finstad, B., and Todd, C. D. 2013. Impact of parasites on salmon recruitment in the Northeast Atlantic Ocean. Proceedings of the Royal Society B, 280: 20122359.

Kruse, I., and Hare, M. P. 2007. Genetic diversity and expanding nonindigenous range of the rhizocephalan *Loxothylacus panopaei* parasitizing mud crabs in the western North Atlantic. Journal of Parasitology, 93: 575–582.

Kuparinen, A., and Merila, J. 2007. Detecting and managing fisheries-induced evolution. Trends in Ecology and Evolution, 22: 652–659.

Kuris, A. M., and Lafferty, K. D. 1992. Modelling crustacean fisheries: effects of parasites on management strategies. Canadian Journal of Fisheries and Aquatic Sciences, 49: 327–336.

Lafferty, K. D. 2004. Fishing for lobsters indirectly increases epidemics in sea urchins. Ecological Applications, 14: 1566–1573.

Lafferty, K., and Ben-Horin, T. 2013. Abalone farm discharges the withering syndrome pathogen into the wild. Frontiers in Microbiology, 4: 373.

Lafferty, K., and Gerber, L. R. 2002. Good medicine for conservation biology: the intersection of epidemiology and conservation theory. Conservation Biology, 16: 1–12.

Lafferty, K., and Kuris, A. 1993. Mass mortality of abalone *Haliotis cracherodii* or the California Channel Islands: tests of epidemiological hypothesis. Marine Ecology Progress Series, 96: 239–248.

Lafferty, K., and Morris, A. 1996. Altered behavior of parasitized killifish increases susceptibility to predation by bird final hosts. Ecology, 77: 1390–1397.

Lafferty, K. D., Harvell, C. D., Conrad, J. M., Friedman, C. S., Kent, M. L., Kuris, A. M., Powell, E. N., et al. 2015. Infectious diseases affect marine fisheries and aquaculture economics. Annual Review of Marine Science, 7: 471–496.

Law, R. 2007. Fisheries-induced evolution: present status and future directions. Marine Ecology Progress Series, 335: 271–277.

Lefebvre, F., and Crivelli, A. J. 2012. Salinity effects on anguillicolosis in Atlantic eels: a natural tool for disease control. Marine Ecology Progress Series, 471: 193-U216.

Lefevre, T., Lebarbenchon, C., Gauthier-Clerc, M., Misse, D., Poulin, R., and Thomas, F. 2009. The ecological significance of manipulative parasites. Trends in Ecology and Evolution, 24: 41–48.

Liu, W.-D., Li, H.-J., Bao, X.-B., Gao, X.-g., Li, Y.-f., He, C.-B., and Liu, Z.-J. 2010. Genetic differentiation between natural and hatchery stocks of japanese scallop (*Mizuhopecten yessoensis*) as revealed by AFLP analysis. International Journal of Molecular Sciences, 11: 3933–3941.

Llarena-Reino, M., Abollo, E., Regueira, M., Rodriguez, H., and Pascual, S. 2015. Horizon scanning for management of emerging parasitic infections in fishery products. Food Control, 49: 49–58.

Llewellyn, M. S., Boutin, S., Hoseinifar, S. H., and Derome, N. 2014. Teleost microbiomes: the state of the art in their characterization, manipulation and importance in aquaculture and fisheries. Frontiers in Microbiology, 5: 207.

Lo, C. M., Morand, S., and Galzin, R. 1998. Parasite diversity/host age and size relationship in three coral-reef fishes from French Polynesia. International Journal for Parasitology, 28: 1695–1708.

MacKenzie, K., and Pert, C. 2018. Evidence for the decline and possible extinction of a marine parasite species caused by intensive fishing. Fisheries Research, 198: 63–65.

Martin, L. B., Hasselquist, D., and Wikelski, M. 2006. Investment in immune defense is linked to pace of life in house sparrows. Oecologia, 147: 565–575.

Martinez Cruz, P., Ibáñez, A. L., Monroy Hermosillo, O. A., and Ramírez Saad, H. C. 2012. Use of probiotics in aquaculture. ISRN Microbiology, 2012: 13.

Martins, E. G., Hinch, S. G., Patterson, D. A., Hague, M. J., Cooke, S. J., Miller, K. M., Robichaud, D., et al. 2012. High river temperature reduces survival of sockeye salmon (*Oncorhynchus nerka*) approaching spawning grounds and exacerbates female mortality. Canadian Journal of Fisheries and Aquatic Sciences, 69: 330–342.

McCallum, H., Barlow, N., and Hone, J. 2001. How should pathogen transmission be modelled? Trends in Ecology & Evolution (Personal edition), 16: 295–300.

McCallum, H., Harvell, D., and Dobson, A. 2003. Rates of spread of marine pathogens. Ecology Letters, 6: 1062–1067.

McCallum, H. I., Kuris, A., Harvell, C. D., Lafferty, K. D., Smith, G. W., and Porter, J. 2004. Does terrestrial epidemiology apply to marine systems? Trends in Ecology & Evolution, 19: 585–591.

McCallum, H., Gerber, L., and Jani, A. 2005. Does infectious disease influence the efficacy of marine protected areas? A theoretical framework. Journal of Applied Ecology, 42: 688–698.

McClanahan, T. R. 1992. Resource utilization, competition, and predation: a model and example from coral reef grazers. Ecological Modelling, 61: 195–215.

McClelland, G. 2002. The trouble with sealworms (*Pseudoterranova decipiens* species complex, Nematoda): a review. Parasitology, 124: S183–S203.

McClenachan, L. 2008. Documenting loss of large trophy fish from the Florida Keys with historical photographs. Conservation Biology, 23: 636–643.

McCollough, C. B., Albright, B. W., Abbe, G. W., Barker, L. S., and Dungan, C. F. 2007. Acquisition and progression of *Perkinsus marinus* infections by specific-pathogen-free juvenile oysters (*Crassostrea virginica* Gmelin) in a mesohaline Chesapeake Bay tributary. Journal of Shellfish Research, 26: 465–477.

Meres, N. J., Ajuzie, C. C., Sikaroodi, M., Vemulapalli, M., Shields, J. D., and Gillevet, P. M. 2012. Dysbiosis in epizootic shell disease of the american lobster (*Homarus americanus*). Journal of Shellfish Research, 31: 463–472.

Miller, K. M., Teffer, A., Tucker, S., Li, S., Schulze, A. D., Trudel, M., Juanes, F., et al. 2014. Infectious disease, shifting climates, and opportunistic predators: cumulative factors potentially impacting wild salmon declines. Evolutionary Applications, 7: 812–855.

Moran, J., Whitaker, D., and Kent, M. 1999. A review of the myxosporean genus Kudoa Meglitsch, 1947, and its impact on the international aquaculture industry and commercial fisheries. Aquaculture, 172: 163–196.

Morton, A., and Routledge, R. 2005. Mortality rates for juvenile pink *Oncorhynchus gorbuscha* and chum *O. keta* salmon infested with sea lice *Lepeophtheirus salmonis* in the Broughton Archipelago. Alaska Fisheries Research Bulletin, 11:146–152.

Murray, A. G., and Peeler, E. J. 2005. A framework for understanding the potential for emerging diseases in aquaculture. Preventive Veterinary Medicine, 67: 223–235.

Naylor, R., Hardy, R., Bureau, D., Chiu, A., Elliott, M., Farrell, A., Forster, I., et al. 2009. Feeding aquaculture in an era of finite resources. Proceedings of the National Academy of Sciences, 106: 15103–15110.

Nell, J. A. 2002. Farming triploid oysters. Aquaculture, 210: 69–88.

Nell, J. 2007. Controlling mudworm in oysters. NSW DPI Primefacts, 590.

NFSC. 2018. 65th Northeast Regional Stock Assessment Workshop (65th SAW) Assessment Report. Northeast Fisheries Science Center, Woods Hole, MA.

Nylund, A., Devold, M., Plarre, H., Isdal, E., and Aarseth, M. 2003. Emergence and maintenance of infectious salmon anaemia virus (ISAV) in Europe: a new hypothesis. Diseases of Aquatic Organisms, 56: 11–24.

OECD/FAO. 2015. OECD–FAO Agricultural Outlook 2015. OECD, Paris.

Oidtmann, B. C., Thrush, M. A., Denham, K. L., and Peeler, E. J. 2011. International and national biosecurity strategies in aquatic animal health. Aquaculture, 320: 22–33.

Pacala, S. W., and Dobson, A. P. 1988. The relation between the number of parasites/host and host age: population dynamic causes and maximum likelihood estimation. Parasitology, 96: 197–210.

Packer, C., Holt, R. D., Hudson, P. J., Lafferty, K. D., and Dobson, A. P. 2003. Keeping the herds healthy and alert: implications of predator control for infectious disease. Ecology Letters, 6: 797–802.

Painter, J. A., Hoekstra, R. M., Ayers, T., Tauxe, R. V., Braden, C. R., Angulo, F. J., and Griffin, P. M. 2013. Attribution of foodborne illnesses, hospitalizations, and deaths to food commodities by using outbreak data, United States, 1998–2008. Emerging Infectious Diseases, 19: 407–415.

Panel on Biological Hazards. 2010. Scientific opinion on risk assessment of parasites in fishery products. EFSA Journal, 8: 1543.

Pauly, D., and Watson, R. 2005. Background and interpretation of the "Marine Trophic Index" as a measure of biodiversity. Proceedings of the Royal Society of London Series B, 360: 415–423.

Pauly, D., Alder, J., and Watson, R. 2005. Global trends in world fisheries: impacts on marine ecosystems and food security. Philosophical Transactions of the Royal Society B: Biological Sciences, 360: 5–12.

Peacock, S., Krkošek, M., Proboszcz, S., Orr, C., and Lewis, M. A. 2013. Cessation of a salmon decline with control of parasites. Ecological Applications, 23: 606–620.

Peacock, S., Connors, B., Krkošek, M., Irvine, J., and Lewis, M. A. 2014. Can reduced predation offset negative effects of sea louse parasites on chum salmon? Proceedings of the Royal Society B, 281: 20132913.

Peacock, S. J., Bouhours, J., Lewis, M. A., and Molnar, P. K. 2018. Macroparasite dynamics of migratory host populations. Theoretical Population Biology, 120: 29–41.

Pernet, F., Lupo, C., Bacher, C., and Whittington, R. 2016. Infectious diseases in oyster aquaculture require a new integrated approach. Philosophical Transactions of the Royal Society B: Biological Sciences, 371: 1689.

Pinnegar, J., Milazzo, M., Polunin, N., Hereu, B., Francour, P., Chemello, R., Harmelin-Vivien, M., et al. 2000.

Trophic cascades in benthic marine ecosystems: lessons for fisheries and protected-area management. Environmental Conservation, 27: 179–200.

Poulin, R. 2000. Variation in the intraspecific relationship between fish length and intensity of parasitic infection: biological and statistical causes. Journal of Fish Biology, 56: 123–137.

Poulin, R. 2007. Are there general laws in parasite ecology? Parasitology, 134: 763–776.

Poulin, R. 2010. Parasite manipulation of host behavior: an update and frequently asked questions. Advances in the Study of Behavior, 41: 151–186.

Poulin, R., Closs, G. P., Lill, A. W. T., Hicks, A. S., Herrmann, K. K., and Kelly, D. W. 2012. Migration as an escape from parasitism in New Zealand galaxiid fishes. Oecologia, 169: 955–963.

Powell, E. N., Ashton-Alcox, K. A., Kraeuter, J. N., Ford, S. E., and Bushek, D. 2008. Long-term trends in oyster population dynamics in Delaware Bay: regime shifts and response to disease. Journal of Shellfish Research, 27: 729–755.

Quinn, R., Metzler, A., Smolowitz, R., Tlusty, N., and Christoserdov, A. 2012. Exposures of Homarus americanus shell to three bacteria isolated from naturally occurring epizootic shell disease lesions. Journal of Shellfish Research, 31: 485–493.

Rasmussen, T., and Randhawa, H. 2018. Host diet influences parasite diversity: a case study looking at tapeworm diversity among sharks. Marine Ecology Progress Series, 605: 1–16.

Richards, G. P. 2014. Bacteriophage remediation of bacterial pathogens in aquaculture: a review of the technology. Bacteriophage, 4: e975540.

Ricker, W. E. 1997. Cycles of abundance among Fraser River sockeye salmon (Oncorhynchus nerka). Canadian Journal of Fisheries and Aquatic Sciences, 54: 950–968.

Sakaguchi, S. 1966. Studies on a trematode parasite of the pearl oyster Pinctada martensii on the trematoda of genus Bucephalus found in the fishes Caranx sexfasciatus and C. ignobilis. Nippon Suisan Gakkaishi, 32: 316–321.

Sala, E. 1998. Fishing, trophic cascades, and the structure of algal assemblages: evaluation of an old but untested paradigm. Oikos, 82: 425–439.

Sala, E., and Knowlton, N. 2006. Global marine biodiversity trends. Annual Review of Environment and Resources, 31: 93–122.

Scheibling, R. E., and Hennigar, A. W. 1997. Recurrent outbreaks of disease in sea urchins Strongylocentrotus droebachiensis in Nova Scotia: evidence for a link with large-scale meteorologic and oceanographic events. Marine Ecology Progress Series, 152: 155–165.

Scheibling, R. E., and Lauzon-Guay, J.-S. 2010. Killer storms: North Atlantic hurricanes and disease outbreaks in sea urchins. Limnology and Oceanography, 55: 2331–2338.

Seed, R. 1969. The ecology of Mytilus edulis L. (Lamellibranchiata) on exposed rocky shores. Oecologia, 3: 277–316.

Shields, J. D. 2012. The impact of pathogens on exploited populations of decapod crustaceans. Journal of Invertebrate Pathology, 110: 211–224.

Shields, J. D. 2013. Complex etiologies of emersging diseases in lobsters (Homarus americanus) from Long Island Sound. Canadian Journal of Fisheries and Aquatic Sciences, 70: 1576–1587.

Shields, J. D. 2017. Collection techniques for the analyses of pathogens in crustaceans. Journal of Crustacean Biology, 37: 753–763.

Shuval, H. 2003. Estimating the global burden of thalassogenic diseases: human infectious diseases caused by wastewater pollution of the marine environment. Journal of Water and Health, 1: 53–64.

Small, H. J. 2012. Advances in our understanding of the global diversity and distribution of Hematodinium spp.—significant pathogens of commercially exploited crustaceans. Journal of Invertebrate Pathology, 110: 234–246.

Sokolow, S., Foley, P., Foley, J., Hastings, A., and Richardson, L. 2009. Disease dynamics in marine metapopulations: modelling infectious diseases on coral reefs. Journal of Applied Ecology, 46: 621–631.

Sonnenholzner, J. I., Lafferty, K. D., and Ladah, L. B. 2011. Food webs and fishing affect parasitism of the sea urchin Eucidaris galapagensis in the Galapagos. Ecology, 92: 2276–2284.

Steneck, R. S., Hughes, T. P., Cinner, J. E., Adger, W. N., Arnold, S. N., Berkes, F., Boudreau, S. A., et al. 2011. Creation of a gilded trap by the high economic value of the Maine lobster fishery. Conservation Biology, 25: 904–912.

Stentiford, G. D., Neil, D. M., Peeler, E. J., Shields, J. D., Small, H. J., Flegel, T. W., Vlak, J. M., et al. 2012. Disease will limit future food supply from the global crustacean fishery and aquaculture sectors. Journal of Invertebrate Pathology, 110: 141–157.

Stevens, B. G. 2009. Effects of epizootic shell disease in American lobster Homarus americanus determined using a quantitative disease index. Diseases of Aquatic Organisms, 88: 25–34.

Stucchi, D. J., Guo, M., Foreman, M. G. G., Czajko, P., Galbraith, M., Mackas, D. L., and Gillibrand, P. A. 2011. Modelling sea lice production and concentrations in the Broughton Archipelago, British Columbia. In Salmon lice: an integrated approach to understanding parasite abundance and distribution. Ed. by S. Jones and R. Beamish. Wiley-Blackwell, Chichester.

Teffer, A. K., Bass, A. L., Miller, K. M., Patterson, D. A., Juanes, F., and Hinch, S. G. 2018. Infections, fisheries capture, temperature, and host responses: multistressor influences on survival and behaviour of adult Chinook salmon. Canadian Journal of Fisheries and Aquatic Sciences, 75: 2069–2083.

Tieleman, B. I., Williams, J. B., Ricklefs, R. E., and Klasing, K. C. 2005. Constitutive innate immunity is a component of the pace-of-life syndrome in tropical birds. Proceedings of the Royal Society B, 272: 1715–1720.

Tranter, D. J. 1958. Reproduction in Australian pearl oysters (Lamellibranchia). I. *Pinctada albina* (Lamarck): gametogenesis. Australian Journal of Marine and Freshwater Research, 9: 135–143.

Urban, D., and Byersdorfer, S. C. 2002. Bitter crab syndrome in tanner crab (*Chionoecetes bairdi*), Alitak Bay, Kodiak, Alaska 1991–2000. University of Alaska Sea Grant, Fairbanks.

Varney, R. L., Watts, J. C., and Wilbur, A. E. 2018. Genetic impacts of a commercial aquaculture lease on adjacent oyster populations. Aquaculture, 491: 310–320.

Villalba, A., Reece, K. S., Camino Ordás, M., Casas, S. M., and Figueras, A. 2004. Perkinsosis in molluscs: a review. Aquatic Living Resoures, 17: 411–432.

Vollset, K. W., Barlaup, B. T., Skoglund, H., Normann, E. S., and Skilbrei, O. T. 2014. Salmon lice increase the age of returning Atlantic salmon. Biology Letters, 10.

Vollset, K., R. Krontveit, P. Jansen, B. Finstad, B. Barlaup, O. Skilbrei, M. Krkošek, A. Romunstad, A. Aunsmo, A. Jansen, and I. Doohoo. 2015. Impacts of parasites on marine survival of Atlantic salmon: a meta-analysis. Fish and Fisheries, 7: 91–113.

Vollset, K. W., Dohoo, I., Karlsen, O., Halttunen, E., Kvamme, B. O., Finstad, B., et al. 2018. Disentangling the role of sea lice on the marine survival of Atlantic salmon. ICES Journal of Marine Science, 75: 50–60.

Wahle, R. A., Gibson, M., and Fogarty, M. 2009. Distinguishing disease impacts from larval supply effects in a lobster fishery collapse. Marine Ecology Progress Series, 376: 185–192.

Wanka, K. M., Damerau, T., Costas, B., Krueger, A., Schulz, C., and Wuertz, S. 2018. Isolation and characterization of native probiotics for fish farming. BMC Microbiology, 18: 119.

Waples, R. S., Hindar, K., Karlsson, S., and Hard, J. J. 2016. Evaluating the Ryman–Laikre effect for marine stock enhancement and aquaculture. Current Zoology, 62: 617–627.

Ward, T. M., Hoedt, F., McLeay, L., Dimmlich, W. F., Jackson, G., Rogers, P. J., and Jones, K. 2001. Have recent mass mortalities of the sardine Sardinops sagax facilitated an expansion in the distribution and abundance of the anchovy Engraulis australis in South Australia? Marine Ecology Progress Series, 220: 241–251.

Waycott, M., Duarte, C. M., Carruthers, T. J., Orth, R. J., Dennison, W. C., Olyarnik, S., Calladine, A., et al. 2009. Accelerating loss of seagrasses across the globe threatens coastal ecosystems. Proceedings of the National Academy of Sciences of the USA, 106: 12377–12381.

Weinersmith, K., and Faulkes, Z. 2014. Parasite manipulation of hosts' phenotype, or how to make a zombie: an introduction to the symposium. Integrative and Comparative Biology, 54: 93–100.

White, V. C., Morado, J. F., Crosson, L. M., Vadopalas, B., and Friedman, C. S. 2013. Development and validation of a quantitative PCR assay for *Ichthyophonus* spp. Diseases of Aquatic Organisms, 104: 69–81.

WHO. 1995. Control of foodborne trematode infections: report of a WHO study group. ICES Document WHO Technical Report Series 849.

Wood, C. L., and Johnson, P. T. J. 2015. A world without parasites: exploring the hidden ecology of infection. Frontiers in Ecology and the Environment, 13: 425–434.

Wood, C. L., and Lafferty, K. D. 2015. How have fisheries affected parasite communities? Parasitology, 142: 134–144.

Wood, C. L., Lafferty, K. D., and Micheli, F. 2010. Fishing out marine parasites? Impacts of fishing on rates of parasitism in the ocean. Ecology Letters, 13: 761–775.

Wood, C. L., Micheli, F., Fernandez, M., Gelcich, S., Castilla, J. C., and Carvajal, J. 2013. Marine protected areas facilitate parasite populations among four fished host species of central Chile. Journal of Animal Ecology, 82: 1276–1287.

Wood, C. L., Sandin, S., Zgliczynski, B., Guerra, A. S., and Micheli, F. 2014. Fishing drives declines in fish parasite diversity and has variable effects on parasite abundance. Ecology, 95: 1929–1946.

Wood, C. L., Baum, J. K., Reddy, S., Trebilco, R., Sandin, S., Zgliczynski, B., Briggs, A., et al. 2015. Productivity and fishing pressure drive variability in fish parasite assemblages of the Line Islands, equatorial Pacific. Ecology, 96: 1383–1398.

Wood, C. L., Zgliczynski, B. J., Haupt, A. J., Guerra, A. S., Micheli, F., and Sandin, S. A. 2018. Human impacts decouple a fundamental ecological relationship—the positive association between host diversity and parasite diversity. Global Change Biology, 24: 3666–3679.

World Bank. 2013. Fish to 2030: Prospects for fisheries and aquaculture. ICES Document World Bank Report Number 83177-GLB.

Yáñez, J. M., Houston, R. D., and Newman, S. 2014. Genetics and genomics of disease resistance in salmonid species. Frontiers in Genetics, 5: 415.

Working with Infectious Diseases

CHAPTER 11

Diagnosing marine diseases

Salvatore Frasca, Jr, Rebecca J. Gast, Andrea L. Bogomolni, and
Steven M. Szczepanek

11.1 Introduction

Detection of infectious agents and their toxins is essential to understanding and managing diseases in marine species. Such diseases have become more prominent in recent years as global climate change and pollution have created conditions that allow pathogenic microbes to thrive. Mass mortality events in marine species are now all too common, and die-offs have become particularly pronounced among corals, crustaceans, fishes, and marine mammals (Burge et al. 2014; Harvell and Lamb Chapter 8, this volume; Raymundo et al. Chapter 9, this volume; Behringer et al. Chapter 10, this volume). Unfortunately, the dynamic environment that is the ocean makes detection and isolation of infectious agents of disease somewhat more challenging in aquatic than in terrestrial environments, as well as in marine as compared to land-dwelling species (Behringer et al. Preface, this volume). Such complications are exacerbated by the practical difficulty of sampling environments that truly utilize three dimensions, as opposed to the mostly two dimensions where many land-dwelling species spend the majority of their existence. The unique environment of the ocean must be taken into careful consideration when researchers attempt to diagnose infectious diseases in marine species.

11.2 Considerations of diagnostic testing

11.2.1 Principles

The use of diagnostic techniques for the identification of disease-causing agents requires an understanding of not only the proper execution of the technique but also proper interpretation of the results. Sample collection, handling, and processing according to established principles and with appropriate controls and quality standards are necessary for results to be considered valid, and national and international agencies, such as the World Organisation for Animal Health (i.e., the OIE), provide recommendations for such when considering testing for certain diseases, e.g., *Manual of Diagnostic Tests for Aquatic Animals* (http://www.oie.int/en/standard-setting/aquatic-manual/access-online/). Diagnosis of disease requires systematic description of the pathogen, detailed descriptions of its relationship with the host, and methodical analyses of its relationship to host populations and the environment (Morse 1995). This is exemplified by the emergence of *Aphanomyces invadans* as a cause of epizootic outbreaks of ulcerative skin disease in fish in southern African aquatic ecosystems (Sibanda et al. 2018), and the rise of the rosette agent *Sphaerothecum destruens* in Europe leading to declines in native fish populations (Combe and Gozlan 2018).

Frasca, Jr, S., Gast, R.J., Bogomolni, A.L., and Szczepanek, S.M., *Diagnosing marine diseases* In: *Marine Disease Ecology*. Edited by: Donald C. Behringer, Kevin D. Lafferty, and Brian. R. Silliman, Oxford University Press (2020).
© Oxford University Press. DOI: 10.1093/oso/9780198821632.003.0011

Diagnostic techniques to determine the etiologic agent of an infectious disease can provide binary data on the presence or absence of an agent, or techniques may be capable of determining specific concentrations of a pathogen in a tissue or environmental sample. Diagnostic assays such as the polymerase chain reaction (PCR), rapid antigen test (RAT), and western blot can be performed relatively quickly and interpreted by individuals with technical understanding to determine the presence or absence of an infectious agent in a short period of time. Metagenomic sequencing requires more time to perform and greater expertise to interpret, yet still provides only binary information as to the presence or absence of a pathogen, although it does so in an unbiased way. Rather than using a specific diagnostic assay to determine the presence or absence of a particular pathogen in an animal or environmental sample, metagenomic sequencing provides information as to the identity of most (if not all) of the microbes that are present in a sample. Interpretation of such data may allow for more rapid identification of the etiologic agent, or it may steer the investigator toward a cause of disease that was not anticipated. One study in Goseong Bay, Korea conducted metagenomic analysis of seawater in different seasons and found that many viruses were only found at certain times of year (Hwang et al. 2017), adding a layer of complexity to the dynamics of infectious disease and the ecology of that region that was previously unappreciated using standard single-plex diagnostic methods.

Sometimes simply detecting the presence or absence of a pathogen is not enough to properly diagnose disease, and quantification of the pathogen is necessary. Such quantification typically requires that a standard curve and quality control reagents be run prior to analysis of samples, making these approaches more time consuming, costly, and difficult to accomplish. One of the most common of such assays is real-time PCR (qPCR), which allows for the detection of a specific number of copies of a pathogen's gene to be estimated in biological and environmental samples. Alternatively, the enzyme-linked immunosorbent assay (ELISA) can be used to both capture and quantify an antigen from a pathogen in a sample, or to detect the antibody titer against a specific pathogen that may be circulating in an infected host. The combination of these ELISA approaches is very powerful, as it allows determination of how much pathogen is circulating in the blood of an infected animal, as well as how vigorously the animal is responding to the pathogen. Proper sampling, test selection, method execution, and result interpretation must be maintained throughout disease investigations.

11.2.2 Detection

Detection of a pathogen in an animal host or the environment is a crucial first step in disease identification, but often more information is needed to specifically identify an unknown agent that may be inciting disease. Gross pathologic investigation of a dead animal (or a biopsy from a living animal) can provide the first clues as to the etiology of disease, and sampling body fluids, such as blood, is often part of this initial investigation. Blood chemistry analyzers have become common in diagnostic laboratories, which may help narrow the list of etiologic agents, depending on the combination of results that are obtained. White blood cell (WBC) total and differential counts can provide insight into the inflammatory response of an infected animal. Furthermore, immunophenotyping can be conducted by flow cytometry on WBCs to determine specific subsets of cells that may be present in the blood of the subject (such as CD8+/CD4+ T-cell ratio). Detailed anatomic descriptions of specific lesions are obtained by histologic examination, complementing descriptions from the gross examination of the infected organism. Special stains can be applied to tissue sections, and these can help further characterize the disease process. Combining histologic techniques with one or more of the aforementioned molecular techniques can provide a very powerful dataset and can establish the presence of a specific pathogen in a lesion, which is one key to developing a causal link. In that vein, certain diagnostic techniques, such as immunohistochemistry (IHC) (Section 11.4.4) and *in situ* hybridization (ISH) (Section 11.4.5) can provide detection of specific molecular features of an infectious agent in the context of a tissue section, thereby providing orientation and association with tracked gross and microscopic lesions.

Marine biologists, ecologists, laboratorians, and health professionals need to exercise caution when embarking on such disease investigations. Organisms can be exposed to a disease agent, yet never become infected; likewise, they may be infected with a disease agent, yet not develop disease. While an infectious agent may be readily detectable given current reagents, it may not be the causative agent of disease in a particular subject. In addition, different members of the same species may be infected with a potential pathogen, yet not show signs of disease. Such subclinical infections are particularly important when one conducts an ecological investigation at a population level, as these organisms may be the source of infection for clinically ill subjects. Adding complexity to disease investigations in marine species, susceptibility to a particular pathogen often is variable between species, and transmission can occur between different populations that share marine habitats. For example, while farmed Atlantic salmon (*Salmo salar*) typically survive infection by viral hemorrhagic septicemia virus, transmission of the virus from farmed Atlantic salmon to wild Pacific herring (*Clupea pallasii*) has been demonstrated (Garver et al. 2013, Lovy et al. 2013), and such transmission can result in epizootic die-offs.

To further complicate disease investigations in marine systems, many micro-organisms serve as opportunistic pathogens in complex interactions between susceptible hosts and stressed aquatic environments. Gram-negative motile bacteria, such as the aeromonads and vibrios, are common inhabitants of the marine environment that cause disease in susceptible hosts subjected to environmental stress (Roberts 2012). Stressed environments can be associated with a variety of interrelated factors that include temperature change, regional hypoxia, changes in food availability, pollution levels, and host population densities. Thus, data from several different approaches, such as PCR and amplicon sequencing, morphological analysis of lesions, and electron microscopy, are often needed to properly diagnose infection or disease. Such an approach was used to identify *Neoparamoeba pemaquidensis* infection in the mass die-off of lobsters in western Long Island Sound in 1999 (Mullen et al. 2004, Mullen et al. 2005). In sum, clinicians need to consider if a

"positive" animal is simply exposed to an agent, infected but asymptomatic, or diseased.

11.2.3 Sampling

Sampling of organisms requires anticipation of downstream morphological and molecular diagnostic testing. While morphometric assessments of grossly apparent features of organisms, such as specimen dimensions and weights, do not inhibit future testing, processing for histologic evaluation by light microscopy or electron microscopy requires sampling and fixation of tissues (Section 11.4.1).

Although molecular detection of pathogens can be achieved using fixed tissues if conditions are favorable, this is often not the case (Mullen et al. 2005). The same is true of tissue storage techniques suitable for molecular testing, e.g., freezing, that limit histologic processing. Tissues sampled for the purpose of histology (whether by light microscopy or electron microscopy) require trimming to suitable dimensions and immersion into cross-linking fixatives. This may include 10 percent neutral buffered formalin or modified Davidson's fixative for light microscopy or fixatives containing glutaraldehyde, e.g., Karnovsky's fixative or McDowell–Trump fixative (McDowell and Trump 1976), for subsequent electron microscopic examination. Immersion in cross-linking fixatives of these types is degradative to nucleic acids, limiting the quality of DNA or RNA that can be extracted from such samples. The effect of formalin on nucleic acid integrity is well described and is dependent on such factors as the type of fixative and length of fixation time before processing (Foss et al. 1994, Frankel 2012). In addition, formalin fixation may mask epitopes that are the target of immunological testing, such as IHC labeling reactions. In comparison, tissues that are unfixed and otherwise unprepared but frozen to a temperature of $-80^{\circ}C$, while well preserved for nucleic acid extraction or enzymatic assays, have artifacts due to the formation of ice crystals that disrupt cellular and histologic architecture (Steu et al. 2008), making them less than ideal for histologic and cellular morphologic assessment. However, tissue freezing procedures do exist using specialized materials and equipment, such as optimal cutting temperature (OCT) compound and a cryostat, and can satisfy the morphologic and molecular testing requirements of investigators and laboratorians in certain circumstances (Steu et al. 2008). Therefore, sample acquisition and preparation should be planned in advance and with much consideration given to anticipated diagnostic procedures, which may require that the same tissue be sampled into several fixatives or compounds.

11.3 Classical microbiology and virology

11.3.1 Cultivation, isolation, and identification of bacterial disease agents

Isolation and identification of bacteria in the course of infectious disease investigations typically rely on culture of the microbe in broth or on agar plates, biochemical assays, and/or PCR and sequencing. Bacterial pathogens of marine species are no exception. Some are easy to culture, and such techniques can help identify the microbe(s) causing infection. One such example is the Gram-positive coccus *Streptococcus iniae*, which infects both freshwater and marine finfish and has been isolated from marine mammals. Culture techniques are useful for identification of this pathogen, as it appears as small, white, umbonate colonies, is beta-hemolytic when grown on blood agar plates, and appears as long chains of cocci when viewed under a microscope in broth culture. Further enrichment of the bacterium can be achieved with heart infusion agar, in conjunction with thallium acetate-oxolinic acid or colistin sulfate-oxolinic acid. However, biochemical assays may be necessary to differentiate this species of *Streptococcus* from other, closely related species such as *Streptococcus pyogenes* and *Streptococcus porcinus* (Agnew and Barnes 2007).

Consideration must be given to those diseases that are associated with multiple pathogens, making interpretation of culture results difficult. Winter-ulcer disease occurs in salmonids when temperatures decrease below 8°C. *Aliivibrio wodanis* and *Moritella viscosa* can be isolated from sick fish, both together or by themselves, but the growth of *A. wodanis* has an inhibitory effect on *M. viscosa* when co-cultured *in vitro* (Hjerde et al. 2015). This may result in a false-negative culture result for *M. viscosa*, making the use of molecular techniques for the detection of

M. viscosa from primary samples an important consideration.

Identifying the etiologic agent of bacterial diseases in marine species can also be complicated by the presence of closely related pathogens in the same sample. Simple culture techniques and biochemical tests may fail to differentiate between them. *Vibrio* spp. tend to fall into this category, but as a common cause of food poisoning from the consumption of shellfish, proper identification of the agent is imperative. DNA-based approaches such as pulsed field gel electrophoresis (Staley and Harwood 2010), specific real-time PCR (Hong et al. 2007, Yu et al. 2010, Section 11.5.3), amplified fragment length polymorphism (AFLP), fluorescence *in situ* hybridization (FISH) (Section 11.4.5), random amplified polymorphic DNA (RAPD), repetitive extragenic palindrome-PCR (rep-PCR), and restriction fragment length polymorphism (RFLP) (Chatterjee and Haldar 2012) can be useful alone or in combination for the proper identification of *Vibrio* species.

Some bacterial pathogens are difficult, or impossible, to culture using standard laboratory equipment. Mycobacteria are notoriously difficult to culture. For example, *Mycobacterium marinum*, an agent of mycobacteriosis in fish, requires 7–14 days before initial growth is typically observed. Great care must be taken during *M. marinum* culture to ensure that plates do not go dry and environmental contaminants, which may harbor bacteria that will outgrow the mycobacteria, are kept to a minimum (Kaattari et al. 2006). Gram staining of mycobacteria is also difficult, making the use of PCR and amplicon sequencing from fresh samples important for identification. Unfortunately, some pathogens have never been successfully cultured *in vitro* under normal laboratory conditions. This is the case with the chlamydial bacteria associated with epitheliocystis in fish (Nowak and LaPatra 2006). Detailed pathologic investigations including morphologic and molecular techniques have been required to characterize the agents of epitheliocystis, such as *Piscichlamydia salmonis*, and have included amplification and sequencing of the 16S rRNA gene and subsequent ISH with bacterial inclusions in histologic sections (Draghi et al. 2004). With the recent proliferation of next-generation sequencing technology

(Section 11.5.4), it is now possible to identify novel infectious agents in a wide variety of fish that previously could not be identified, such as the Chlamydiales (Taylor-Brown et al. 2017).

11.3.2 Cultivation, isolation, and identification of fungal disease agents

Proper sample collection for successful culture of fungi requires knowledge of the physical environment in which the sampling will occur, e.g., the field setting or controlled laboratory space, together with anatomical or biological site preferences for sampling, and growth characteristics of the fungus or fungi of concern. Isolation of fungi in culture requires utilization of multiple media that together maximize the potential to grow the fungus or fungi that are relevant to the investigation (de Hoog et al. 2000). The objective in selecting the media to use should be to include those with features that complement one another in a logical selective scheme. Larone (2002) has provided a useful summary of media for primary isolation of fungi that are applicable in a variety of scenarios.

A wide range of samples are suitable for culture to isolate fungi and include specimens sourced from animal tissues as well as from the environment, such as water, sediment, and vegetation. It is recommended that, whenever possible, specimens submitted for fungal culture be subsampled and examined by direct microscopy for fungal hyphae or yeasts using techniques such as the potassium hydroxide technique, India ink technique, or Calcofluor White staining method; this examination should be made in addition to, and not *in lieu* of, culture (Larone 2002). Isolation of environmental opportunistic fungi in culture from external anatomical sites of marine animals, such as the skin and gill, requires complementary histologic or microscopic evidence that the fungus is in the lesioned tissue in order to be clinically relevant (Guarner and Brandt 2011).

Identification of fungal isolates is best accomplished by combining morphological and molecular features (de Hoog et al. 2000). Sequences of fungal ribosomal operons, in particular those of the 18S rRNA gene, internal transcribed spacers (ITS), and the 5′ end of the 28S rRNA gene, have been exploited extensively for this purpose (Khot et al. 2009). However, *in silico* analyses of ITS primers have reported the potential for biased PCR results (Bellemain et al. 2010). A potentially more useful approach for characterization of fungal isolates is that of multi-locus sequence typing (MLST) (Taylor and Fisher 2003). Not only can MLST provide more substantive, and accurate, identification of fungal isolates, but also the genetic information from MLST can provide information relevant to the occurrence of mycoses within aquatic populations or environments, which has been applied in cases of fusariosis in sharks caused by members of the *Fusarium solani* species complex (Desoubeaux et al. 2018).

11.3.3 Cultivation, isolation, and identification of protistan disease agents

In general, obligate parasitic protists can be difficult to recover and maintain in culture due to the need for co-culture with their host, multiple host life cycles, and the fastidious nature associated with their growth requirements (Visvesvara and Garcia 2002). Conversely, opportunistic parasites, or those that have free-living stages, are more likely to be recovered *in vitro*. Protistan parasites involved in marine diseases have nonetheless been successfully isolated and are useful for understanding the survival, diversity, infectivity, existence of co-infections, and distribution potential of the parasites (e.g., Gachon et al. 2009, Bockelmann et al. 2012, Arzul and Carnegie 2015, FioRito et al. 2016, Sullivan et al. 2017).

Parasitic protists are diverse and examples can be found in almost every known lineage (Bateman et al. Chapter 1 this volume); the most common in the marine environment include amoebae, flagellates (nanoflagellate stramenopiles, dinoflagellates, kinetoplastids), ciliates, and cercozoa (Schweikert 2015, Yokoyama et al. 2015). They can be isolated/cultivated from most sample types, including seawater, sediment, homogenized tissue, and gross lesions, and the media of choice depends largely on the organism being sought. Protists can often be identified to some level by morphologic features, e.g., presence of flagella or cilia, but difficulty can arise if the organism has multiple life stages or is within a host, as those features may not be present, e.g., amoebae that form flagellate stages, formation of spores, or dormant cysts. Isolation of the organism in culture can be very helpful in identifying these stages (e.g., Schweikert and Schnepf 1997, Dykova et al. 2000). Histology is a widely used microscopy-based method for detecting protistan parasites in tissues. It allows for the recognition of non-normal cells and structures, with their morphology then permitting a general identification of the type of organism present. Indeed, such techniques have been used to identify a parasome-containing amoeba associated with the mass die-off of American lobsters in Long Island Sound (USA) in 1999 (Mullen et al. 2004), which was later identified by molecular means to be a strain of *Neoparamoeba pemaquidensis* (Mullen et al. 2005). More specific identification can then be accomplished by methods that target the genetic or protein components of the protist.

11.3.4 Cultivation, isolation, and identification of viral disease agents

Viruses are obligate intracellular infectious agents (Bateman et al. Chapter 1, this volume). As such, their replication requires living cells within which they can utilize host organelles and biochemical pathways to replicate the viral genome, transcribe viral mRNA, translate viral proteins, and assemble viral particles. Viral culture is therefore predicated on the need for growing host cells, i.e., cell culture. Isolation of virus by culture is commonly performed by inoculating a monolayer cell culture with a tissue homogenate or filtered environmental sample. The inoculated cell monolayer is observed by light microscopy for evidence of replication of the virus, which is commonly expressed as forms of cytopathic effect (CPE) and often involves rounding, detachment, and lysis of cells or syncytia formation (Smail and Munro 2012). Culture attempts are often limited by the availability of species-relevant cell lines or the permissiveness of cells to infection by the virus in the test sample.

The isolation of virus in cell culture must always be confirmed by a second diagnostic modality (Smail and Munro 2012). While light microscopy is used to detect CPE, electron microscopy performed on fixed preparations of cells from cultures displaying CPE is necessary to visualize and describe virus particles. Cellular lysates or supernatants from cell cultures displaying CPE may be assessed for the presence of proteins specific to a particular virus by using fluorescein- or enzyme-conjugated antibody raised against that virus, e.g., fluorescent antibody test (FAT) or ELISA. Viral nucleic acids can be detected by PCR for DNA viruses and reverse transcriptase PCR for RNA viruses, either of which can be applied to nucleic acids extracted from cells suspected of being infected. PCRs may be directed toward a specific virus or may be directed toward broader groups of phylogenetically related viruses, i.e., consensus PCRs. In either case, sequence determination of amplification products is a necessity, and sequencing of amplicons from consensus PCRs can allow for characterization of the virus isolated in cell culture by strain, genotype, or subgenotype.

11.4 Morphologic approaches to detection and diagnosis

11.4.1 Histology

Microscopic examination of tissue sections prepared by standard histologic processing techniques affords investigators two major advantages in the diagnosis and understanding of disease processes: (1) identification of an infectious agent in a histologic context, and (2) recognition of morphologic changes or features that are representative of disease. This can be particularly advantageous in the identification and characterization of new or emerging pathogens, where molecular approaches alone are not sufficient. The histopathologic evaluation of tissues can be critical to characterizing the inflammatory reaction of the host to infection, assessing lesioned tissues for particular classes of agents using special histochemical stains (Bateman et al. Chapter 1, this volume), understanding the spatial relationship between agents and the host tissue reaction, and inferring causal relationships (Procop and Wilson 2001). Acute inflammatory reactions typically are characterized by the infiltration of granulocytes in mammalian and non-mammalian vertebrates, whereas a granulomatous inflammatory reaction is a chronic response (Montali 1988). Such insight can be harder to obtain from the infiltrative responses seen in marine invertebrates in which a single circulating cell may be present. However, multiple variants may exist for that circulating cell, which is the case for the hemocyte of the American lobster *Homarus americanus*. Differential hemocyte counts have been useful in assessing and characterizing inflammatory responses (Battison et al. 2003), and histopathology has been integral to characterizing infectious disease in a number of diagnostic investigations, e.g., lobster paramoebiasis (Mullen et al. 2004) and epizootic shell disease of lobster (Smolowitz et al. 2005; Behringer et al. Chapter 10, this volume).

In order to provide consistent and accurate microscopic interpretations of histologic sections, tissues need to be prepared with fundamental histotechnological principles in mind. Tissues must be preserved in a fixative that inactivates enzymatic processes which would cause degradation and provides cross-linking of proteins for subsequent microtome sectioning. The most common fixative is 10 percent neutral buffered formalin, although fixatives such as modified Davidson's fixative provide advantages of concomitant dehydration and decalcification in addition to fixation (Fournie et al. 2000). After fixation, tissues must be processed for infiltration and subsequent embedment in paraffin or plastics. Preparation of histologic sections of appropriate thickness, e.g., 3 or 4 μm, without artifacts such as folds and tears, is technically challenging and requires proper training and supervised practice to master.

Histochemical staining of tissue sections can highlight particular features of tissues or cells or infectious agents (Procop and Wilson 2001, Bateman et al. Chapter 1, this volume). Routine histochemical staining typically involves the use of hematoxylin, a basic stain that binds to acidic moieties in cells, such as nucleic acids, and imparts a blue color, and eosin, an acidic stain that binds to basic moieties in cells, such as basic residues of proteins, and imparts a pink color. A wide range of specialized histochemical stains exist to highlight components of the extracellular matrix, such as collagen or elastin, e.g., Masson's trichrome and van Gieson stains, respectively, compounds deposited within cells or tissues, such as iron or mineral salts, e.g., Prussian blue reaction and von Kossa stain, respectively, and particular infectious agents, such as bacteria and fungi. Tissue Gram stains can be applied to provide insight into the Gram-staining characteristics of bacteria in histologic sections, while Grocott methenamine silver and periodic acid-Schiff stains can be used to highlight fungal hyphae and yeasts (Luna 1968).

11.4.2 Laser capture microdissection

Microscopic evaluation of tissue sections for the identification of infectious agents can lead to other diagnostic investigative techniques, such as laser capture microdissection (LCM) (Curran et al. 2000). The LCM technique allows the precise excision of a region-of-interest from a histologically prepared tissue section, which can then be used for nucleic acid extraction for PCR testing. The technique is predicated on the ability of the investigator to interpret the features of the histologic section accurately and guide the laser excision to the correct region of the section. Although the equipment can be expensive, LCM can afford distinct advantages in certain instances, such as allowing for molecular analysis of a specific pathogen in the presence of a mixed infection.

11.4.3 Electron microscopy

Electron microscopic techniques consist of transmission electron microscopy (TEM) and scanning electron microscopy (SEM). TEM allows one to resolve the ultrastructure of cells and organelles, while SEM allows resolution of the surface structures and surface topography of tissues and cells. Electron microscopic studies often follow histologic investigations, provide significantly higher resolution, and allow detailed examination of structures within histologic sections. TEM is used to examine the cellular and subcellular features of diseased cells and to characterize pathogenic micro-organisms. For example, TEM has been used to resolve the structure and cytopathology of intracellular bacteria such as chlamydia and rickettsia (Avakyan and Popov 1984). Additionally, TEM has been utilized to resolve the structure of virus particles that comprise viral inclusion bodies observed in histologic sections (e.g., Weber et al. 2009). TEM can be combined with labeling using antibody conjugated to gold particles, i.e., immunogold labeling (Gerard et al. 2005, Qian and Lloyd 2006) to demonstrate the presence of clinically or taxonomically relevant antigens particular to an infectious agent (e.g., Draghi et al. 2004).

11.4.4 Immunohistochemistry

IHC exploits the binding specificity of antibody to antigen and can be a particularly useful diagnostic technique when the antigen is a molecular marker of a disease process or pathogen (Taylor 1994). Antibody is typically of the IgG class, although occasionally IgM is used (Adams et al. 1995), and antigenic sites are three-dimensional configurations on macromolecules such as proteins and polysaccharides. Antibody is used as a probe to detect antigen within a histologic section, localizing it to cells that define a lesion or to a pathogenic micro-organism within a lesion (Taylor 1994, Adams et al. 1995, Tu et al. 2014). Monoclonal antibodies provide practical advantages in this regard, e.g., defined specificity, excessive production potential, and molecular uniformity. Monoclonal antibodies have been used to detect a range of fish pathogens such as bacteria, e.g., *Renibacterium salmoninarum*, *Mycobacterium* spp., and myxozoa, e.g., *Tetracapsuloides bryosalmonae*, the agent of proliferative kidney disease in salmonids (Adams et al. 1995). A monoclonal antibody against the capsid protein (ORF72) of koi herpesvirus isolated from diseased koi in Taiwan has been used to detect the virus by IHC (Tu et al. 2014). IHC techniques have employed fluorescent compounds and enzymes in order to permit visualization of the antibody–antigen complex in histologic section, and different methods of immunostaining exist, such as the direct conjugate method, indirect conjugate method, peroxidase–antiperoxidase method, and avidin–biotin conjugate method, to name a few (Taylor 1994).

11.4.5 *In situ* hybridization

Detection of specific nucleic acids in a histologic context

ISH is a diagnostic technique used to detect a specific nucleic acid sequence within cells or tissues. This allows for detection and anatomical interpretation of infectious agents, and expression patterns of genes in injured or activated cells (Qian and Lloyd 2006). ISH uses labeled nucleic acid sequences specific for a target to detect the presence of complementary sequences. Detection can be accomplished in frozen or formalin-fixed paraffin-embedded tissue samples and cells (Wilcox 1993, Levsky and Singer 2003). Probes for ISH include RNA probes (cRNA or riboprobes), synthetic oligonucleotide probes (e.g., peptide nucleic acids (PNA), locked nucleic acid (LNA), single-stranded DNA (ssDNA), and cDNA (created from mRNA)), and can be small (20–40 base pairs) or large (up to 1,000 base pairs). Successful hybridization by the probe is detected by a signal from either a radioactive (^{35}S, ^{131}I, ^{32}P, ^{33}P, ^{3}H) or non-radioactive (biotin, digoxigenin (DIG), or fluorescent) label (Corthell 2014). Gold particles coupled to antibodies (immunogold staining (IGS)) can be used for visualization of ISH targets within tissues and cells via electron microscopy (Qian and Lloyd 2006). Labeling by IGS can be done pre- and post-embedding preparation, and immuno-gold silver staining (IGSS) has improved sensitivity and detection efficiency (Gu and Hacker 2012). With regard to improved sensitivity and detection, an advanced ultrasensitive ISH technology, RNAscope®, employs a unique probe design strategy to simultaneously amplify hybridization signal and suppress background, while maintaining tissue morphology (Wang et al. 2012) (Figure 11.1).

Detection of nucleic acids of infectious agents

Visualization of pathogens through ISH can aid in determining the etiology of infection at the location of associated pathology. It is especially useful for diagnosing latent infections where virus particles are no longer produced, yet genomic material remains present in the infected cell (Segales et al. 1999). ISH has been used successfully to identify bacterial, viral, parasitic, and fungal agents in a wide variety of aquatic species, and it is often used in combination with other diagnostics methods (Griffitt et al. 2011, Fernandez-Lopez et al. 2012). Bodewes et al. (2013) used ISH to detect a novel b19-like parvovirus of the genus *Erythrovirus* in a rehabilitated harbor seal with clinical signs of central nervous system disease. Weli et al. (2017) identified the

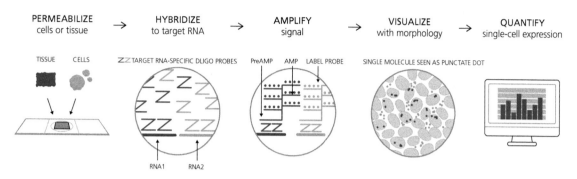

Figure 11.1 RNAscope® assay procedure. The unique double Z probe system allows specific hybridization while supporting a platform for signal amplification and maintaining tissue morphology. The system can be used with chromogenic or fluorescent dyes and achieves single-molecule visualization that can be quantified in a histologic context. (Image courtesy of Bio-Techne Corporation.)

etiology of an unknown disease in farmed Atlantic salmon (*Salmo salar* L.) using ISH. Although microsporidian infection of tissues was observed, routine diagnostics could not reveal the cause of a disease outbreak characterized by chronic proliferative gill inflammation and peritonitis. Using a combination of histopathology, ISH, CFW staining, and real-time PCR, the disease progression and visualization of *Desmozoon lepeophtherii* stages were possible *in situ*. ISH can succeed in identifying pathogens where routine diagnostics fail.

Location of infectious agents by ISH can also be used to detect and elucidate host–parasite life cycles and biology. Harbor porpoises (*Phocoena phocoena*) and harbor seals (*Phoca vitulina*) from German waters are infected by six species of lungworms (Metastrongyloidea). Using sequencing, phylogenetic analysis, histology, and ISH, Lehnert et al. (2010) identified the potential wild intermediate hosts of lungworms by identifying larval nematodes via ISH in tissues of prey fish.

Detection of mRNA expression in injured or activated cells

ISH is also used to reveal the location of specific nucleic acid sequences on chromosomes or in tissues, a crucial step for understanding the organization, regulation, and function of genes. Kawabe et al. (2012) used anti-sense cRNA probes to demonstrate the presence of lymphocyte-specific protein tyrosine kinase gene (*lck*) mRNA-positive cells during

early T-cell and thymus development to better understand immune system development in the Japanese eel (*Anguilla japonica*). Exposure to environmental contaminants can cause changes in gene expression, and these changes can be located and monitored using ISH. Expression of metallothionein (MT) mRNA in mussel (*Mytilus galloprovincialis*) was measured from contaminated and control sites. Mussels collected from sites with higher Cd, Pb, and Cr had higher basal rates of MT-10 mRNA in their gills than those from control sites, indicating that in stressed mussels, the defensive processes increase when exposed to metal pollution (Fasulo et al. 2008). Terminal deoxynucleotidyl transferase mediated dUTP nick end labeling (TUNEL) is another ISH method that detects DNA fragmentation by labeling 3′- hydroxyl termini in the double-stranded DNA breaks generated during apoptosis and necrosis. This technique, in combination with morphologic characterization of dying cells, can help to understand disease progression by identifying the specific types of cell death (e.g., apoptotic or necrotic) that occur within a lesion. Galimany and Sunila (2008) identified more apoptotic neoplastic cells in the early stage of disseminated neoplasia disease in mussels (*Mytilus edulis*) than later stages. The presence and decrease of these cells during disease progression indicated that proliferation of disseminated neoplasia in mussels is genetically regulated and could result in differences in susceptibility (Galimany and Sunila 2008).

Immunogold ISH to detect nucleic acids in a cellular context

Immunogold ISH can be used to detect nucleic acids with light microscopy, TEM, and SEM (Gerard et al. 2005). When immunogold is combined with silver staining (immunogold-silver staining (IGSS)) labeling efficiency and visibility are maximized (de Graaf et al. 1991, Humbel and Biegelmann 1992). Lignot et al. (1999) used immunofluorescence microscopy and immunogold electron microscopy to demonstrate that enhanced expression of the Na⁺, K⁺-ATPase under low salinity was localized to the basolateral infolding of the epipodite and specifically the branchiostegite epithelia, indicating that these structures may be important for osmoregulation in the European lobster *Homarus gammarus*. Hepatopancreatic parvovirus (HPV) infection in cultured shrimp has been linked to chronic mortalities during early larval and post-larval stages (Lightner 1999), and Pantoja and Lightner (2001) used an HPV-specific DIG-labeled probe detected by anti-DIG gold conjugate and silver enhancement to follow the infection and development cycle of the virus in cells of the penaeid shrimp (*P. vannamei*). Penaeidins are antimicrobial peptides produced and stored in the hemocytes of penaeid shrimp; ISH analyses using *pen-3* antisense and sense RNA probes and immunocytochemistry have been used to establish the origin of penaeidin mRNA and to localize the peptide in shrimp tissues (Munoz et al. 2002).

11.5 Nucleic acid-based approaches for detection and diagnosis

Nucleic acid-based approaches make use of the genomic and transcriptomic information in cells for determining the identity, presence, abundance, and virulence of pathogens and parasites.

11.5.1 Polymerase chain reaction

PCR amplification is a rapid process (< 2 hours) whereby many copies of a DNA target are made. This allows very small amounts of sample to be assayed and can range from phylogenetically conserved target detection (e.g., pan-eukaryotic 18S

rRNA genes) to a unique sequence present in only the target organism. These characteristics have made PCR a widely applied method for the detection of disease agents in tissues and environmental samples. Although conceptually straightforward, development of a reliable disease agent PCR assay requires careful planning and testing. Design of the primers used in the amplification is key to high specificity and sensitivity of the resulting assay. One needs to not only know the sequence of the target in the disease agent, but also take into account other organisms that may be present in the sample or environment to design primers that will only amplify the target (e.g., Bower et al. 2004). There is also the issue of low target DNA copy numbers in the sample, which are challenging to detect in a background of abundant non-target nucleic acids. A great variety of contaminants in sample preparations can also inhibit PCR amplification (Schrader et al. 2012).

11.5.2 Endpoint PCR

Detection of a product at the end of the amplification process without estimating quantity or abundance is called endpoint PCR. The process can be accomplished with one round of PCR, but reactions with primer sets that are "nested" relative to the first amplification primers are used to increase sensitivity and specificity. Gast et al. (2006, 2008) developed a nested amplification strategy to detect the thraustochytrid parasite of hard clams called Quahog Parasite Unknown (QPX) in samples of seawater, sediment, macroalgae, and other local invertebrates. Primer sets specific for *Leptospira interrogans* and *L. kirschneri* were used to detect these bacteria in tissue and urine samples from captive, wild, and stranded pinnipeds along the coasts of California, Washington, and British Columbia (Cameron et al. 2008). Bass et al. (2015) provide a diverse collection of studies on disease agents in both aquatic and terrestrial environments in their review on DNA methods in parasitology.

11.5.3 Quantitative PCR

qPCR, or real-time PCR, refers to the method whereby target DNA production is monitored during the amplification process in order to quantify

Before DNA polymerase:

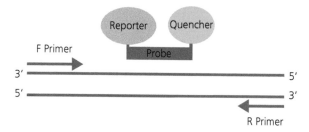

After DNA polymerase:

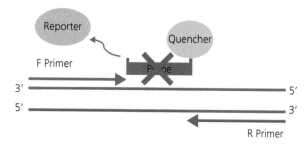

Figure 11.2 Probe-based qPCR. In this technique, a probe is constructed with nucleotide sequence identity to a region of the target sequence. At one end of the probe is a fluorescent reporter molecule, the signal from which is suppressed by a quencher molecule located at the opposite end of the intact probe. Hybridization of the probe to the target sequence during annealing positions the probe for subsequent cleavage by the endogenous exonuclease activity of DNA polymerase during extension, thereby releasing the reporter from the quencher and generating fluorescent signal that is detectable at each cycle and specific to amplification of the target sequence. (Diagram courtesy of Abigail S. Clark.)

the abundance of the disease agent in tissue or environmental samples. Two approaches are commonly used for the detection of products in qPCR: one uses non-specific fluorescent dyes that bind any double-stranded DNA in the reaction, and the other uses sequence-specific probes that are labeled with a fluorescent tag that is only detectable after the probe hybridizes with its target sequence and is subsequently cleaved by the exonuclease activity of DNA polymerase (Figure 11.2).

In both cases, the amount of fluorescence detected at each cycle is used to estimate the abundance of the target based upon a standard curve generated by a dilution series of known target quantity. The dilution series can be created using cloned target sequences, or from extracts of known cell abundance if cultures are available. While knowing that a disease agent is present can be useful for diagnosis or for following disease distribution, knowing how many pathogens are present is essential to being

able to estimate risk of infection or stage of infection. Furthermore, expression of disease-related genes can be assessed by reverse transcriptase qPCR of RNA from infected organisms, providing information on the activity of the agent or production of toxins.

Probe-based qPCR is considered to be more sensitive and specific than the intercalating fluorescent dye method, but the dye method can be cheaper and more flexible for adaptation to many different target assays. Probe-based qPCR does allow the development of multiplex assays for the detection of multiple targets within a single reaction, but it is possible to obtain data on multiple targets using dye-based qPCR by analyzing fragment melt curves at the end of the run (Lalonde et al. 2013). This has been accomplished for a wide range of organisms and targets, but ones relevant to marine diseases are the detection and discrimination of *Vibrio* and *Brucella* species (Winchell et al. 2010, Kim et al. 2012, Sidor et al. 2013).

A recent advance in qPCR is digital droplet PCR, which relies on the dilution of the sample prior to amplification into a large number of subsamples, resulting in an individual nucleic acid per subsample (Baker 2012). These subsamples can be in micro-well plates, capillaries, or oil emulsions. The number of positive versus negative samples is used to estimate the number of targets present based upon a Poisson distribution, thereby eliminating the need for standard curves. At the time this chapter was written, only one publication reported on digital PCR in a marine disease context. Jia et al. (2018) evaluated the effectiveness of Singapore grouper iridovirus inhibitory compounds using an *in vitro* cell viability assay with digital droplet PCR to confirm the inhibition of viral replication.

In contrast, a variety of studies have used traditional qPCR to examine diseases in marine organisms. Amoebic gill disease (AGD) has largely impacted farmed salmon, but it has also been identified in non-salmonid finfish and is a growing threat to aquaculture of these species. Bridle et al. (2010) developed a SYBR®-based qPCR assay to detect *Neoparamoeba perurans*, the causative agent of AGD in Atlantic salmon, in water samples collected from around farming and non-farming sites in Tasmania, Australia, which could be potentially very useful for assessing infection risk. *Eurychasma dicksonii* is an oomycete that parasitizes most brown algal species and has the potential to significantly disrupt cold and temperate rocky coastal marine ecosystems where brown algae comprise the majority of the biomass. Gachon et al. (2009) developed a fluorescent dye qPCR assay to detect *E. dicksonii* to facilitate disease monitoring. They were able to detect *E. dicksonii* infections in a range of filamentous brown algae from around the world and suggest that the method could be expanded to include hyphochytrid (heterokont) pathogens as well. This could be accomplished by using melt curve analysis (Lalonde et al. 2013), or by converting to a probe-based method. Disseminated neoplasia in bivalves is a transmissible leukemia characterized by multiple cell lineages specific for different bivalve species (Carballal et al. 2015, Metzger et al. 2016). Arriagada et al. (2014) used a reverse transcriptase qPCR to determine that expression levels of the retroelement *Steamer* were strongly correlated with disease status

in soft-shelled clams; the DNA copy number of *Steamer* was remarkably high in leukemic cells, constituting a potential diagnostic target. Bogomolni et al. (2015) utilized the Taqman probe approach to develop a duplex reverse transcriptase qPCR method to detect RNA from both phocine distemper virus (PDV) and glyceraldehyde-3-phosphate dehydrogenase (a positive control for RNA quality). Detection of the virus was possible in tissues from several seal species and has potential as a tool to monitor disease status and outbreak progression. Researchers were also able to use the method to assess *in vitro* effects of the toxins PCB Aroclor 1260 (Bogomolni et al. 2016a) and saxitoxin (Bogomolni et al. 2016b) using pinniped immune cell cultures. qPCR is currently one of the most useful, important, and cost-effective tools for marine disease research.

11.5.4 Next Generation Sequencing

NGS, also called high-throughput sequencing (HTS), refers to several different technologies that generate extremely large numbers of sequences without cloning individual targets. While most current methods produce sequence reads of < 500 base pairs, there are some that generate sequences averaging several thousand base pairs. NGS has been used by marine disease scientists in a number of productive applications (Figure 11.3).

It has been used to discover the distribution of disease agents in tissue or environmental surveys (e.g., Pagenkopp Lohan et al. 2016), to sequence whole genomes of disease agents from cultured isolates or from metagenomes (e.g., Katharios et al. 2015, Pérez-Pascual et al. 2017), and in transcriptomic analyses to understand virulence and the activity of host–pathogen interactions (e.g., Jang et al. 2012, Guo and Ford 2016, Sato et al. 2017).

11.5.5 Short-read high-throughput methods

Short-read high-throughput methods generate millions of short sequence reads per sample, and include 454 pyrosequencing (no longer supported), Ion Torrent, Illumina, and SOLiD sequencing (van Dijk et al. 2014). Depending on the sample, these reads can be computationally assembled as genomes or transcriptomes, or examined as shorter fragments

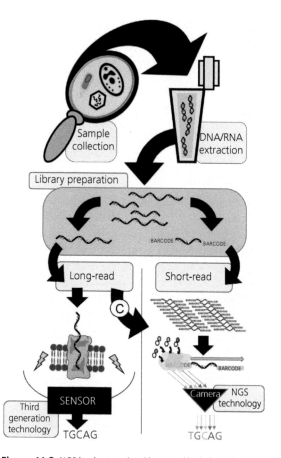

Figure 11.3 NGS by short-read and long-read high-throughput methods. Tissues, water samples, or other nucleotide-containing substances could be samples for DNA/RNA extraction. Extracts require preparation for use with high-throughput sequencing technologies. The short-read method produces high quantities of read data in the range of ~ 100–500 bp and requires addition of a barcode and primer ("BARCODE"). NGS then uses fluorescent tags of different colors to identify the base addition via a DNA polymerase. Color bursts are imaged by a camera to provide the output as a string of nucleotides. The long-read method makes use of a similar chemistry (black arrow with "C") but has a lower average data output despite longer read availability. The most recent development includes third-generation sequencing using protein channels embedded in an electro-conductive membrane. As strings of nucleotides pass through the protein channel, an electric current passes over the membrane and is translated by a sensor into a base call. (Image and figure legend provided by Jamie Bojko.)

sequenced using 454 pyrosequencing to generate reads with an average length of 479 bp from the diseased sample, and Ion Torrent to generate reads with an average length of 118 bp, but in greater numbers, from the healthy sample. After assembly of the metagenomes and elimination of oyster genomic sequences, the healthy and diseased assemblies were compared, and sequences found only in the diseased state were considered disease condition specific. While additional confirmation of the particular sequence type as the causative agent is certainly necessary, the use of metagenomics and amplicon sequencing helped to identify potential taxa of interest. A metatranscriptomic approach was used by Rosales and Thurber (2015) to investigate whether viral or bacterial agents were involved in seven harbor seal deaths. RNA was recovered and sequenced from samples of brain, along with tissue from seven other harbor seals that died from known causes. After assembly and removal of seal and human sequences, bacterial and viral-like transcripts were identified and their abundances compared within the unknown cause of death (UCD) group, and between the UCD group and the comparative group.

11.5.6 Long-read high-throughput methods

Long-read high-throughput methods are relatively new and not yet widely used in disease diagnosis, particularly in marine organisms. Also known as third-generation sequencing platforms, they were developed to help complete genome assemblies, sequence through complex regions like repetitive elements, or to effectively sequence phage genomes. The Pacific Biosciences (PacBio; Rhoads and Au 2015) single-molecule real-time method utilizes sequencing by light pulses during DNA synthesis; the association of fluorescently labeled nucleotides with the template at the active site of the DNA polymerase results in a pulse of fluorescence that is recorded to indicate the base at that sequence position. The Oxford Nanopore Technologies minION (Jian et al. 2016) identifies DNA bases by a change in electric conductivity as the nucleotide strand passes through the protein nanopore. The minION can also be used for rapid sequencing of shorter reads such as amplicons and transcriptomes. Both methods currently have higher error rates than short-read technology,

in amplicon-based diversity studies. Matsuyama et al. (2017) used both metagenomic and amplicon approaches to identify a potential agent of pearl oyster disease. DNA from hemolymph samples collected from diseased and healthy oysters was

but this will likely change as the technology is improved or correction algorithms are developed. As both short- and long-read NGS technology become more widely used in clinical situations, the technology will transfer more effectively to studies of marine organisms (Motro and Moran-Gilad 2017, Besser et al. 2018).

Currently, the few marine disease-related studies using long-read NGS have reported on the sequencing of pathogen genomes (Dong et al. 2017, Liu et al. 2017). The approach uses both short reads from Illumina and long reads from PacBio to sequence bacterial pathogen genomes and the plasmids that are also often found. Virulence and toxin genes, novel and known antibiotic resistance genes, and genes used by the bacterial pathogen for growth have all been found, and provide information on how the pathogen functions. It is reasonable to expect that, as long-read technology becomes more main stream, we will see more investigations like these in the field of marine diagnostics.

11.6 Summary

- Managing diseases of marine species and disease investigations in marine environments requires knowledge of the capabilities and limitations of a wide variety of diagnostic techniques.
- Utilization of multiple approaches in parallel is suggested to properly identify the etiologic agents of infectious diseases.
- Detection of a pathogen in an animal host or the environment does not necessarily indicate disease. Causal links to disease require spatial and temporal association of pathogens with lesions in host tissues.
- Cultivation, isolation, and identification of bacteria, fungi, protists, and viruses from aquatic hosts have been accomplished in marine disease investigations and require an understanding of proper sampling, microbial growth requirements, and morphologic and molecular diagnostic techniques.
- Microscopic tissue examination through histology, histochemical staining, electron microscopy, IHC, and ISH affords investigators the major advantages of (1) identifying the infectious agent

in the context of the host tissue, (2) recognizing the morphologic changes representative of disease, and (3) providing morphologic and molecular characterization of pathogens, which taken together are essential to the investigation of infectious disease.
- ISH can be used to detect specific nucleic acid sequences of infectious agents and of the host, e.g., injured or active cells, providing important insights into host–pathogen interactions in disease.
- Nucleic acid-based diagnostic techniques such as PCR and NGS make use of the genomic and transcriptomic information in cells to determine the identity, presence, abundance, and virulence of pathogens and parasites. These techniques provide genetic and genomic sequence data that can be essential to the identification of infectious agents in disease scenarios, as well as characterizing host responses to infection.

Acknowledgments

R.J.G.'s contribution to this chapter was supported by a subcontract on a Northeast Regional Aquaculture Grant to Roxanna Smolowitz at Roger Williams University.

References

Adams, A., Thompson, K.D., Morris, D., Farias, C., and Chen, S.C. 1995. Development and use of monoclonal antibody probes for immunohistochemistry, ELISA, and IFAT to detect bacterial and parasitic fish pathogens. *Fish & Shellfish Immunology* 5: 537–47

Agnew, W., and Barnes, A.C. 2007. *Streptococcus iniae*: an aquatic pathogen of global veterinary significance and a challenging candidate for reliable vaccination. *Veterinary Microbiology* 122: 1–15

Arriagada, G., Metzger, M.J., Muttray, A.F., Sherry, J., Reinisch, C., Street, C., Lipkin, W.I., and Goff, S.P. 2014. Activation of transcription and retrotransposition of a novel retroelement, *Steamer*, in neoplastic hemocytes of the mollusk *Mya arenaria*. *Proceedings of the National Academy of Sciences* 111: 14175–80

Arzul, I., and Carnegie, R.B. 2015. New perspective on the haplosporidian parasites of molluscs. *Journal of Invertebrate Pathology* 131: 32–42

Avakyan, A.A., and Popov, V.L. 1984. Rickettsiae and chlamydiae: comparative electron microscopic studies. *Acta Virologica* 28: 159–73

Baker, M. 2012. Digital PCR hits its stride. *Nature Methods* 9: 541–4

Bass, D., Stentiford, G.D., Littlewood, D.T.J., and Hartikainen, H. 2015. Diverse applications of environmental DNA methods in parasitology. *Trends in Parasitology* 31: 499–513

Battison, A., Cawthorn, R., and Horney, B. 2003. Classification of *Homarus americanus* hemocytes and the use of differential hemocyte counts in lobsters infected with *Aerococcus viridans* var. *homari* (Gaffkemia). *Journal of Invertebrate Pathology* 84: 177–97

Bellemain, E., Carlsen, T., Brochmann, C., Coissac, E., Taberlet, P., and Kauserud, H. 2010. ITS as an environmental DNA barcode for fungi: an *in silico* approach reveals potential PCR biases. *BioMedical Central Microbiology* 10: 189–98

Besser, J., Carleton, H.A., Gerner-Smidt, P., Lindsey, R.L., and Trees, E. 2018. Next-generation sequencing technologies and their application to the study and control of bacterial infections. *Clinical Microbiology and Infection* 24: 335–41

Bockelmann, A.-C., Beining, K., and Reusch, T.B.H. 2012. Widespread occurrence of endophytic *Labyrinthula* spp. in northern European eelgrass *Zostera marina* beds. *Marine Ecology Progress Series* 445: 109–16

Bodewes, R., Garcia, A.R., Wiersma, L.C.M., Getu, S., Beukers, M., Schapendornk, C.M.E., van Run, P.R.W.A., van de Bildt, M.W.G., Poen, M.J., Osinga, N., Sanchez Contreras, G.J., Kuiken, T., Smits, S.L., and Osterhaus, A.D.M.E. 2013. Novel B19-like parvovirus in the brain of a harbor seal. *PLoS One* 8: e79259

Bogomolni, A.L., Frasca, Jr., S., Matassa, K.A., Nielsen, O., Rogers, K., and De Guise, S. 2015. Development of a one-step duplex RT-qPCR for the quantification of phocine distemper virus. *Journal of Wildlife Diseases* 51: 454–65

Bogomolni, A., Frasca, Jr., S., Levin, M., Matassa, K., Nielsen, O., Waring, G., and De Guise, S. 2016a. In vitro exposure of harbor seal immune cells to Aroclor 1260 alters phocine distemper virus replication. *Archives of Environmental Contamination and Toxicology* 70: 121–32

Bogomolni, A.L., Bass, A.L., Fire, S., Jasperse, L., Levin, M., Nielsen, O., Waring, G., and De Guise, S. 2016b. Saxitoxin increases phocine distemper virus replication upon *in-vitro* infection in harbor seal immune cells. *Harmful Algae* 51: 89–96

Bower, S.M., Carnegie, R.B., Goh, B., Jones, S.R.M., Lowe, G.J., and Mak, M.W.S. 2004. Preferential PCR amplification of parasitic protistan small subunit rDNA from metazoan tissues. *Journal of Eukaryotic Microbiology* 51: 325–32

Bridle, A.R., Crosbie, P.B.B., Cadoret, K., and Nowak, B.F. 2010. Rapid detection and quantification of *Neoparamoeba perurans* in the marine environment. *Aquaculture* 309: 56–61

Burge, C.A., Mark Eakin, C., Friedman, C.S., Froelich, B., Hershberger, P.K., Hofmann, E.E., Petes, L.E., Prager, K.C., Weil, E., Willis, B.L., Ford, S.E., and Harvell, C.D. 2014. Climate change influences on marine infectious diseases: implications for management and society. *Annual Review of Marine Science* 6: 249–77

Cameron, C.E., Zuerner, R.L., Raverty, S., Colegrove, K.M., Norman, S.A., Lambourn, D.M., Jeffries, S.J., and Gulland, F.M. 2008. Detection of pathogenic *Leptospira* bacteria in pinniped populations via PCR and identification of a source of transmission for zoonotic leptospirosis in the marine environment. *Journal of Clinical Microbiology* 46: 1728–33

Carballal, M.J., Barber, B.J., Iglesias, D., and Villalba, A. 2015. Neoplastic diseases of marine bivalves. *Journal of Invertebrate Pathology* 131: 83–106

Chatterjee, S., and Haldar, S. 2012. *Vibrio* related diseases in aquaculture and development of rapid and accurate identification methods. *Journal of Marine Science: Research and Development* S1: 002. doi:10.4172/2155-9910.S1-002

Combe, M., and Gozlan, R.E. 2018. The rise of the rosette agent in Europe: an epidemiological enigma. *Transboundary and Emerging Diseases* 65: 1474–81

Corthell, J.T. 2014. *In situ* hybridization. In: J.T. Corthell (ed.), *Basic Molecular Protocols in Neuroscience: Tips, Tricks, and Pitfalls*, pp. 105–11. San Diego: Academic Press

Curran, S., McKay, J.A., McLeod, H.L., and Murray, G.I. 2000. Laser capture microscopy. *Molecular Pathology* 53: 64–8

de Graaf, A., van Bergen en Henegouwen, P.M., Meijne, A.M., van Driel, R., and Verkleij, A.J. 1991. Ultrastructural localization of nuclear matrix proteins in HeLa cells using silver-enhanced ultra-small gold probes. *Journal of Histochemistry and Cytochemistry* 39: 1035–45

de Hoog, G.S., Guarro, J., Gené, J., and Figueras, M.J. 2000. General techniques. In: G.S. de Hoog, J. Guarro, J. Gené, and M.J. Figueras (eds), *Atlas of Clinical Fungi*, Second Edition, pp. 39–53. Utretcht: Centraalbureau voor Schimmelcultures

Desoubeaux, G., Debourgogne, A, Wiederhold, N., Zaffino, M., Sutton, D., Burns, R.E., Frasca, Jr., S., Hyatt, M.W., and Cray, C. 2018. Multi-locus sequence typing provides epidemiological insights for diseased sharks infected with fungi belonging to the *Fusarium solani* species complex. *Medical Mycology* 56: 591–601

Dong, X., Wang, H.L., Zou, P.Z., Chen, J.Y., Liu, Z., Wang, X.P., and Huang, J. 2017. Complete genome sequence of *Vibrio campbellii* strain 20130629003S01 isolated from shrimp with acute hepatopancreatic necrosis disease. *Gut Pathogens* 9: 31. doi: 10.1186/s13099-017-0180-2

Draghi 2nd, A., Popov, V.L., Kahl, M.M., Stanton, J.B., Brown, C.C., Tsongalis, G.J., West, A.B., and Frasca, Jr., S. 2004. Characterization of *"Candidatus piscichlamydia salmonis"* (order *Chlamydiales*), a chlamydia-like bacterium associated with epitheliocystis in farmed Atlantic salmon (*Salmo salar*). *Journal of Clinical Microbiology* 42: 5286–97

Dykova, I., Figueras, A., and Peric, Z. 2000. *Neoparamoeba* Page, 1987: light and electron microscopic observations on six strains of different origin. *Diseases of Aquatic Organisms* 43: 217–23

Fasulo, S., Mauceri, A., Giannetto, A., Maisano, M., Bianchi, N., and Parrino, V. 2008. Expression of metallothionein mRNAs by *in situ* hybridization in the gills of *Mytilus galloprivincialis*, from natural polluted environments. *Aquatic Toxicology* 88:62–8

Fernandez-Lopez, B., Villar-Cervino, V., Valle-Maroto, S., Barreiro-Iglesias, A., Anadon, R., and Rodicio, M.C. 2012. The glutamatergic neruons in the spinal cord of the sea lamprey: an *in situ* hybridization and immuno-histochemical study. *PLoS One* 10: e47898

FioRito, R., Leander, C., and Leander, B. 2016. Characterization of three novel species of Labyrinthulomycota isolated from ochre sea stars (*Piaster ochraceus*). *Marine Biology* 163: 170–80

Foss, R.D., Guha-Thakurta, N., Conran, R.M., and Gutman, P. 1994. Effects of fixative and fixation time on the extraction and polymerase chain reaction amplification of RNA from paraffin-embedded tissue. Comparison of two housekeeping gene mRNA controls. *Diagnostic Molecular Pathology* 3: 148–55

Fournie, J.W., Krol, R.M., and Hawkins, W.E. 2000. Fixation of fish tissues. In: G.K. Ostrander (ed.), *The Laboratory Fish*, pp. 569–77. San Diego: Academic Press

Frankel, A. 2012. Formalin fixation in the "-omics" era: a primer for the surgeon-scientist. *ANZ Journal of Surgery* 82: 395–402

Gachon, C.M.M., Strittmatter, M., Müller, D.G., Kleinteich, J., and Küpper, F.C. 2009. Detection of differential host susceptibility to the marine oomycete pathogen *Eurychasma dicksonii* by real-time PCR: not all algae are equal. *Applied and Environmental Microbiology* 75: 322–8

Galimany, E., and Sunila, I. 2008. Several cases of disseminated neoplasia in mussels *Mytilus edulis* (L.) in western Long Island Sound. *Journal of Shellfish Research* 27: 1201–7

Garver, K.A., Traxler, G.S., Hawley, L.M., Richard, J., Ross, J.P., and Lovy, J. 2013. Molecular epidemiology of viral haemorrhagic septicaemia virus (VHSV) in British Columbia, Canada, reveals transmission from wild to farmed fish. *Diseases of Aquatic Organisms* 104: 93–104

Gast, R.J., Cushman, E., Moran, D.M., Uhlinger, K.R., Leavitt, D., and Smolowitz, R. 2006. DGGE-based detection method for Quahog Parasite Unknown (QPX). *Diseases of Aquatic Organisms* 70: 115–22

Gast, R.J., Moran, D.M., Audemard, C., Lyons, M.M., DeFavari, J., Reece, K.S., Leavitt, D., and Smolowitz, R. 2008. Environmental distribution and persistence of Quahog Parasite Unknown (QPX). *Diseases of Aquatic Organisms* 81: 219–29

Gerard, E., Guyot, F., Philippot, P., and Lopez-Garcia, P. 2005. Fluorescence *in situ* hybridisation coupled to ultra small immunogold detection to identify prokarytic cells using transmission and scanning electron microscopy. *Journal of Microbiological Methods* 63: 20–8

Griffitt, K.J., Noriea III, N.F., Jonson, C.N., and Grimes, D.J. 2011. Enumeration of *Vibrio parahaemolyticus* in the viable but nonculturable state using direct plate counts and recognition of individual gene fluorescence *in situ* hybridization. *Journal of Microbiological Methods* 85: 114–18

Gu, J., and Hacker, G.W. 2012. *Modern Methods in Analytical Morphology*. New York: Springer

Guarner, J., and Brandt, M.E. 2011. Histopathologic diagnosis of fungal infections in the 21st century. *Clinical Microbiology Reviews* 24: 247–80

Guo, X., and Ford, S.E. 2016. Infectious diseases of marine molluscs and host responses as revealed by genomic tools. *Philosophical Transactions of the Royal Society B* 371: 20150206

Hjerde, E., Karlsen, C., Sørum, H., Parkhill, J., Willassen. N.P., and Thomson, N.R. 2015. Co-cultivation and transcriptome sequencing of two co-existing fish pathogens *Moritella viscosa* and *Aliivibrio wodanis*. *BioMedical Central Genomics* 16: 447–59

Hong, G.E., Kim, D.G., Bae, J.Y., Ahn, S.H., Bai, S.C., and Kong, I.S. 2007. Species-specific PCR detection of the fish pathogen, *Vibrio anguillarum*, using the amiB gene, which encodes N-acetylmuramoyl-L-alanine amidase. *FEMS Microbiology Letters* 269: 201–6

Humbel, B.M., and Biegelmann, E. 1992. Preparation protocol for postembedding immunoelectron microscopy of *Dictyostelium discoideum* cells with monoclonal antibodies. *Scanning Microscopy* 6: 817–25

Hwang, J., Park, S.Y., Park, M., Lee, S., and Lee, T.K. 2017. Seasonal dynamics and metagenomic characterization of marine viruses in Goseong Bay, Korea. *PLoS One* 12: e0169841

Jang, J.B., Kim, Y.K., del Castillo, C.S., Nho, S.W., Cha, I.S., Park, S.B., Ha, M.A., Hikima, J., Hong, S.J., Aoki, T., and Jung, T.S. 2012. RNA-Seq-based metatranscriptomic and microscopic investigation reveals novel metalloproteases of *Neobodo* sp. as potential virulence factors for soft tunic syndrome in *Halocynthia roretzi*. *PLoS One* 7: e52379

Jia, K.T., Yuan, Y.M., Liu, W., Liu, L., Qin, Q.W. and Yi, M.S. 2018. Identification of inhibitory compounds against Singapore grouper iridovirus infection by cell viability-based screening assay and droplet digital PCR. *Marine Biotechnology* 20: 35–44

Jian, M., Olsen, J.E., Paten, B., and Akeson, M. 2016. The Oxford Nanopore MinION: delivery of nanopore sequencing to the genomics community. *Genome Biology* 17: 239

Kaattari, I.M., Rhodes, M.W., Kaattari, S.L., and Shotts, E.B. 2006. The evolving story of *Mycobacterium tuberculosis* clade members detected in fish. *Journal of Fish Diseases* 29: 509–20

Katharios, P., Seth-Smith, H.M.B., Fehr, A., Mateos, J.M., Qi, W., Richter, D., Nufer, L., Ruetten, M., Soto, M.G., Ziegler, U., Thomson, N.R., Schlapbach, R., and Vaughan, L. 2015. Environmental marine pathogen isolation using mesocosm culture of sharpsnout seabream: striking genomic and morphological features of novel *Endozoicomonas* sp. *Scientific Reports* 5: 17609. doi: 10.1038/srep17609

Kawabe, M., Suetake, H., Kikuchi, K., and Suzuki, Y. 2012. Early T-cell and thymus development in Japanese eel *Anguilla japonica*. *Fisheries Science* 78: 539–47

Khot, P.D., Ko, D.L., and Fredricks, D.N. 2009. Sequencing and analysis of fungal rRNA operons for development of broad-range fungal PCR assays. *Applied and Environmental Microbiology* 75: 1559–65

Kim, H.-J., Lee, H.-J., Lee, K.-H., and Cho, J.-C. 2012. Simultaneous detection of pathogenic *Vibrio* species using multiplex real-time PCR. *Food Control* 23: 491–8

Lalonde, L.F., Reyes, J., and Gajadhar, A.A. 2013. Application of a qPCR assay with melting curve analysis for detection and differentiation of protozoan oocysts in human fecal samples from Dominican Republic. *American Journal of Tropical Medicine and Hygiene* 89: 892–8

Larone, D.H. 2002. Laboratory procedures. In: *Medically Important Fungi: A Guide to Identification*, Fourth Edition, pp. 293–312. Washington, DC: American Society for Microbiology Press

Lehnert, K., von Samson-Himmelstjerna, G., Schaudien, D., Bleidorn, C., Wohlsein, P., and Siebert, U. 2010. Transmission of lungworms of harbour porpoises and harbour seals: molecular tools determine potential vertebrate intermediate hosts. *International Journal for Parasitology* 40: 845–53

Levsky, J.M., and Singer, R.H. 2003. Fluorescence *in situ* hybridization: past, present and future. *Journal of Cell Science* 116: 2833–8

Lightner, D.V. 1999. The penaeid shrimp viruses TSV, IHHNV, WSSV, and YHV: current status in the Americas, available diagnostic methods, and management strategies. *Journal of Applied Aquaculture* 9: 27–52

Lignot, J.-H., Charmantier-Daures, M., and Charmantier, G. 1999. Immunolocalization of Na$^+$, K$^+$-ATPase in the organs of the branchial cavity of the European lobster *Homarus gammarus* (Crustacea, Decapoda). *Cell and Tissue Research* 296: 417–26

Liu, J.X., Ahao, Z., Deng, Y.Q., Shi, Y., Liu, Y.P., Wu, C., Luo, P., and Hu, C.Q. 2017. Complete genome sequence of *Vibrio campbellii* LmB 29 isolated from Red Drum with four native megaplasmids. *Frontiers in Microbiology* 8: 2035. doi: 10.3389/fmicb.2017.02035

Lovy, J., Piesik, P., Hershberger, P.K., and Garver, K.A. 2013. Experimental infection studies demonstrating Atlantic salmon as a host and reservoir of viral hemorrhagic septicemia virus type IVa with insights into pathology and host immunity. *Veterinary Microbiology* 166: 91–101

Luna, L.G. 1968. *Manual of Histologic Staining Methods of the Armed Forces Institute of Pathology*, Third Edition, pp. 2–277. New York: McGraw-Hill

Matsuyama, T., Yasuike, M., Fujiwara, A., Nakamura, Y., Takano, T., Takeuchi, T., Satoh, N., Adachi, Y., Tsuchihashi, Y., Aoki, H., Odawara, K., Iwangaga, S., Kurita, J., Kamaishi, T., and Nakayasu, C. 2017. A spirochaete is suggested as the causative agent of Akoya oyster disease by metagenomic analysis. *PLoS One* 12: e0182280

McDowell, E.M., and Trump, B.F. 1976. Histologic fixatives suitable for diagnostic light and electron microscopy. *Archives of Pathology and Laboratory Medicine* 100: 405–14

Metzger, M.J., Villalba, A., Carballal, M.J., Iglesias, D., Sherry, J., Reinisch, C., Muttray, A.F., Baldwin, S.A., and Goff, S.P. 2016. Widespread transmission of independent cancer lineages within multiple bivalve species. *Nature* 534: 705–9

Montali, R.J. 1988. Comparative pathology of inflammation in the higher vertebrates (reptiles, birds and mammals). *Journal of Comparative Pathology* 99: 1–26

Morse, S.S. 1995. Factors in the emergence of infectious diseases. *Emerging Infectious Diseases* 1: 7–15

Motro, Y., and Moran-Gilad, J. 2017. Next-generation sequencing application in clinical bacteriology. *Biomolecular Detection and Quantification* 14: 1–6

Mullen, T.E., Russell, S., Tucker, M.T., Maratea, J.L., Koerting, C., Hinckley, L., De Guise, S., Frasca, Jr., S., French, R.A., Burrage, T.G., and Perkins, C. 2004. Paramoebiasis associated with mass mortality of American lobster *Homarus americanus* in Long Island Sound, USA. *Journal of Aquatic Animal Health* 16: 29–38

Mullen, T.E., Nevis, K.R., O'Kelly, C.J., Gast, R.J., and Frasca, Jr., S. 2005. Nuclear small-subunit ribosomal RNA gene-based characterization molecular phylogeny and PCR detection of the *Neoparamoeba* from western Long Island Sound lobster. *Journal of Shellfish Research* 24: 719–31

Munoz, M., Vandenbulcke, F., Saulnier, D., and Bachere, E. 2002. Expression and distribution of penaeidin antimicrobial peptides are regulated by haemocyte reactions in microbial challenged shrimp. *European Journal of Biochemistry* 269: 2687–9

Nowak, B.F., and LaPatra, S.E. 2006. Epitheliocystis in fish. *Journal of Fish Diseases* 29: 573–88

Pagenkopp Lohan, K.M., Fleischer, R.C., Carney, K.J., Holzer, K.K., and Ruiz, G.M. 2016. Amplicon-based pyrosequencing reveals high diversity of protistan parasites in ships' ballast water: implications for biogeography and infectious diseases. *Microbiology of Aquatic Systems* 71: 530–42

Pantoja, C.R., and Lightner, D.V. 2001. Detection of hepatopancreatic parvovirus (HPV) of penaeid shrimp by *in situ* hybridization at the electron microscope level. *Diseases of Aquatic Organisms* 44: 87–96

Pérez-Pascual, D., Lunazzi, A., Magdelenat, G., Rouy, Z., Roulet, A., Lopez-Roques, C., Larocque, R., Barbeyron, T., Gobet, A., Michel, G., Bernardet, J.-F., and Duchaud, E. 2017. The complete genome sequence of the fish pathogen *Tenacibaculum maritimum* provides insights into virulence mechanisms. *Frontiers in Microbiology* 8: 1542

Procop, G.W., and Wilson, M. 2001. Infectious disease pathology. *Clinical Infectious Diseases* 32: 1589–601

Qian, X., and Lloyd, R.V. 2006. *In situ* hybridization. In: A. Lorincz (ed.), *Nucleic Acid Testing for Human Disease*, p. 113. New York: CRC Press

Rhoads, A., and Au, K.F. 2015. PacBio sequencing and its applications. *Genomics Proteomics Bioinformatics* 13: 278–89

Roberts, R.J. 2012. The bacteriology of teleosts. In: R.J. Roberts (ed.), *Fish Pathology*, Fourth Edition, pp. 339–82. Chichester: Wiley-Blackwell

Rosales, S.M., and Thurber, R.V. 2015. Brain metatranscripts from harbor seals to infer the role of the microbiome and virome in a stranding event. *PLoS One* 10: e0143944

Sato, Y., Ling, E.Y.S., Turaev, D., Laffy, P., Weynberg, K.D., Rattei, T., Willis, B.L., and Bourne, D.G. 2017. Unraveling the microbial processes of black band disease in corals through integrated genomics. *Scientific Reports* 7: 40455. doi: 10.1038/srep40455

Schrader, C., Schielke, A., Ellerbroek, L., and Johne, R. 2012. PCR inhibitors—occurrence, properties and removal. *Journal of Applied Microbiology* 113: 1014–26

Schweikert, M. 2015. Biology of heterotrophic nanoflagellates: parasitoids of diatoms. In: S. Ohtsuka, T. Suzaki, T. Horiguchi, N. Suzuki, and F. Not (eds), *Marine Protists*, pp. 519–30. Tokyo: Springer

Schweikert, M., and Schnepf, E. 1997. Light and electron microscopical observations on *Pirsonia punctigerae* spec.

nov., a nanoflagellate feeding on the marine centric diatom *Thassiosira punctigera*. *European Journal of Protistology* 33: 168–77

Segales, J., Ramos-Vara, J.A., Duran, C.O., and Porter, A. 1999. Diagnosing infectious diseases using *in situ* hybridization. *Swine Health and Production* 7: 125–8

Sibanda, S., Pfukenyi, D.M., Barson, M., Hang'ombe, B., and Matope, G. 2018. Emergence of infection with *Aphanomyces invadans* in fish in some main aquatic ecosystems in Zimbabwe: a threat to national fisheries production. *Transboundary and Emerging Diseases* 65: 1648–56

Sidor, I.F., Dunn, J.L., Tsongalis, G.J., Carlson, J., and Frasca, Jr., S. 2013. A multiplex real-time polymerase chain reaction assay with two internal controls for the detection of *Brucella* species in tissues, blood, and feces from marine mammals. *Journal of Veterinary Diagnostic Investigation* 25: 72–81

Smail, D.A., and Munro, E.S. 2012. The virology of teleosts. In: R.J. Roberts (ed.), *Fish Pathology*, Fourth Edition, pp. 186–291. Chichester: Wiley-Blackwell

Smolowitz, R., Chistoserdov, A.Y., and Hsu, A. 2005. A description of the pathology of epizootic shell disease in the American lobster, *Homarus americanus*, H. Milne Edwards 1837. *Journal of Shellfish Research* 24: 749–56

Staley, C., and Harwood, V.J. 2010. The use of genetic typing methods to discriminate among strains of *Vibrio cholerae*, *Vibrio parahaemolyticus*, and *Vibrio vulnificus*. *Journal—Association of Official Analytical Chemists International* 93: 1553–69

Steu, S., Baucamp, M., von Dach, G., Bawohl, M., Dettwiler, S., Storz, M., Moch, H., and Schraml, P. 2008. A procedure for tissue freezing and processing applicable to both intra-operative frozen section diagnosis and tissue banking in surgical pathology. *Virchows Archiv: An International Journal of Pathology* 452: 305–12

Sullivan, B.K., Robinson, K.L., Trevathan-Tackett, S.M., Lije, E.S., Gleason, F.H., and Lije, O. 2017. The first isolation and characterisation of the protist *Labyrinthula* sp. in southeastern Australia. *Eukaryotic Microbiology* 64: 504–13

Taylor, C.R. 1994. Principles of immunomicroscopy. In: C.R. Taylor and R.J. Cote (eds), *Immunomicroscopy: A Diagnostic Tool for the Surgical Pathologist*, Second Edition, pp. 1–20. Philadelphia: W.B. Saunders Company

Taylor, J.W., and Fisher, M.C. 2003. Fungal multilocus sequence typing—it's not just for bacteria. *Current Opinion in Microbiology* 6: 351–6

Taylor-Brown, A., Spang, L., Borel, N., and Polkinghorne, A. 2017. Culture-independent metagenomics supports discovery of uncultivable bacteria within the genus *Chlamydia*. *Scientific Reports* 7: 10661. doi: 10.1038/s41598-017-10757-5

Tu, C., Lu, Y.-P., Hsieh, C.-Y., Huang, S.-M., Chang, S.-K., and Chen, M.-M. 2014. Production of monoclonal antibody against ORF72 of koi herpesvirus isolated in Taiwan. *Folia Microbiologica* 59: 159–65

van Dijk, E.L., Auger, H., Jaszczyszyn, Y., and Thermes, C. 2014. Ten years of next-generation sequencing technology. *Trends in Genetics* 30: 418–26

Visvesvara, G.S., and Garcia, L.S. 2002. Culture of protozoan parasites. *Clinical Microbiology Reviews* 15: 327–8

Wang, F., Flanagan, J., Su, N., Wang, L.C., Bui, S., Nielson, A., Wu, X., Vo, H.T., Ma, X.J., and Luo, Y. 2012. RNAscope: a novel *in situ* RNA analysis platform for formalin-fixed, paraffin-embedded tissues. *Journal of Molecular Diagnostics* 14: 22–9

Weber 3rd, E.S., Waltzek, T.B., Young, D.A., Twitchell, E.L., Gates, A.E., Vagelli, A., Risatti, G.R., Hedrick, R.P., and Frasca, Jr., S. 2009. Systemic iridovirus infection in the Banggai cardinalfish (*Pterapogon kauderni* Koumans 1933). *Journal of Veterinary Diagnostic Investigation* 21: 306–20

Weli, S.C., Dale, O.B., Hansen, H., Gjessing, M.C., Ronneberg, L.B., and Falk, K. 2017. A case study of *Desmozoon lepeophtherii* infection in farmed Atlantic salmon associated with gill disease, peritonitis, intestinal infection, stunted growth, and increased mortality. *Parasites and Vectors* 10: 370. doi: 10.1186/s13071-017-2303-5

Wilcox, J.N. 1993. Fundamental principles of *in situ* hybridization. *Journal of Histochemistry and Cytochemistry* 41: 1725–33

Winchell, J.M., Wolff, B.J., Tiller, R., Bowen, M.D., and Hoffmaster, A.R. 2010. Rapid identification and discrimination of *Brucella* isolates by use of real-time PCR and high-resolution melt analysis. *Journal of Clinical Microbiology* 48: 697–702

Yokoyama, H., Itoh, N., and Ogawa, K. 2015. Fish and shellfish diseases caused by marine protists. In: S. Ohtsuka, T. Suzaki, T. Horiguchi, N. Suzuki, and F. Not (eds), *Marine Protists*, pp. 533–49. Tokyo: Springer

Yu, S., Chen, W., Wang, D., He, X., Zhu, X., and Shi, X. 2010. Species-specific PCR detection of the food-borne pathogen *Vibrio parahaemolyticus* using the irgB gene identified by comparative genomic analysis. *FEMS Microbiology Letters* 307: 65–71

CHAPTER 12

Modelling marine diseases

Tal Ben-Horin, Gorka Bidegain, Giulio de Leo, Maya L. Groner,
Eileen Hofmann, Hamish McCallum, and Eric Powell

12.1 Introduction

In late March 1995, Australian pilchards began wash-ing ashore in mass near South Australia's Anxious Bay. Fish strandings happen regularly, especially in shoaling fish such as pilchards, but beyond its mag-nitude this event seemed bizarre. Pilchards were washing ashore with strangely enlarged lesion-covered gills, and dead fish were found with their mouths open and their opercula closed (Whittington et al. 1997). No longer confined to Anxious Bay, it became common to see kilometer-sized rafts of dead pil-chards floating at sea. Within 4 months this massive mortality event spread across Australia's entire southern coastline, often against prevailing ocean currents (Figure 12.1), with impacts extending out to New Zealand (Whittington et al. 1997; Gaughan et al. 2000). This fish kill is thought to be the largest ever recorded, in both numbers impacted and geo-graphic range, and pilchard stocks along a coastline of more than 5,000 km were reduced by over 70 per-cent (Whittington et al. 1997). Also, because pil-chards are forage fish that dominate food chain links between plankton and larger fish, birds, and mammals, impacts extended to entire coastal food webs. Penguins, for instance, had increased death rates and failed to breed the following summer due to the food shortage (Dann et al. 2000). To what

extent could epidemiological models help explain this historic mass mortality event?

The wave of fish kills observed outward from Anxious Bay was associated with a novel herpes-virus (pilchard herpesvirus (PHV)) (Hyatt et al. 1997). However, because herpesviruses also lead to latent infections once initial infections clear (Virgin et al. 2009), it remained a mystery as to whether PHV was a recent introduction or something that had been present all along, only to be reactivated with an ecological or environmental trigger. The answer to this question had major regulatory impli-cations. Large amounts of frozen pilchards had been imported into Anxious Bay as feed for sea-farmed tuna, and it was alternatively hypothesized that PHV was introduced into Australian waters through bait imports or ballast water (Jones et al. 1997). There was clearly a need to understand how PHV caused such devastation and the distinctive mortal-ity wave provided an opportunity to solve this mystery with epidemiological models (Murray et al. 2001a; Murray et al. 2001b). Models fitting the observed data required an infection rate over 90 percent and a diffusion coefficient consistent with that of a swimming pilchard school, providing a hypothesis for how PHV spread against prevailing currents (Murray et al. 2001a). Furthermore, epi-demiological models showed that the PHV epidemic

Ben-Horin, T., Bidegain, G., de Leo, G., Groner, M.L., Hofmann, E., McCallum, H., and Powell, E., *Modelling marine diseases*
In: *Marine Disease Ecology*. Edited by: Donald C. Behringer, Kevin D. Lafferty, and Brian. R. Silliman,
Oxford University Press (2020). © Oxford University Press. DOI: 10.1093/oso/9780198821632.003.0012

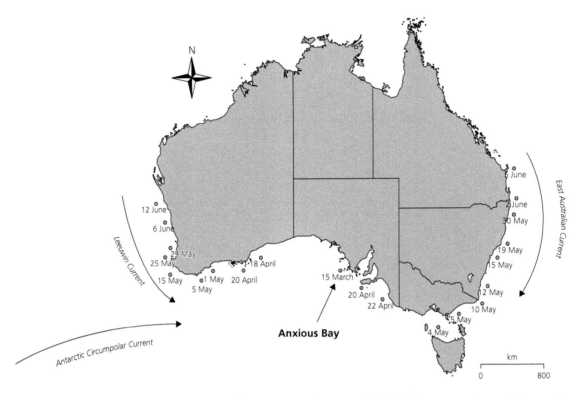

Figure 12.1 The Australian pilchard die-off in spring 1995. Beginning in March, reports of dead pilchards were reported progressively outward from Anxious Bay. Points on the map show dates mortalities were first observed and curved arrows indicate the major ocean surface currents, which generally run counter to the observed pilchard mortality wave. (Map redrawn from Whittington et al. 1997)

was host-density independent and generated from a single point source (Murray et al. 2001a; Murray et al. 2001b). Multiple infection sources, or the presence of vectors such as seabirds, would not have generated the observed wave. This modelling work supports the hypothesis that PHV was introduced to Anxious Bay and demonstrates the unique and important insights into marine disease dynamics provided by epidemiological models.

This chapter reviews modelling approaches that have been used to capture marine disease dynamics. The basic principles and foundations of epidemiological modelling apply whether disease outbreaks occur on land or in the sea (McCallum et al. 2004), but marine ecosystems have unique attributes. Many of the most familiar transmission modes, for example biting vectors, are rare in the ocean. Instead, transmission more commonly occurs when hosts contact or eat free-living parasite stages (i.e., environmental

transmission) such as spores and cysts (Ben-Horin et al. 2015). These stages occur in both marine and terrestrial environments, but because they often do not disperse far on land, models of terrestrial disease can subsume their dynamics into simplified transmission functions (Lafferty et al. 2015; Lafferty 2017; McCallum et al. 2017). Seawater's physical attributes, such as boundary layers and low Reynolds numbers, prolong parasite survival outside hosts, and these often ciliated or flagellated parasite stages use ocean currents to spread over spatial scales exceeding tens to hundreds of kilometers (McCallum et al. 2003). This incessant parasite rain increases interactions with other hosts and potential reservoirs, reducing the importance of local transmission in host population dynamics. For this reason, unlike classic infectious disease models that focus on transmission from host to host, many models of diseases in marine hosts assume infection has already occurred and focus on parasite dynamics within hosts (Hofmann et al. 1995; Powell and Hofmann 2015; Caldwell et al. 2016).

This chapter begins by reviewing such models, which often require the complexity required to quantify and forecast how parasite dynamics respond to environmental conditions (Section 12.2). We then describe compartmental marine disease models used to consider host demographic processes and how these processes influence epidemiological outcomes (Section 12.3), and then use this foundation to build models that consider transmission pathways and ecological dynamics seen in several notable marine host-parasite systems (Sections 12.4 and 12.5). We next consider cases where marine disease models have been applied to understand disease impacts on fished populations (Section 12.6) and conclude by overviewing methods to model how the ocean environment affects parasite dispersal and spread (Section 12.7).

12.2 Modelling and forecasting parasites in and on hosts

A typical liter of seawater collected from coastal seas can contain millions of algal cells, billions of bacteria and microbial eukaryotes, and up to 100 billion viruses (Breitbart 2012; Ben-Horin et al. 2015). Although bacteria–bacteriophage interactions dominate this microbial soup (McLaughlin et al. Chapter 2, this volume), this abundance confronts hosts in open marine ecosystems with many disease-causing parasites, which often minimizes the importance of local host to host transmission to marine disease dynamics. For this reason, understanding parasite dynamics assuming exposure has already occurred has been a common target for modelling disease impacts (Calvo et al. 2001; Powell et al. 2011; Powell et al. 2012; Paillard et al. 2014). Such models have been applied to questions of how environmental conditions influence host immune defenses and parasite proliferation and growth, and disease outcomes such as host mortality and infectiousness, including what causes hosts to have susceptibility or resistance to infection (Hofmann et al. 1995; Powell and Hofmann 2015).

Modelling how environmental conditions drive parasite dynamics relies on either well-established statistical relationships between abiotic variables and disease outcomes (Miller and Richardson 2014; Caldwell et al. 2016; Maynard et al. 2016) or linkages between environmental conditions and disease processes in terms of the host energy budget (Hofmann et al. 1995; Powell et al. 1999, Paillard et al. 2014). The latter require host processes such as ingestion, nutrient assimilation, and somatic and reproductive growth, and parasite processes such as within-host proliferation to be explicitly described and parameterized, as well as the response of both to changing abiotic conditions. Models developed for dermo disease in oysters (Hofmann et al. 1995) and brown ring disease in clams (Paillard et al. 2014) highlight the value of attaining this level of complexity. In the case of dermo disease, models demonstrate that temperature and salinity drive *in vivo* proliferation of the disease agent *Perkinsus marinus* (Powell et al. 2011; Powell et al. 2012). On the other hand, the host immune system response, as moderated by temperature, primarily controls the colonization and proliferation of the brown ring disease agent *Vibrio tapetis* (Paillard et al. 2014).

For diseases and disease agents with strong links to environmental drivers, given reliable environmental data, forecasts may be made to quantify marine disease risks into the future. This is important; in most cases, marine diseases are only detectable during or after their epidemic peak (Berkelman et al. 1994; Groner et al. 2016), while interventions taken before epidemics occur are often the ones most likely to succeed (e.g., Culver and Kuris 2000). Marine disease forecasts can inform interventions in near real-time (days to weeks), seasonally (typically 1–6 months), and over the long term (years to decades), each with different applications. Near real-time forecasts are based on recently modelled or ongoing surveillance data and are most often used for rapid response management actions such as closures to tourist activities or alterations to fisheries management (Maynard et al. 2009). Seasonal forecasts may be useful for informing managers, decision makers, and the public about potential disease risks, planning for management actions, or creating targeted surveillance programs to better track disease outbreaks (Liu et al. 2018). Long-term forecasts are developed using climate models established by national and international agencies such as the Intergovernmental Panel on Climate Change (IPCC). Forecasts on all these timescales exist for a variety of temperature-dependent coral diseases (Maynard et al. 2015; Caldwell et al. 2016) and are valuable for predicting how disease risks may change under different climate scenarios. These latter forecasts also serve as communication and outreach tools for generating public awareness and political willpower to manage marine disease outbreaks (Hooidonk et al. 2015).

Marine disease forecasting is still in its infancy; however, models have been developed for white syndrome, which is caused by a suite of different pathogens in primarily *Acropora* corals (Heron et al. 2010; Maynard et al. 2011; Maynard et al. 2015) and

are in development for epizootic shell disease in American lobsters (Maynard et al. 2016). Both white syndrome and epizootic shell disease are strongly associated with temperature (Brown 1997; Glenn and Pugh 2006; Bruno et al. 2007; Groner et al. 2018b), but even with tight temperature associations, forecasting biological complexity is a challenge (Box 12.1). Reliable marine disease forecasts become even more challenging with less reliable environmental associations (Groner et al. 2016); however, Maynard et al. (2016) identified infectious diseases of lobster, corals, oysters, abalone, salmonids, and seagrasses where known temperature-dependencies may make initial forecasting efforts possible.

Box 12.1 What would it take to forecast epizootic shell disease in lobsters?

Epizootic shell disease, caused by a dysbiotic suite of chitinolytic bacteria, first appeared in southern New England lobsters in 1996 (Castro and Somers 2012). It has since become established throughout the mid-Atlantic and southern New England, with peak prevalence exceeding 80 percent in male lobsters in some years (Groner et al. 2018a). Survival of severely diseased lobsters (with lesions on more than 10 percent of their carapace) is 70 percent lower than in healthy lobsters (Hoenig et al. 2017). This disease, along with several other stressors, including warming temperatures, has been associated with the collapse of the lobster fishery in Long Island Sound and southern New England (Howell 2012; Hoenig et al. 2017; Groner et al. 2018a).

While opportunities to manage epizootic shell disease may be limited in the southern New England lobster stock, where the impacts are already severe, it isn't too late to consider interventions to manage this disease further north in the Gulf of Maine. The Gulf of Maine lobster fishery represents about 75 percent of the state's commercial fishery value and annual dockside revenues frequently exceed US $500 million. Recognizing the ecological and cultural importance of the Gulf of Maine lobster stock, Maynard et al. (2016) created a simple "beta" program to forecast epizootic shell disease in the near-term, seasonally, and over the long term under different climate change scenarios. Previous studies had identified a threshold benthic water temperature of 12°C required for epizootic shell disease to emerge. Focusing on the year 2014, near-term projections

showed that the southern New England stock have already exceeded this threshold, while the Gulf of Maine stock did not. Seasonal projections showed that the Gulf of Maine could occasionally exceed this threshold, while long-term outlooks based on the fossil fuels aggressive emissions scenario (IPCC projection RCP8.5) revealed that warming in the Gulf of Maine could lead to summer temperatures exceeding the 12°C threshold regularly by 2020. Indeed, 2017 observations show increasing prevalence of shell disease in the Gulf of Maine (Reardon et al. 2018).

Although this model is simple, it aligns well with past and current observations. These temperature projections suggest that the Gulf of Maine may be on the verge of an epidemic. Future model iterations may be able to improve forecasts by adding new biological information. For example, recent work suggests that lobster molting phenology is shifting with changing winter and spring temperatures (Groner et al. 2018a). Summer molting is intrinsically linked to shell disease because it allows lobsters to discard their diseased carapace, essentially resetting their diseased state. In years with warmer spring temperatures, lobsters molt earlier, which increases the intermolt period in the summer, allowing shell disease to infect more lobsters (Groner et al. 2018a). Climate data from the Gulf of Maine suggest that winter temperatures are warming, but not nearly as rapidly as spring and summer temperatures, which are warming faster than 99 percent of the ocean (Thomas et al. 2017). This could alter molting phenology and increase disease progression, neither of which is good news for lobsters (Figure 12.2).

Population models that can project lobster populations under different climate and management scenarios can be used to forecast disease to advise fishing practices. For example, lobsters are less likely to be diseased after summer molting, and they will be a larger size and therefore be more valuable. Models that can predict the timing of summer molting based on spring temperatures can be used to identify the best time of year to fish. If these models incorporate shell disease, they can predict when prevalence is so high that fishing will no longer be lucrative, because diseased lobsters only sell for a fraction of the price of healthy lobsters. Mark-recapture programs and their associated statistical models (*sensu* Groner et al. 2018b) can be used to estimate the effect of temperature on disease mortality and disease progression. These functions, along with previously modelled impacts of temperature on molting, can be used in lobster population models to create an informative shell disease forecast benefitting both lobster populations and lobster fishing communities.

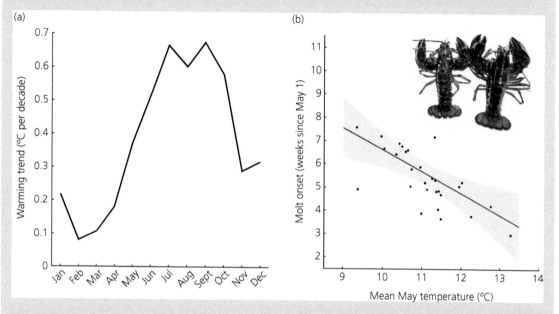

Figure 12.2 (A) Sea surface temperature warming trends in the Gulf of Maine, 1982–2014 (redrawn from Thomas et al. 2017). The Gulf of Maine region has exceeded the global average warming rate, and spring and summer temperatures have warmed faster than 99.9 percent of the global ocean. (B) Relationship between spring temperatures and lobster molting onset in Long Island Sound (redrawn from Groner et al. 2018a). Molting earlier in spring leads to increased epizootic shell disease in summer.

12.3 Compartmental marine host–parasite models

Parasite proliferation and forecast models are necessarily complex to quantify and predict how parasite levels and their associated disease outcomes respond to external drivers such as temperature. However, at the population level, individual hosts show demographic turnover, where deaths remove individuals (and their parasites) from the population and births bring new susceptible ones. These demo-graphic processes invariably shape how infections sweep through host populations (Anderson and May 1978; Anderson and May 1992). Alternative approaches are therefore required to address questions such as how epidemics rise and fall and understand outcomes such as the proportion of host populations experiencing infection.

Compartmental models simplify within-host processes to address host demography and its epidemiological outcomes. These include microparasite models, which consider all infected hosts to be equivalent with respect to the level of infection (Anderson and May 1979), and macroparasite models, which consider infection levels but do not allow infections to develop within hosts (Anderson and May 1978). Both microparasite and macroparasite models divide host populations into epidemiologically relevant compartments in which individuals reside. For example, a microparasite model describing an environmentally transmitted parasite that causes latent infections that eventually confer lifetime immunity would divide the host population (X) into susceptible (X_S), exposed (X_E), infected (or infective; X_I), and resistant (or recovered; X_R) host classes, and the model may also consider the environmental parasite stages (P) exposed, infective, and/or resistant host stages produce (a susceptible-exposed-infective-resistant-parasite model ($SEIRP$)). Systems where hosts become infective soon after infection and infection does not confer immunity, such as abalone (*Haliotis* spp.) and the Rickettsiales-like parasite causing abalone withering syndrome (WS-RLO) (Friedman et al. 2000; Moore et al. 2000), require fewer compartments and therefore allow simpler representations. In the case of abalone and WS-RLO, a susceptible-infected-parasite (SIP) model can

represent the system by dividing the total abalone population (X) into two compartments: one that is susceptible to but has yet to experience infection (X_S), and a second class that becomes infected (X_I) when contacting free-living parasite stages (P) in the local environment (Figure 12.3). It is relatively straightforward to consider host and parasite demography in compartmental models such as this SIP example, here by removing susceptible and infected hosts through natural and disease-induced mortality and bringing new susceptible hosts into the local population through reproduction. Larval abalone and environmental parasite stages can also emigrate into the local environment, with the system's openness modelled by increasing host and parasite recruitment relative to local production. The basic mathematics of this SIP model are presented in Box 12.2 and methods to apply the model to estimate the conditions under which marine parasites can invade host populations are presented in Box 12.3.

Kuris and Lafferty (1992) modelled possible outcomes arising when hosts and parasites interact in open marine ecosystems, where both larval hosts and parasite stages can emigrate into the local

---

Box 12.2 Mathematical formulation of the *SIP* model

The *SIP* model can be created by separating a host population X into a susceptible, uninfected class and a second class infected with a parasite that causes additive disease-induced mortality rate μ_I. The natural mortality rate in the absence of infection is μ_x. Here, X_s is the susceptible host density and X_I is the density of infected hosts ($X = X_s + X_I$). This simple model assumes that susceptible and infected hosts contact parasite stages (P) in their local environment (Figure 12.3), where contact dynamics are simplified by the coefficient c, which describes a host's capacity to contact parasites in the local environment per unit time. Contacts lead to new infections in susceptible hosts at the per parasite host susceptibility γ, and for simplicity we summarize new transmissions as the product of contacts and host susceptibility, where the coefficient of transmission $\beta = \gamma c$. The model allows host (B_x) and parasite stages (B_p) to recruit from outside the local

environment, whereas b represents the local production of hosts (b_1) and parasite stages shed from infected hosts (b_2). Parasite stages can then be lost from the local environment through inactivation and export at the rate (μ_p). Parasite stages can also be lost when encountering susceptible and infected hosts, although this can be ignored when $cX << \mu_p$. The dynamics of this system can be described by the following linked differential equations:

$$\frac{dX_S}{dt} = B_x + b_1 X - \mu_x X_S - \beta X_S P \quad (12.1)$$

$$\frac{dX_I}{dt} = \beta X_S P - (\mu_x + \mu_I) X_I \quad (12.2)$$

$$\frac{dP}{dt} = B_P + b_2 X_I - \mu_P P - cXP \quad (12.3)$$

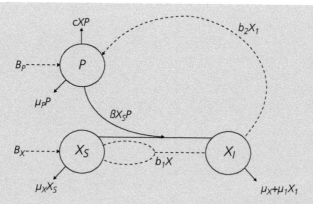

Figure 12.3 Flow diagram representing the *SIP* model for host species *X*. Hosts are separated into susceptible (X_s) and infected (X_I) classes that interact with parasite stages (*P*) in the local environment. Solid lines show changes to the host and parasite populations and dashed lines show host and parasite recruitment into the local environment.

Box 12.3 Estimating the basic reproductive number R_0

Whether or not an invading parasite can trigger an epidemic is contingent on the basic reproductive number (R_0), or the expected number of new infections occurring when an infected host, along with the parasite stages it produces, encounters an unexploited (i.e., entirely susceptible) host population. Arguably the most widely used quantity in wild-life epidemiology, when $R_0 > 1$, each infected individual is expected to produce more than one other case, leading to an epidemic outbreak. Alternatively, when $R_0 < 1$, each infected individual is expected not even to replace itself in the infected population, and disease-causing parasites will eventually be lost from host populations.

Various methods have been developed to estimate R_0 (see Anderson and May 1992; Dietz 1993; Diekmann et al. 2010; Diekmann et al. 2013). The next-generation matrix approach is particularly well suited for systems such as the one described by the *SIP* model presented in Box 12.2, where the number of discrete host and parasite classes is finite (Diekmann et al. 2010). A next-generation matrix of the *SIP* model quantifies the expected number of new infections arising from a single infected host contacting an entirely susceptible host population (i.e., $X = X_s$). If host and parasite

recruitment from outside the local population is similar to export ($B_X = B_P = 0$), a next-generation matrix may be constructed by linearizing the equations involved in the local production, transmission, and loss of parasite stages (equations 12.2 and 12.3) around X for X_I and $P \ll X$. The linearized system can be decomposed by the Jacobian matrix, which is decomposed into two matrices, T and Σ, to facilitate epidemiological interpretations. The transmission matrix T describes new infections and the production of parasite stages, and the transition matrix Σ describes the loss of infected hosts and parasite stages. All epidemiological events that lead to new infections, such as successful host contact and the release of parasite stages into the local environment, are described by T, where the first row corresponds to gain terms for X_I and the second row corresponds to gain terms for P:

$$T = \begin{bmatrix} 0 & \beta X \\ b_2 & 0 \end{bmatrix} \qquad (12.4)$$

All events leading to the loss of hosts and parasites, such as infected host mortality and the inactivation of parasite stages, are described by the matrix Σ. Here, the first row

continued

Box 12.3 *Continued*

corresponds to loss terms for X_i and the second row corresponds to loss terms for P:

$$\Sigma = \begin{bmatrix} -(\mu_x + \mu_i) & 0 \\ 0 & -(\mu_p + cX) \end{bmatrix} \quad (12.5)$$

The multiplicative inverse of Σ represents the expected lifespan of an infected host and the parasite stages it produces. The product of $-T$ and Σ^{-1}, K_L, gives the average number of new infections produced by a single infected host contacting an entirely susceptible host population:

$$K_L = \begin{bmatrix} 0 & \dfrac{\beta X}{\mu_p + cX} \\ \dfrac{b_2}{\mu_x + \mu_i} & 0 \end{bmatrix} \quad (12.6)$$

The dominant eigenvalue of K_L (λ) then defines R_0, where:

$$\lambda = R_0 = \sqrt{\frac{\beta X b_2}{(\mu_x + \mu_i)(\mu_p + cX)}} \quad (12.7)$$

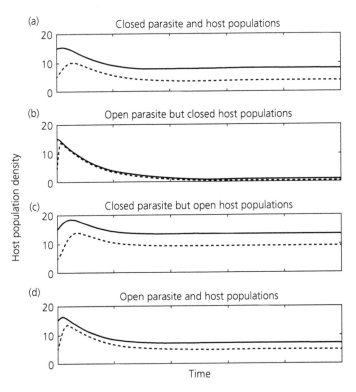

Figure 12.4 Four epidemiological outcomes produced from the *SIP* model presented in Box 13.2, illustrating four scenarios occurring when the relative openness of host and parasite populations are varied (following Kuris and Lafferty 1992). Solid line shows density of total host population (X) and dashed line shows density of infected hosts (X).

environment (illustrated in Figure 12.4 with the *SIP* model). When both host and parasite populations are closed ($B_p = 0$ & $B_X = 0$), introducing an infected host into the local population leads to an initial epidemic as parasites pass through the host

population and convert susceptible to infected hosts (Figure 12.4A). Fewer susceptible hosts leads to a decline in new infections, and infected host mortality slows parasite stage production. This allows the epidemic to settle down to an endemic equilibrium that balances the recruitment of new susceptible hosts into the population with infected host mortality. Host extinction can occur when the parasite population is open ($B_P \gg b_2 X_I$) but the host population remains closed ($B_X = 0$; Figure 12.4B). Here, parasite recruitment into the local environment is decoupled from local production, permitting infection to spread and be maintained even as susceptible hosts decline. Alternatively, when hosts recruit primarily from outside the local population ($B_X \gg b_1 X$), infection spreads because the susceptible host population is maintained (Figure 12.4C), even if parasite stages are produced only from local hosts ($B_P = 0$). In this latter scenario, introducing parasite stages produced from outside the local host population ($B_P \gg b_2 X_I$) will increase infection prevalence (Figure 12.4D).

The *SIP* model outlined in Box 12.2 reproduces several behaviors seen in disease-impacted marine populations. The following section uses this framework to develop models to capture transmission modes and infer ecological dynamics in notable marine host-parasite systems.

12.4 Ecological dynamics from marine disease models

12.4.1 Phocine distemper virus outbreaks in harbor seal colonies

Phocine distemper virus (PDV) was first identified throughout the North Sea in 1988 following a massive die-off of European harbor seals (*Phoca vitulina vitulina*) and a smaller number of sympatric gray seals (*Halicheroerus grypus*). This epidemic is among the best studied in marine mammals (Heide-Jøgensen and Härkönen 1992; Duignan et al. 2014) and was followed in 2002 by a second epidemic of similar temporal and geographic range (Härkönen et al. 2006).

PDV is a morbillivirus in the family Paramyxoviridae, closely related to the viruses that cause canine distemper, measles, and rinderpest. As with many Paramyxoviridae, virions are quickly inactivated outside the host environment (Jo et al. 2018). Transmission is likely to occur directly by contact between infected and susceptible seals, or via

aerosolized viral particles when infected and susceptible seals are in close contact on land, which likely occurs when seals haul out at high densities during the breeding season (Duignan et al. 2014; Jo et al. 2018). Outbreaks can lead to significant mortality in infected seals, although animals that do not die recover within 2 weeks and appear to be immune for life. Phocids such as harbor and gray seals reproduce only once a year and, as a result, PDV outbreaks spread rapidly through seal colonies but then die out before new susceptibles are born (Grenfell et al. 1992). Grenfell et al. (1992) developed a simple model representing these short-term PDV dynamics by ignoring recruitment (R_x and $r1$) and natural mortality (μ_x) of the local seal population, treating the rise and fall of PDV infection as a simple and single epidemic. By eliminating the contribution of free-living parasite stages to PDV transmission in favor of direct host-to-host transmission, and adding an additional class that has recovered from infection (X_R) at rate ρ and is immune to future PDV infection, equations 12.1–12.3 can be revised as:

$$\frac{dX_S}{dt} = -\beta X_S X_I \tag{12.8}$$

$$\frac{dX_I}{dt} = \beta X_S X_I - (\alpha + \rho) X_I \tag{12.9}$$

$$\frac{dX_R}{dt} = \rho X_I \tag{12.10}$$

This simple model captures the characteristic rise and fall of a PDV outbreak in a small, local population (Figure 12.5). It fails, however, to capture the broader epidemiological patterns observed across the North Sea ecosystem. For both the 1988 and 2002 PDV epidemics, seal die-offs were first observed in harbor seals hauled out on the Danish island of Anholt in the Kattegat, and immediately spread to adjacent seal colonies (Härkönen et al. 2006). But rather than continuing to spread like a wave, new centers of infection commonly appeared far from known PDV-infected populations during both epidemics. Harbor seals remain relatively sedentary at haul-outs, but gray seals, which occur in mixed colonies with harbor seals, move long distances and likely contribute to PDV spread and persistence. Capturing the broader PDV dynamics therefore requires accounting for the full host range of the

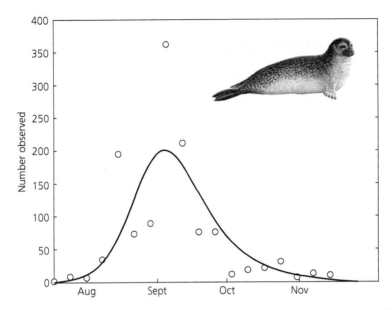

Figure 12.5 Rise and fall of a local PDV epidemic. Weekly seal mortalities observed in the Wash, East Anglia, UK, during the 1988 North Sea epidemic (open points), with the fit of the simple model (solid line). Pupping in harbor seals occurs from late winter to early summer, and this local PDV outbreak appears to have died out before new susceptible hosts arrived (after Grenfell et al. 1992).

virus and the interactions among these sympatric species within and between haul-out sites (Swinton et al. 1998).

12.4.2 Demographic resistance to disease in novel communities

Much attention has focused on global losses in coral cover and its consequences on reef function, including ecological shifts toward increased macroalgal dominance (Jackson 1997; Knowlton 2001; Hughes et al. 2017). Less discussed have been the subtle shifts toward smaller-bodied, faster-growing coral species (Green et al. 2008), which may lead to important outcomes for disease in coral assemblages (Yakob and Mumby 2011). Caribbean corals were once dominated by large, long-lived species (e.g., *Acropora cervicornis* and *Montastraea annularis*) but have come to be dominated by small-bodied "weedy" species (i.e., *Porites astreoides* and *Agaricia agaricites*; Gardner et al. 2003; Green et al. 2008), characterized by high population turnover owing to their short longevity and high fecundity. Increasing the average population turnover of corals makes it more difficult for epidemics to occur, as an epidemic requires that a

colony survive long enough to become infected, and then infect, on average, more than one additional colony (Box 12.3).

Yakob and Mumby (2011) developed a simple epidemiological model to consider the outcome of changes in the turnover of coral assemblages on the prevalence of white plague type II in the Florida Keys, USA. Here, the life-history shift toward increased population turnover can be incorporated into equations 12.1–12.3 by increasing recruitment rates at lower coral densities and increasing the natural mortality rate at higher coral densities. White plague type II is thought to be caused by the heterotrophic bacterium *Aurantimonas coralicida* (Denner et al. 2003) and spreads from infected colonies to infect susceptible corals. Although the survival ability of *A. coralicida* outside the host environment is not well known, Yakob and Mumby (2011) made the simplifying assumption that transmission occurs directly between infected and susceptible coral colonies, subsuming the contribution of free-living parasite stages to transmission into the coefficient β. White plague type II dynamics can therefore be generalized as:

$$\frac{dX_S}{dt} = b_1 X \left(1 - \frac{X}{X_K}\right) - \mu_X X_S \left(1 - \frac{X}{K}\right) - \beta X_S X_I \quad (12.11)$$

$$\frac{dX_I}{dt} = \beta X_S X_I - \mu_X X_I \left(1 - \frac{X}{K}\right) - \alpha X_I \quad (12.12)$$

Here, K is the carrying capacity of the host population, or the level of coral coverage for which recruitment and natural mortality are effectively zero.

Considering dynamic coral population turnover in epidemiological models accurately captures patterns of white plague type II infection observed at the Florida Keys (Figure 12.6) and suggests that novel coral assemblages may become less prone to disease outbreaks in a changing ocean. This result, however, is not universal (Burge and Hershberger Chapter 5, this volume). In the Indo-Pacific, coral diseases have been observed across the slow–fast life-history continuum (Sutherland et al. 2004). Higher transmission rates, which may occur when parasites better survive and spread outside the host environment, will counter any demographic resistance to disease levied by rapid population turnover (Figure 12.3B; Sokolow et al. 2009). Understanding how disease dynamics vary across systems will therefore require a better accounting of epidemiological processes occurring within host populations as well as outside the host environment.

12.4.3 Allee effects in colonizing ectoparasites

Allee effects may dampen parasite colonization (i.e., infection) and spread to uninfected host populations (May and Woolhouse 1993). Case studies have documented Allee effects arising when infection establishment is dose-dependent (Regoes et al. 2002), and theory predicts that mate limitation may also give rise to Allee effects that restrict the infection spread in dioecious parasites that reproduce sexually (May 1977; May and Woolhouse 1993).

Krkošek et al. (2012) tested whether mate limitation can dampen parasite spread in coastal marine ecosystems by quantifying how salmon lice (*Oncorhynchus keta*) infection intensity on wild juvenile Pacific salmon (*Oncorhynchus* spp.) influenced pair formation, and then integrated these data in a compartmental disease model. Salmon lice are directly transmitted parasitic copepods that are dioecious and reproduce sexually while attached to host salmonids (Pike and Wadsworth 2000). Mated adult females produce free-living, non-feeding nauplii, which molt into copepodites in the water column before attaching to host fish. The

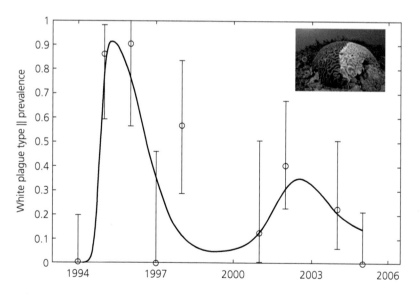

Figure 12.6 Demographic resistance to disease in novel coral assemblages. Observed prevalence (open circles, error bars show 95 percent confidence intervals) of white plague type II in corals sampled in Florida Keys National Marine Sanctuary and surrounding reefs from 1994 to 2005 (data redrawn from Sokolow et al. 2009). Solid line shows the fitted model described by Yakob and Mumby (2011), illustrated in equations 12.11 and 12.12. Prevalence of coral disease fluctuates with transient changes in the abundance of corals, and falls with increased coral turnover, indicating demographic resistance to epidemics with the emergence of novel, "weedy" corals.

attached copepodites then develop through a series of chalimus followed by motile pre-adult and adult stages (Johnson and Albright 1991), feeding on host surface tissues which leads to sublethal behavioral and physiological changes and host mortality at high infection intensities (Pike and Wadsworth 2000; Krkošek et al. 2011). Mating occurs with the formation of a pre-copular pair between an adult male and pre-adult female, completing the parasite life cycle.

The microparasite model described in Box 12.1 may be retrofitted as a macroparasite model to consider salmon louse life history and spread, first by assuming that salmon density (X) is constant over the louse colonization period ($dX/dt \rightarrow 0$) and then by considering the dynamics of adult female salmon lice attached to hosts (L) and louse stages free-living in the environment (P). The first assumption allows equations 12.1 and 12.2 to be ignored when estimating parasite spread, and the latter consideration permits evaluating how parasite growth and spread respond to colonizing parasite densities. Following Krkošek et al. (2012), the function describing free-living parasite dynamics (equation 12.3) can be modified to consider the life history of salmon lice:

$$\frac{dP}{dt} = b_2 XL\Phi_L - \mu_P P - \beta XP \quad (12.13)$$

where, as in equation 12.3, b_2 is production of free-living nauplii stages produced per adult female, Φ_L is the density-dependent mating probability function estimated by Krkošek et al. (2012), μ_P is the mortality rate of free-living parasite stages, and β is the transmission coefficient assuming that all host contacts lead to successful host colonization ($\gamma = 1$). The dynamics of adult female sea lice colonizing hosts can be described as:

$$\frac{dL}{dt} = \beta XP - \mu_L L - \left(\alpha + \frac{\alpha}{k} L \right) L \quad (12.14)$$

where μ_L is the mortality rate of adult female salmon lice and, as lice also die with their hosts, α is a per-parasite host mortality rate. The quantity k comes from the negative binomial distribution, which accounts for the overdispersion of parasites on their hosts (Shaw and Dobson 1995), which has

been observed in both male and female salmon lice infecting juvenile salmon (Krkošek et al. 2012). The lifespan of the free-living stage is short ($\sim$ 5 days) relative to the lifespan of attached stages (several weeks or months; Pike and Wadsworth 2000), and Krkošek et al. (2012) therefore assumed that densities of larval lice will quickly equilibrate and follow the dynamics of attached lice. Assuming $dP/dt \rightarrow 0$ and solving and substituting P into equation 12.14 gets:

$$\frac{dL}{dt} = \frac{\beta b_2 XL\Phi_L}{\mu_P + \beta X} - \left(\mu_L + \alpha + \frac{\alpha}{k} L \right) L \quad (12.15)$$

The expression for R_0 that comes from equation 12.15 is:

$$R_0 = \frac{\beta b_2 X}{(\mu_L + \alpha)(\mu_P + \beta X)} \quad (12.16)$$

In the system modelled here, R_0 describes the expected number of adult female offspring produced by a mated female when introduced into an entirely susceptible host population. Substituting the expression for R_0 into equation 12.15 gives the following model for the dynamics of the colonizing louse population:

$$\frac{dL}{dt} = \left[(R_0 \Phi_L - 1)\left(\mu_L + \alpha + \frac{\alpha}{k} L \right) \right] L \quad (12.17)$$

By integrating mating probabilities observed when salmon lice infect wild salmon with an epidemiological model describing louse population spread, Krkošek et al. (2012) suggest that louse populations may decline when parasite densities fall below approximately one adult female louse per juvenile salmon (Figure 12.7). This result has influenced sea louse control in salmon aquaculture, where a portfolio of management interventions can be used to keep sea louse abundance below this threshold, including chemical treatments, co-culture of sea louse predators, and fallowing. Further work (Stormoen et al. 2013; Groner et al. 2014) suggests that when the target louse abundance is just below the mate limitation threshold, better control can be achieved at the lowest cost, highlighting insights into disease dynamics and control provided by epidemiological models.

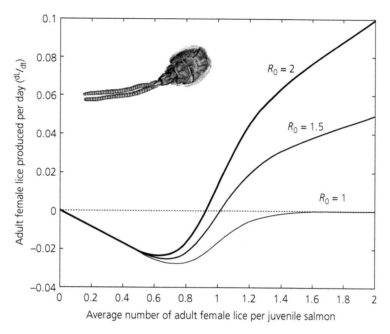

Figure 12.7 Demographic Allee effects in colonizing salmon lice, adapted from Krkošek et al. (2012). Predicted rate of change in adult female salmon lice populations colonizing juvenile salmon as a function of louse population density for values of $R_0 = 1$, $R_0 = 1.5$, and $R_0 = 2$. The horizontal dotted line shows zero net louse population growth ($\frac{dL}{dt} = 0$ adult female lice day^{-1}). For all values of R_0, louse populations decline at densities below approximately 0.8 adult female louse per juvenile salmon. Breakpoints in louse population dynamics, or density permitting population growth, occur at lower parasite densities with increasing values of R_0.

12.5 Considering parasite proliferation in compartmental models

A critical challenge to understanding host and parasite population dynamics is bridging the gap between proliferation and compartmental models. For microparasites such as PDV, disease intensity and pathology, and often infectiousness, tend to be determined simply by whether hosts are infected. That is, because microparasites generally have fast life histories relative to their hosts (Poulin 1995), models often assume that all infected hosts are equivalent with respect to the level of infection. This assumption implies that the number of microparasites per host is not as important as it is for macroparasites, such as salmon lice. However, as explained in Section 12.2, this can be problematic for several marine diseases, including brown ring disease in clams and dermo disease in oysters, where pathology and infectiousness depend on variation among and within hosts in parasite growth and proliferation (Powell and Hoffman 2015).

One way to bridge the gap between microparasite and macroparasite models is to use integral projection modelling, which can help put continuous traits like parasite burden and its effects on host survival and fecundity into compartment models. Doing so requires estimates for the growth (or decay) function of parasite burden within an individual host (i.e., how the probability distribution of burden at time $t +1$ depends on burden at time t). A further complication unique to host–parasite integral projection models is the need to estimate infectiousness and transmission to other hosts as a function of parasite burden (Metcalf et al. 2016; Wilber et al. 2016).

Bruno et al. (2011) used an integral projection approach to model the impact of aspergillosis, a fungal disease, on populations of sea fans (*Gorgonia ventalina*) in the Caribbean. Aspergillosis causes partial or whole colony mortality of sea fans. A 7-year epidemic commenced in 1994, which peaked at 50 percent prevalence in 1997 and led to substantial mortality and near complete reproductive

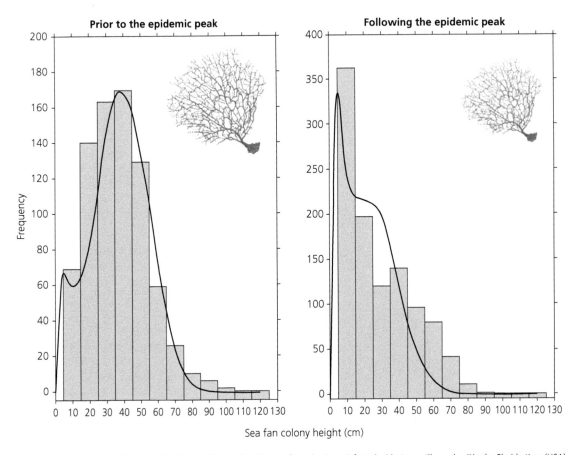

Figure 12.8 Observed size frequency distributions (bars) of healthy sea fan colonies uninfected with *Aspergillus sydowii* in the Florida Keys (USA) prior to the peak of the Caribbean epidemic (1997; left panel) and following the epidemic peak (2004; right panel; after Bruno et al. 2011). Solid lines show size frequency distributions predicted from the integral projection model of Bruno et al. (2011).

failure. Recovery was essentially complete by 2005, where the disease remained but at low prevalence.

The model considered two disease states—infected and uninfected—and, as a continuous variable, parasite burden measured as the total area of diseased tissue. Field data from the Florida Keys, USA, were used to estimate survival and fecundity as a function of disease status and healthy tissue area, and to estimate changes in healthy tissue area between years. A simplifying feature of this system is that incidence and prevalence of aspergillosis is predominantly independent of host density, suggesting that infection occurs from external sources, as is often the case with marine pathogens, avoiding the need to estimate transmission as a function of disease burden. The model accurately captured observed changes in

the size distribution of sea fan colonies through the aspergillosis outbreak (Figure 12.8).

12.6 Modelling disease in fished populations

Over three decades ago, Dobson and May (1987) proposed the idea of fishing out marine diseases, arguing that reducing host population densities and therefore parasite transmission efficiencies would curtail parasite spread and reduce disease impacts to fished populations. Their idea never gained traction, and the decades since have revealed complex effects of fishing on parasite dynamics (Kuris and Lafferty 1992; Wood et al. 2010; Wood et al. 2013; Wood and Lafferty 2015), both in harvested fish species and the broader ecological communities where harvested fish reside (Behringer et al. Chapter 10, this volume). Fishing out parasites is an unlikely target for future fisheries management; nevertheless, considering fishery harvest in compartmental disease models highlights systems where a fuller consideration of parasite dynamics may help optimize management.

Fisheries management focuses on maximum sustainable yield (MSY), which in theory represents the largest catch or harvest that can be taken from a fished stock over an indefinite period. This concept reflects a balance between maximizing recruitment into fished stocks and minimizing losses from density-dependent mortality (Brooks 2013; Mangel et al. 2013). The goal of managing for MSY is to maintain fished stocks at population levels that maximize their growth potential by harvesting fish at the rate they are added to the population (i.e., surplus production). MSY occurs when fished stocks are harvested to a fraction of the biomass or population density supported by the environment (i.e., the environmental carrying capacity; Hilborn and Walters 1992; Maunder 2003); assuming populations follow logistic growth (as in equations 12.11 and 12.12), MSY occurs at half the environmental carrying capacity. More refined fisheries models that consider complexity such as species' interactions, life-history variability, and age- or size-specific fishing selectivity estimate that MSY occurs at 20–40 percent of the unexploited population size (Botsford et al. 1997; Pikitch et al. 2004; Thorpe et al. 2015). It is straightforward to consider fishery harvest as an additive mortality source in compartmental disease models, for example rewriting equations 12.1–12.3 as:

$$\frac{dX_S}{dt} = b_1 X \left(1 - \frac{X}{K}\right) - \left(\mu_X + f\right)X_S - \beta X_S P \quad (12.18)$$

$$\frac{dX_I}{dt} = \beta X_S P - \left(\mu_X + \mu_I + f\right)X_I \quad (12.19)$$

$$\frac{dP}{dt} = b_2 X_I - \mu_P P \quad (12.20)$$

where f is the per capita fishery harvest rate. In this example the decline in recruitment follows a logistic pattern with increasing population density, where K is the population abundance or density where recruitment is effectively zero. Without

parasites ($P = 0$ and $X_S = X$), equations 12.18–12.20 simplify to:

$$\frac{dX_S}{dt} = \frac{dX}{dt} = b_1 X \left(1 - \frac{X}{K}\right) - \left(\mu_X + f\right)X \quad (12.21)$$

Here, population growth and surplus production available to the fishery is a concave function of population density, and following logistic growth, MSY occurs at half the environmental carrying capacity (illustrated in Figure 12.9 with parameter estimates from disease-free abalone populations).

When parasites increase mortality in their hosts, they compete with fisheries for surplus production and therefore reduce MSY. Combining equations 12.18 and 12.19, surplus production in disease-impacted stocks can be approximated as:

$$\frac{dX}{dt} = b_1 X \left(1 - \frac{X}{K}\right) - \left(\mu_X + \rho\mu_I + f\right)X \quad (12.22)$$

where ρ is the infection prevalence. Figure 12.9 shows how surplus production and MSY decline when parasites are introduced to host populations, using parameter estimates from abalone populations impacted by WS-RLO (from Ben-Horin et al. 2016).

Dobson and May (1987) applied similar compartmental models to a hypothetical fishery, demonstrating that fishing could reduce parasite transmission and even eliminate a parasite ($R_0 < 1$) without causing fishery collapse so long as transmission thresholds remains above MSY. The expression for $R0$ from equations 12.18–12.20:

$$R_0 = \sqrt{\frac{b_2 \beta X}{\mu_P \left(\mu_X + \mu_I + f\right)}} \quad (12.23)$$

illustrates that R_0 decreases with increasing fishery harvests, so long as fished stocks can support increasing harvest. In the case of abalone and WS-RLO, the threshold abalone density required to eliminate the parasite is far below densities supporting MSY (Figure 12.9).

The example of abalone fisheries impacted by WS-RLO as well as the hypothetical fisheries investigated by Dobson and May (1987) assume that all individuals are harvested with equal probability. But fisheries are almost always selective; that is, fish and shellfish of different age, size, sex, and spatial distribution are targeted by fisheries and their fishing methods (Myers and Hoenig 1997; Pikitch et al.

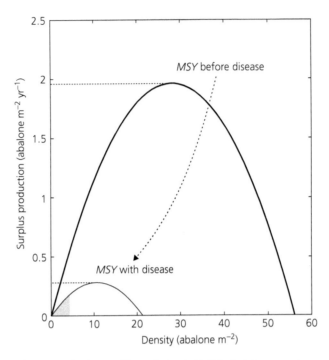

Figure 12.9 Surplus production of abalone in the absence of (heavy line) and with (light line) disease-causing parasites. Parasites causing mortality in their hosts compete with fisheries for surplus production, decreasing available yield. Host densities below the threshold for disease invasion ($R_0 < 1$) are shown in the shaded area of the surplus production curve with parasite introduction. Host densities supporting parasite extirpation fall below densities supporting MSY, making this management strategy unlikely.

2004; Ying et al. 2011; Garcia et al. 2012). Selectivity can have profound consequences for managing disease-impacted fisheries. Ben-Horin et al. (2016) found that dive fisheries directly targeting abalone infected with WS-RLO (measured through a concurrent abalone tagging and parasite sampling program) in their catch can both provide fishery yield and increase abalone densities by reducing parasite impacts. Similarly, Kuris and Lafferty (1992) modelled crustacean fisheries impacted by nemertean egg predators and parasitic castrators, which are common and affect only females or feminize males. Crustacean fisheries often self-impose catch restrictions on reproductive females to better maximize stocks' reproductive potential, but this practice may inadvertently increase parasite impacts by increasing the abundance of parasitized hosts. Kuris and Lafferty (1992) found that retaining females in the catch can reduce disease impacts, but only when host and parasite populations are closed. Females

should be treated for infection (if possible) when parasites recruit from outside populations. This result is supported by empirical results from Wood et al. (2013), finding the strongest negative effects of fishing to occur in parasite species with short dispersal distances. That is, low dispersal parasite populations are less likely to be rescued when fishing reduces transmission efficiencies in local host populations. In this same vein, McCallum et al. (2005) modelled how conservation measures such as no-fishing zones might increase host densities and thereby the transmission of endemic parasites, increasing disease impacts to fished host populations nearby.

Managing fisheries requires understanding how impacts to fished stocks are structured in space as well as by age, size, and sex. Epidemiological models considering selectivity and structure of fish stocks will guide the next major breakthroughs in managing fisheries impacted by disease.

12.7 Integrating disease models with physical ocean models

Physical ocean models such as the Regional Ocean Modelling System (ROMS) (Haidvogel et al. 2008) have transformed applied research in marine fisheries and conservation. These models allow particles to be tracked in simulated three-dimensional time-evolving circulation distributions, or Lagrangian particle tracking. Particle tracking simulations have been used to predict the transport and dispersion of salmon louse larvae colonizing Atlantic salmon (*Salmo salar*) from point sources in a Scottish loch (Amundrud and Murray 2009) and throughout the Broughton Archipelago in Canada (Cantrell et al. 2018) for a range of environmental conditions. In both examples the passive particles matured and died following a set of statistical rules that represented the planktonic stages of salmon louse life history.

Wang et al. (2012) used an implementation of ROMS for Delaware Bay to investigate environmental factors that affect the prevalence of infection with the agent of MSX (*Haplosporidium nelsoni*) in oysters. Passive particles were released at locations that correspond to the major oyster reefs, and environmental conditions along the trajectories were correlated with observed space and time variations in MSX infection preva-lence. The particle trajectories also provided insights into the movement patterns and potential source regions of MSX infective stages. Kough et al. (2015) used a Lagrangian particle tracking model of the Caribbean basin to test whether transmission of the *Panulirus argus* virus 1 (PaV1) to Caribbean spiny lobsters (*Panulirus argus*) occurs strictly through seawater or via infected lobster post-larvae. Their results suggest that if viral transmission is strictly waterborne, then PaV1 is unlikely to impact both the eastern and northwestern Caribbean, as the virus is only viable in seawater for a few days (Box 12.4). However, PaV1 transport by infected post-larvae allows the virus to spread throughout much of the wider Caribbean, confirming the present distribution, prevalence, and genetic diversity of the virus (Moss et al. 2012; Moss et al. 2013). Similarly, Ferreria et al. (2014) used passive particle tracking simulations to consider the potential for disease spread between offshore and inshore shellfish aquaculture sites off the coast of Portugal. The risk exposure map developed from the particle tracking trajectories illustrated the importance of understanding the on- and offshore connectivity provided by the circulation to disease occurrence.

Particle tracking simulations and epidemiological models each provide insights about processes, trends, and patterns in marine disease dynamics

---

Box 12.4 Modelling waterborne *Panulirus argus* virus 1

PaV1 infects Caribbean spiny lobsters (*Panulirus argus*) throughout much of the wider Caribbean (Shields and Behringer 2004; Moss et al. 2013). A large, non-enveloped DNA virus, PaV1 is highly pathogenic to juvenile lobsters (Butler et al. 2008), challenging the management of the US $1 billion lobster industry in the region. Virions can be infectious in seawater, and, like other non-enveloped viruses, PaV1 is presumed to remain viable outside hosts for 1 week or more (Shields and Behringer 2004). However, population genetic studies suggest that an additional vector or dispersal mechanism must be required. High levels of PaV1 genetic diversity are observed throughout the Caribbean Sea and viral alleles are shared even among distant locations (Moss et al. 2012; Moss et al. 2013). Moss et al. (2012) discovered PaV1 infecting *P. argus* post-larvae arriving in the Florida Keys, USA, leading to the hypothesis that the virus may disperse throughout the Caribbean along with its hosts' planktonic stages. Given the extended duration of pelagic *P. argus* larvae and post-larvae (5–9 months), an association with these host stages represents a novel mechanism for virus dispersal in the sea.

Kough et al. (2015) addressed probable mechanisms of PaV1 dispersal by coupling particle tracking models of the Caribbean basin with a stepping-stone model of PaV1 spread. Their results indicate that if transmission occurs strictly between juvenile and adult lobsters via waterborne virions, then PaV1 is unlikely to impact both the eastern and northwest Caribbean, which are separated by strong dispersal barriers. However, if PaV1 can be transported between locations by infected post-larvae, the entire Caribbean becomes linked, with higher viral delivery and PaV1 prevalence in the north. This analysis identified the subregions of the Caribbean where PaV1 is most likely to spread and highlighted the highly connected locations that may serve as dispersal "gateways," permitting rapid PaV1 spread into otherwise isolated areas (Figure 12.10).

continued

Box 12.4 *Continued*

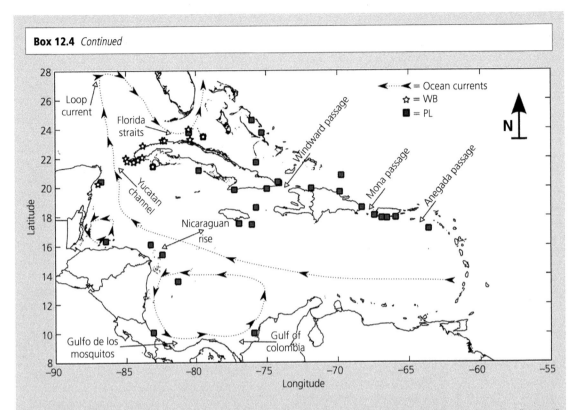

Figure 12.10 Locations serving as "gateways" from which PaV1 may spread most rapidly to other locations. White stars show "gateway" locations if transmission occurs strictly from waterborne virions (WB). Gray squares show "gateway" locations if transmission is mediated by pelagic lobster post-larvae (PL). Due to physical dispersal barriers, PaV1 spread to the eastern Caribbean is unlikely if transmission occurs only from waterborne virions. (Reprinted with permission from Kough et al. 2015)

that are not possible to obtain through experimental and observational programs. The approaches and techniques used in each are well established and have been tested in a wide range of applications in marine environments. Dynamically coupling Lagrangian simulations with epidemiological models provides the next major challenge to understanding feedbacks and controls on marine diseases, which is needed to forecast marine disease dynamics and inform effective mitigation strategies.

12.8 Synthesis

Although not wholly unique to marine systems, characteristics such as strong links to environmental conditions, environmental transmission, and the more open nature of host and parasite recruitment have pushed investigators to refine well-tested and widely applied epidemiological methods that were developed in terrestrial and aquatic systems. We concur with the conclusion made by McCallum et al. (2004) over a decade ago: the basic principles and foundations of epidemiological modelling apply whether disease outbreaks occur on land or in the sea. It will be the application and broad integration of approaches drawing on the fields of epidemiology, oceanography, and fisheries science that will provide the next breakthroughs in our understanding of the feedbacks and controls on marine diseases and refine our ability to predict and forecast disease dynamics to better inform disease management and mitigation.

12.9 Summary

- The foundations of epidemiological modelling apply whether disease outbreaks occur on land or in the sea, but the marine ecosystem's unique characteristics have pushed investigators to refine well-tested and widely applied epidemiological methods.

- The abundance of microbial diversity in open marine ecosystems can minimize the importance of local host-to-host transmission to marine disease dynamics. As a result, understanding parasite dynamics assuming exposure has already occurred has been a common target for modelling disease impacts, particularly in response to changing environmental conditions.

- When parasites and disease impacts respond strongly to environmental conditions, and reliable environmental data are available, forecasting models can quantify marine disease risks into the future.

- Epidemiological outcomes drive but are also driven by host demographic processes. Compartmental models simplify within-host processes to address host demography and its influence on disease outcomes.

- Compartmental models are also well suited to consider additional processes that influence host demography, such as fishery and aquaculture activities.

- Managing fisheries requires understanding how impacts to fished stocks are structured in space as well as by age, size, and sex. Epidemiological models considering selectivity and structure of fish stocks will guide the next major breakthroughs in managing fisheries impacted by disease.

- Disease dynamics often vary across local host populations. Coupling epidemiological models with physical ocean models such as ROMS will bring new insights toward understanding epidemiological processes within and between local host populations.

References

Amundrud, T.L. and Murray, A.G., 2009. Modelling sea lice dispersion under varying environmental forcing in a Scottish sea loch. *Journal of Fish Diseases* 32: 27–44

Anderson, R.M. and May, R.M. 1978. Regulation and stability of host–parasite population interactions: I. Regulatory processes. *Journal of Animal Ecology* 47: 219–47

Anderson, R.M. and May, R.M. 1979. Population biology of infectious diseases: Part I. *Nature* 280: 361–7

Anderson, R.M. and May, R.M. 1992. Infectious Diseases in Humans: Dynamics and Control. Oxford University Press, Oxford

Ben-Horin, T., Bidegain, G., Huey, L., Narvaez, D.A., and Bushek, D. 2015. Parasite transmission through suspension feeding. *Journal of Invertebrate Pathology* 131:155–76

Ben-Horin, T., Lafferty, K.D., Bidegain, G., and Lenihan, H.S. 2016. Fishing diseased abalone to promote yield and conservation. *Philosophical Transactions of the Royal Society B* 371: 20150211

Berkelman, R.L., Bryan, R.T., Osterholm, M.T., LeDuc, J.W., and Hughes, J.M. 1994. Infectious disease surveillance: a crumbling foundation. *Science* 264: 368–70

Botsford, L.W., Castilla, J.C., and Peterson, C.H. 1997. The management of fisheries and marine ecosystems. *Science* 277: 509–15

Breitbart, M. 2012. Marine viruses: truth or dare. *Annual Review of Marine Science* 4: 425–48

Brooks, E.N. 2013. Effects of variable reproductive potential on reference points for fisheries management. *Fisheries Research* 138: 152–8

Brown, B.E. 1997. Coral bleaching: causes and consequences. *Coral Reefs* 16: S129–38

Bruno, J.F., Selig, E.R., Casey, K.S., Page, C.A., Willis, B.L., Harvell, C.D., Sweatman, H., and Melendy, A.M. 2007. Thermal stress and coral cover as drivers of coral disease outbreaks. *PLoS Biology* 5: e124

Bruno, J.F., Ellner, S.P., Vu, I., Kim, K., and Harvell, C.D. 2011. Impacts of aspergillosis on sea fan coral demography: modelling a moving target. *Ecological Monographs* 81: 123–39

Butler, M.J., Behringer, D.C., and Shields, J.D. 2008. Transmission of *Panulirus argus* virus 1 (PaV1) and its effect on the survival of juvenile Caribbean spiny lobster. *Diseases of Aquatic Organisms* 79: 173–82

Caldwell, J.M., Heron, S.F., Eakin, C.M., and Donahue, M.J. 2016. Satellite SST-based coral disease outbreak predictions for the Hawaiian Archipelago. *Remote Sensing* 8: 93

Calvo, L.M.R., Wetzel, R.L., and Burreson, E.M. 2001. Development and verification of a model for the population dynamics of the protistan parasite, *Perkinsus marinus*, within its host, the eastern oyster, *Crassostrea virginica*, in Chesapeake Bay. *Journal of Shellfish Research* 20: 231–41

Cantrell, D.L., Rees, E.E., Vanderstichel, R., Grant, J., Filgueira, R., and Revie, C.W. 2018. The use of kernel density estimation with a bio-physical model provides a method to quantify connectivity among salmon farms: spatial planning and management with epidemiological relevance. *Frontiers in Veterinary Science* 5: 269

Castro, K.M. and Somers, B.A. 2012. Observations of epizootic shell disease in American lobsters, *Homarus americanus*, in southern New England. *Journal of Shellfish Research* 31: 423–30

Culver, C.S. and Kuris, A.M. 2000. The apparent eradication of a locally established marine pest. *Biological Invasions* 2: 245–53

Dann, P., Norman, F.I., Cullen, J.M., Neira, F.J., and Chiaradia, A. 2000. Mortality and breeding failure of little penguins, *Eudyptula minor*, in Victoria 1995–1996, following a widespread mortality of pilchard, *Sardinops sagax*. *Marine and Freshwater Research* 51: 355–62

Denner, E.B.M., Smith, G.W., Busse, H.-J., Schumann, P., Narzt, T., Polson, S.W., Lubitz, W., and Richardson, L.L. 2003. *Aurantimonas coralicida* gen. nov., sp. nov. the causative agent of white plague type II on Caribbean scleractinian corals. *International Journal of Systematic and Evolutionary Microbiology* 53: 1115–22

Diekmann, O., Heesterbeek, J.A.P., and Roberts, M.G. 2010. The construction of next-generation matrices for compartmental epidemic models. *Journal of the Royal Society Interface* 7:873–85

Diekmann, O., Heesterbeek, J.A.P., and Britton, T. 2013. Mathematical Tools for Understanding Infectious Disease Dynamics. Princeton University Press, Princeton, NJ

Dietz, K. 1993. The estimation of the basic reproduction number for infectious diseases. *Statistical Methods in Medical Research* 2: 23–41

Dobson, A.P. and May, R.M. 1987. The effects of parasites on fish populations—theoretical aspects. *International Journal for Parasitology* 17: 363–70

Duignan, P.J., Van Bressem, M.-F., Baker, J.D., Barbieri, M., Colegrove, K.M., De Guise, S., de Swart, R.L., Di Guardo, G., Dobson, A., Duprex, W.P., Early, G., Fauquier, D., Goldstein, T., Goodman, S.J., Grenfell, B., Groch, K.R., Gulland, F., Hall, A., Jensen, B.A., Lamy, K., Matassa, K., Mazzariol, S., Morris, S.E., Nielsen, O., Rotstein, D., Rowles, T.K., Saliki, J.T., Siebert, U., Waltzek, T., and Wellehan, J.F.X. 2014. Phocine distemper virus: current knowledge and future directions. *Viruses* 6: 5093–134

Ferreira, J.G., Saurel, C., Silva, J.L., Nunes, J.P., and Vazquez, F. 2014. Modelling of interactions between inshore and offshore aquaculture. *Aquaculture* 426: 154–64

Friedman, C.S., Andree, K.B., Beuchamp, K., Moore, J.D., Robbins, T.T., Shields, J.D., and Hedrick, R.P. 2000. *Candidatus Xenohaliotis californiensis* gen. nov., sp. nov. a pathogen of abalone, *Haliotis* spp., along the west coast of North America. *International Journal of Systematic and Evolutionary Microbiology* 50: 847–55

Garcia, S.M., Kolding, J., Rice, J., Rochet, M.-J., Zhou, S., Arimoto, T., Beyer, J.E., Borges, J.E., Bundy, A., Dunn, D., Fulton, E.A., Hall, M., Heino, M., Law, R., Makino, M., Rijnsdorp, AD, Simard, F., and Smith, A.D.M. 2012. Reconsidering the consequences of selective fisheries. *Science* 335: 1045–7

Gardner, T.A., Côté, I.M., Gill, J.A., Grant, A., and Watkinson, A.R. 2003. Long-term region-wide declines in Caribbean corals. *Science* 301: 958–60

Gaughan, D.J., Mitchell, R.W., and Blight, S.J. 2000. Impact and mortality, possibly due to herpesvirus, on pilchard *Sardinops sagax* stocks along the south coast of western Australia in 1988–1999. *Marine and Freshwater Research* 51: 601–12

Glenn, R.P. and Pugh, T.L. 2006. Epizootic shell disease in American lobster (*Homarus americanus*) in Massachusetts coastal waters: interactions of temperature, maturity, and intermolt duration. *Journal of Crustacean Biology* 26:639–45

Green, D.H., Edmunds, P.J., and Carpenter, R.C. 2008. Increasing relative abundance of *Porites astreoides* on Caribbean reefs mediated by an overall decline in coral cover. *Marine Ecology Progress Series* 359: 1–10

Grenfell, B.T., Lonergran, M.E., and Harwood, J. 1992. Quantitative investigations of the epidemiology of phocine distemper virus (PDV) in European common seal populations. *Science of the Total Environment* 115: 15–29

Groner, M.L., Gettinby, G., Stormoen, M., Revie, C.W., and Cox, R. 2014. Modelling the impact of temperature-induced life history plasticity and mate limitation on the epidemic potential of a marine ectoparasite. *PLoS One* 9: e88465

Groner, M.L., Maynard, J., Breyta, R., Carnegie, R.B., Dobson, A., Friedman, C.S., Froelich, B., Garren, M., Gulland, F. M.D., Heron, S.F., Noble, R.T., Revie, C.W., Shields, J.D., Vanderstichel, R., Weil, E., Wyllie-Echeverria, S., and Harvell, C.D. 2016. Managing marine disease emergencies in an era of rapid change. *Philosophical Transactions of the Royal Society B* 371: 20150364

Groner M.L., Shields J.D., Landers Jr, D.F., Swenarton, J., and Hoenig, J. 2018a. Rising temperature causes phenological mismatch between molting and epizootic shell disease in the American lobster. *American Naturalist* 192: E163–77

Groner, M.L., Hoenig, J.M., Pradel, R., Choquet, R., Vogelbein, W.K., Gauthier, D.T., and Friedrichs, M.A.M. 2018b. Dermal mycobacteriosis and warming sea surface temperatures are associated with elevated mortality of striped bass in Chesapeake Bay. *Ecology and Evolution* 8: 9384–97

Haidvogel, D.B., Arango, H., Budgell, W.P., Cornuelle, B.D., Curchitser, E., Di Lorenzo, E., Fennel, K., Geyer, W.R., Hermann, A.J., Lanerolle, L. and Levin, J. 2008.

Ocean forecasting in terrain-following coordinates: formulation and skill assessment of the Regional Ocean Modelling System. *Journal of Computational Physics* 227: 3595–624

Härkönen, T., Dietz, R., Reijnders, P. Teilmann, J., Harding, K., Hall, A., Brasseur, S., Siebert, U., Goodman, S.J., Jepson, P.D., Rasmussen, T.D., and Thompson, P. 2006. The 1998 and 2002 phocine distemper virus epidemics in European harbour seals. *Diseases of Aquatic Organisms* 68: 115–30

Heide-Jørgensen, M.-P. and Härkönen, T. 1992. Epizootiology of the seal disease in the eastern North Sea. *Journal of Applied Ecology* 29: 99–107

Heron, S.F., Willis, B.L., Skirving, W.J., Eakin, C.M., Page, C.A., and Miller, I.R. 2010. Summer hot snaps and winter conditions: modelling white syndrome outbreaks on Great Barrier Reef corals. *PLoS One* 5: e12210

Hilborn, R. and Walters, C.J. 1992. Quantitative Fisheries Stock Assessment. Choice Dynamics and Uncertainty. Chapman & Hall, New York

Hoenig, J.M., Groner, M.L., Smith, M.W., Vogelbein, W.K., Taylor, D.M., Landers Jr, D.F., Swenarton, J.T., Gauthier, D.T., Sadler, P., Matsche, M.A., Haines, A.N., Small, H.J., Pradel, R., Choquet, R., and Shields, J.D. 2017. Impact of disease on the survival of three commercially fished species. *Ecological Applications* 27: 2116–27

Hofmann, E.E., Powell, E.N., Klinck, J.M., and Saunders, G. 1995. Modelling diseased oyster populations. I. Modelling *Perkinsus marinus* infections in oysters. *Journal of Shellfish Research* 14: 121–51

Hooidonk, R., Maynard, J.A., Liu, Y., and Lee, S.K. 2015. Downscaled projections of Caribbean coral bleaching that can inform conservation planning. *Global Change Biology* 21: 3389–401

Howell, P. 2012. The status of the southern New England lobster stock. *Journal of Shellfish Research* 31: 573–9

Hughes, T.P., Barnes, M.L., Bellwood, D.R., Cinner, J.E., Cumming, G.S., Jackson, J.BC, Kleypas, J., van de Leemput, I.A., Lough, J.M., Morrison, T.H., Palumbi, S.R., van Nes, E.H., and Scheffer, M. 2017. Coral reefs in the Anthropocene. *Nature* 546: 82–90

Hyatt, A.D., Hine, P.M., Jones, J.B., Whittington, R.J., Kearnes, C., Wise, T.G., Crane, M.S., and Williams, L.M. 1997. Epizootic mortality in the pilchard *Sardinops sagax neopilchardardus* in Australia and New Zealand in 1995. 2. Identification of a herpesvirus within the gill epithelium. *Diseases of Aquatic Organisms* 28: 17–29

Jackson, J.B.C. 1997. Reefs since Columbus. *Coral Reefs* 16: S23–32

Jo, W.K., Osterhaus, A.D.M.E., and Ludlow, M. 2018. Transmission of morbilliviruses within and among marine mammal species. *Current Opinion in Virology* 28: 133–41

Johnson, S.C. and Albright, L.J. 1991. The development stages of *Lepeophtheirus salmonis* (Kroyer, 1837) (Copepode, Caligidae). *Canadian Journal of Zoology* 69: 929–50

Jones, J.B., Hyatt, A.D., Hine, P.M., Whittington, R.J., Griffin, D.A., and Bax, N.J. 1997. Australian pilchard mortalities. *World Journal of Microbiology & Biotechnology* 13: 383–92

Knowlton, N. 2001. The future of coral reefs. *Proceedings of the National Academy of Sciences USA* 98: 5419–25

Kough, A.S., Paris, C.B., Behringer Jr, D.C., and Butler IV, M.J. 2015. Modelling the spread and connectivity of waterborne marine pathogens: the case of PaV1 in the Caribbean. *ICES Journal of Marine Science* 72: i139–46

Krkošek, M., Connors, B., Mages, P., Peacock, S., Ford, H., Ford, J.S., Morton, A., Volpe, J.P., Hilborn, R., Dill, L.M., and Lewis, M.A. 2011. Fish farms, parasites, and predators: implications for salmon population dynamics. *Ecological Applications* 21: 897–914

Krkošek, M., Connors, B.M., Lewis, M.A., and Poulin, R. 2012. Allee effects may slow the spread of parasites in a coastal marine ecosystem. *The American Naturalist* 179: 401–12

Kuris, A.M. and Lafferty, K.D. 1992. Modelling crustacean fisheries: effects of parasites on management strategies. *Canadian Journal of Fisheries and Aquatic Sciences* 49: 327–36

Lafferty, K.D. 2017. Marine infectious disease ecology. *Annual Review of Ecology, Evolution, and Systematics* 48: 473–96

Lafferty, K.D., DeLeo, G., Briggs, C.J., Dobson, A.P., Gross, T., and Kuris, A.M. 2015. A general consumer-resource population model. *Science* 349: 854–7

Liu, G., Eakin, C.M., Chen, M., Kumar, A., De La Cour, J.L., Heron, S.F., Geiger, E.F., Skirving, W.J., Tirak, K.V., and Strong, A.E. 2018. Predicting heat stress to inform reef management: NOAA Coral Reef Watch's 4-month coral bleaching outlook. *Frontiers in Marine Science* 5: 57

Mangel. M., MacCall, A.D., Brodziak, J., Dick, E.J., Forrest, R.E., Pourzand, R., and Ralston, S. 2013. A perspective on steepness, reference points, and stock assessment. *Canadian Journal of Fisheries and Aquatic Sciences* 70: 930–40

Maunder, M.N. 2003. Is it time to discard the Schaefer model from the stock assessment scientist's toolbox? *Fisheries Research* 61: 145–9

May, R.M. 1977. Togetherness among schistosomes: its effects on the dynamics of the infection. *Mathematical Biosciences* 35: 301–43

May, R.M. and Woolhouse, M.E.J. 1993. Biased sex ratios and parasite mating probabilities. *Parasitology* 107: 287–95

Maynard, J.A., Johnson, J.E., Marshall, P.A., Eakin, C.M., Goby, G., Schuttenberg, H., and Spillman, C.M. 2009. A strategic framework for responding to coral bleaching events in a changing climate. *Environmental Management* 44: 1–11

Maynard, J.A., Anthony, K.R.N., Harvell, C.D., Burgman, M.A., Beeden, R., Sweatman, H., Heron, S.F., Lamb, J.B., and Willis, B.L. 2011. Predicting outbreaks of a climate-driven coral disease in the Great Barrier Reef. *Coral Reefs* 30: 485–95

Maynard, J., Van Hooidonk, R., Eakin, C.M., Puotinen, M., Garren, M., Williams, G., Heron, S.F., Lamb, J., Weil, E., Willis, B., and Harvell, C.D. 2015. Projections of climate conditions that increase coral disease susceptibility and pathogen abundance and virulence. *Nature Climate Change* 5: 688–94

Maynard, J., Van Hooidonk, R., Harvell, C.D., Eakin, C.M., Liu, G., Willis, B.L., Williams, G.J., Groner, M.L., Dobson, A., Heron, S.F., and Glenn, R. 2016. Improving marine disease surveillance through sea temperature monitoring, outlooks and projections. *Philosophical Transactions of the Royal Society B* 371: 20150208

McCallum, H.D., Harvell, C.D., and Dobson, A.P. 2003. Rates of spread of marine pathogens. *Ecology Letters* 6: 1062–7

McCallum, H.D., Kuris, A., Harvell, C.D., Lafferty, K.D., Smith, G.W., and Porter, J. 2004. Does terrestrial epidemiology apply to marine systems? *Trends in Ecology and Evolution* 19: 585–91

McCallum, H.D., Gerber, L., and Jani, A. 2005. Does infectious disease influence the efficacy of marine protected areas? A theoretical framework. *Journal of Applied Ecology* 42: 688–98

McCallum, H.D., Fenton, A., Hudson, P.J., Lee, B., Levick, B., Norman, R., Perkins, S.E., Viney, M., Wilson, A.J., and Lello, J. 2017. Breaking beta: deconstructing the parasite transmission function. *Philosophical Transactions of the Royal Society B* 372: 20160084

Metcalf, C.J.E., Graham, A.L., Martinez-Bakker, M., and Childs, D.Z. 2016. Opportunities and challenges of integral projection models for modelling host–parasite dynamics. *Journal of Animal Ecology* 85: 343–55

Miller, A.W. and Richardson, L.L. 2014. Emerging coral diseases: a temperature-driven process? *Marine Ecology* 36: 1–14

Moore, J.D., Robbins, T.T., and Friedman, C.S. 2000. Withering syndrome in farmed red abalone, *Haliotis rufescens*: thermal induction and association with a gastrointestinal Rickettsiales-like prokaryote. *Journal of Aquatic Animal Health* 12: 26–34

Moss, J., Butler IV, M.J., Behringer, D.C., and Shields, J.D. 2012. Genetic diversity of the Caribbean spiny lobster virus, *Panulirus argus* virus 1 (PaV1), and the discovery of PaV1 in lobster postlarvae. *Aquatic Biology* 14: 223–32

Moss, J., Behringer, D., Shields, J.D., Baeza, A., Aguilar-Perera, A., Bush, P.G., Dromer, C., Herrara-Moreno, A.,

Gittens, L., Matthews, T.R., McCord, M.R., Schärer, M.T., Reynal, L., Truelove, N., and Butler IV, M.J. 2013. Distribution, prevalence, and genetic analysis of *Panulirus argus* virus 1 (PaV1) from the Caribbean Sea. *Diseases of Aquatic Organisms* 104: 129–40

Murray, A.G., O'Callaghan, M., and Jones, B. 2001a. A model of transmission of a viral epidemic among schools within a shoal of pilchards. *Ecological Modelling* 144: 245–59

Murray, A.G., O'Callaghan, M., and Jones, B. 2001b. Simple models of massive epidemics of herpesvirus in Australian (and New Zealand) pilchards. *Environment International* 27: 242–8

Myers, R.A. and Hoenig, J.M. 1997. Direct estimates of gear selectivity from multiple tagging experiments. *Canadian Journal of Fisheries and Aquatic Sciences* 54: 1–9

Paillard, C., Jean, F., Ford, S.E., Powell, E.N., Klinck, J.M., Hofmann, E.E., and Flye Sainte Marie, J. 2014. A theoretical individual-based model of brown ring disease in Manila clams, *Venerupis philippinarum*. *Journal of Sea Research* 91: 15–34

Pike, A.W. and Wadsworth, S.L. 2000. Sea lice on salmonids: their biology and control. *Advances in Parasitology* 44: 233–337

Pikitch, E.K., Cantora, C., Babcock, E.A., Bonfil, R., Conover, D.O., Dayton, P., Doukakis, P., Fluharty, D., Heneman, B., Houde, E.D., Link, J., Livingston, P.A., Mangel, M., McAllister, M.K., Pope, J., and Sainsbury, K.J. 2004. Ecosystem-based fisheries management. *Science* 305: 346–7

Poulin, R. 1995. Evolution of parasite life history traits: myths and reality. *Parasitology Today* 11: 342–5

Powell, E.N. and Hofmann, E.E. 2015. Models of marine molluscan diseases: trends and challenges. *Journal of Invertebrate Pathology* 131: 212–25

Powell, E.N., Klinck, J.M., Ford, S.E., Hofmann, E.E., and Jordan, S.J. 1999. Modelling the MSX parasite in eastern oyster (*Crassostrea virginica*) populations. III. Regional application and the problem of transmission. *Journal of Shellfish Research* 18: 517–38

Powell, E.N., Klinck, J.M., Guo, X., Ford, S.E., and Bushek, D. 2011. The potential for oysters, *Crassostrea virginica*, to develop resistance to Dermo disease in the field: evaluation using a gene-based population dynamics model. *Journal of Shellfish Research* 30: 685–712

Powell, E.N., Klinck, J.M., Guo, X., Hofmann, E.E., Ford, S.E., and Bushek, D. 2012. Can oysters *Crassostrea virginica* develop resistance to Dermo disease in the field: the impediment posed by climate cycles. *Journal of Marine Research* 70: 309–55

Reardon, K.M., Wilson, C.J., Gillevet, P.M., Sikaroodi, M., and Shields, J.D. 2018. Increasing prevalence of epizootic shell disease in American lobster from the nearshore Gulf of Maine. *Bulletin of Marine Science* 94: 903–21

Regoes, R.R., Ebert, D., and Bonhoeffer, S. 2002. Dose-dependent infection rates of parasites produce the Allee effect in epidemiology. *Proceedings of the Royal Society B: Biological Sciences* 269: 271–9

Shaw, D.J. and Dobson, A.P. 1995. Patterns of macroparasite abundance and aggregation in wildlife populations: a quantitative review. *Parasitology* 111: S111–33

Shields, J.D. and Behringer Jr, D.C. 2004. A new pathogenic virus in the Caribbean spiny lobster *Panulirus argus* from the Florida Keys. *Diseases of Aquatic Organisms* 59: 109–18

Sokolow, S.H., Foley, P., Foley, J.E., Hastings, A., and Richardson, L.L. 2009. Disease dynamics in marine metapopulations: modelling infectious diseases on coral reefs. *Journal of Applied Ecology* 46: 621–31

Stormoen, M., Skjerve, E., and Aunsmo, A. 2013. Modelling salmon lice, *Lepeophtheirus salmonis*, reproduction on farmed Atlantic salmon, *Salmo salar*. *Journal of Fish Diseases* 36: 25–33

Sutherland, K.P., Porter, J.W., and Torres, C. 2004. Disease and immunity in Caribbean and Indo-Pacific zooxanthellate corals. *Marine Ecology Progress Series* 266: 283–302

Swinton, J., Harwood, J., Grenfell, B.T., and Gilligan, C.A. 1998. Persistence thresholds for phocine distemper infection in harbour seal *Phoca vitulina* metapopulations. *Journal of Animal Ecology* 67: 54–68

Thomas, A.C., Pershing, A.J., Friedland, K.D., Nye, J.A., Mills, K.E., Alexander, M.A., Record, N.R., Weatherbee, R., and Henderson, M.E. 2017. Seasonal trends and phenology shifts in sea surface temperature on the North American northeastern continental shelf. *Elementa Science of the Anthropocene* 5: 48

Thorpe, R.B., Le Quesne, W.J.F., Luxford, F., Collie, J.S., and Jennings, S. 2015. Evaluation and management implications of uncertainty in a multispecies size-structured model of population and community responses to fishing. *Methods in Ecology and Evolution* 6: 49–58

Virgin, H.W., Wherry, E.J., and Ahmed, R. 2009. Redefining chronic viral infection. *Cell* 138: 30–50

Wang, Z., Haidvogel, D.B., Bushek, D., Ford, S.E., Hofmann, E.E., Powell, E.N., and Wilkin, J. 2012. Circulation and water properties and their relationship to the disease, MSX, in Delaware Bay. *Journal of Marine Research* 70: 279–308

Whittington, J.R., Jones, J.B., and Hyatt, A.D. 1997. Pilchard herpesvirus in Australia 1995–1999. *Diseases in Asian Aquaculture* 5: 137–40

Wilber, M.Q., Weinstein, S.B., and Briggs, C.J. 2016. Detecting and quantifying parasite-induced host mortality from intensity data: method comparisons and limitations. *International Journal for Parasitology* 46: 59–66

Wood, C.L. and Lafferty, K.D. 2015. How have fisheries affected parasite communities? *Parasitology* 142: 134–44

Wood, C.L., Lafferty, K.D., and Micheli, F. 2010. Fishing out marine parasites? Impacts of fishing on rates of parasitism in the ocean. *Ecology Letters* 13: 761–75

Wood, C.L., Micheli, F., Fernandez, M., Gelich, S., Castilla, J.C., and Carvajal, J. 2013. Marine protected areas facilitate parasite populations among four fished host species in central Chile. *Journal of Animal Ecology* 82: 1276–87

Yakob, L. and Mumby, P.J. 2011. Climate change induces demographic resistance to disease in novel coral assemblages. *Proceedings of the National Academy of Sciences USA* 108: 1967–9

Ying, Y., Chen, Y., Lin, L., and Gao, T. 2011. Risks of ignoring fish population spatial structure in fisheries management. *Canadian Journal of Fisheries and Aquatic Sciences* 68: 2101–20

CHAPTER 13

Future directions for marine disease research

Rebecca Vega Thurber

13.1 The intersectional nature of disease

In 2002, my father officially died of kidney failure. I was at his bedside. One week earlier, while I was attending the American Society for Cell Biology Meeting in San Francisco, he was diagnosed with double lobe pneumonia and admitted to the hospital for monitoring and IV antibiotics. Two years earlier, he could not remember his assistant's name and accurately diagnosed himself as having glioblastoma, a severe form of brain cancer that is associated with exposure to formaldehyde, a compound my pathologist father was persistently exposed to in his line of work.

When people ask me from what disease my father died, I often hesitate because I am not sure myself. And when people ask why it is so hard for us to define the etiological agents of disease in marine systems, I often think of the complex nature of my father's disease and death. Had I only been provided evidence on the day he died, I would have surmised that he died of multi-organ failure driven by a loss of kidney function. Had I amplified the 16S rRNA gene from within his lungs one week before he succumbed, I could have hypothesized he was suffering from an acute *Pseudomonas* bacterial infection. Had I used metagenomics, I may have guessed, more likely, that his lungs were filled with a diverse commensal community of bacterial and fungal taxa

most of which are not associated with pneumonia. Should I have been the surgeon who removed the baseball-sized tumor from his frontal lobe, I would have said he died of cancer. And lastly, had I been the environmental health and safety individual at his clinical laboratory, I might have assumed he died of excessive inhalation exposure to a known carcinogen. So, which is it then? Answering this question is hard enough in a well-studied human patient under continuous physician care, but can be even more elusive when the host is a marine organism, and the deathbed is underwater.

We struggle to define the true root causes of illness, whether it be in our own hominid species or in basal invertebrates. This is because context and surveillance are essential for accurate inference. The complex nature of disease is driven by nuanced interactions of host genotype, host phenotype, the pathogen state and permissibility, and the past or prevailing environmental conditions that create a myriad set of possible disease scenarios. To borrow a term from sociology, all diseases are likely "intersectional," referring to the complexity of factors that every individual and every population and its pathogens experience. The "intersectional" context of every host and every disease outbreak matter to epidemiologists, and they should matter to every marine disease ecologist as well. Therein lies the issue of modern-day marine disease ecology; context

Thurber, R.V., *Future directions for marine disease research* In: *Marine Disease Ecology*. Edited by: Donald C. Behringer, Kevin D. Lafferty, and Brian. R. Silliman, Oxford University Press (2020). © Oxford University Press.
DOI: 10.1093/oso/9780198821632.003.0013

and timing, we now know, are everything. As presented in this textbook, it is clear that we have remarkably sophisticated tools to track shifts in host physiology, microbial community dynamics, and ecosystem variation. However, what we often lack when new diseases emerge is basic surveillance-style data of the hosts, the pathogens, and even the food-web dynamics of an ecosystem that are necessary to interpret context and define the origins and mechanisms of marine disease. For instance, there is a large diversity of marine pathogens and parasites (chapter 1—Bateman et al.), and many occur naturally in functional marine ecosystems (chapter 2—McLaughlin et al.), where they can alter marine community structure (chapter 3—Morton et al.), or interact to drive parasite virulence (chapter 4—Little et al.). Yet the central challenges in our field are: (1) how do we know when changes to marine systems interact to cause some of these diseases (chapter 5—Burge and Hershberger; chapter 6—Bojko et al.; chapter 7—Pagenkopp Lohan et al.; chapter 8—Harvell and Lamb; chapter 9—Raymundo et al.; chapter 10—Behringer et al.); (2) how do we accurately identify and define all the factors and their interactions responsible for these problems (chapter 11—Frasca et al.; chapter 12—Ben-Horin et al.); and lastly (3) how do we synthesize these data in order to minimize these issues and preserve marine species in the future (chapter 9—Raymundo et al; chapter 10—Behringer et al.; chapter 12—Ben-Horin et al.)?

13.2 The context dependency of disease

Interactions between a host and potential pathogens, while sufficient, are not necessary for disease. Host–microbe relationships can be contextual, where changes in environmental conditions can cause a non-permissive host to become susceptible to infection and/or a commensal parasite to switch from a non-virulent state to a pathogenic one (chapter 4—Little et al.; chapter 5—Burge and Hershberger; chapter 6—Bojko et al.). Thus, disease is at the intersection of (1) factors influencing the virulence of the potential pathogen(s), (2) the variable environmental conditions, and (3) host susceptibility. This context dependency is often the reason that asymptomatic carriers of potential pathogens do not dis-

play any visual signs or symptoms of disease. These phenomena also exist in humans, for example, with the re-emergence of varicella zoster virus to create a shingles outbreak (Arvin 2005), the transition of *Clostridium difficile* from a low abundance and non-virulent member of the human microbiome to a deadly opportunistic pathogen (Loo et al. 2011), and the clinical manifestation of meningococcal disease in school-aged children and young adults, but the absence of the disease in older carriers (Rouphael & Stephens 2012).

In marine organisms, identifying the proper alignment of host–pathogen–environment remains challenging due to a paucity of models that incorporate the environment, food-web dynamics, and the genotypes and phenotypes of both the host and pathogen (chapter 12—Ben-Horin et al.). Yet, as we have read in this book, disease is a major cause of the loss of critical foundational marine species such as seagrass, corals, marsh grass, and oysters (chapter 8—Harvell and Lamb). Corals represent an extreme case, where disease in either one or a few species has caused not only extensive habitat loss, but also entire ecosystem phase shifts (chapter 8—Harvell et al.; chapter 3—Morton et al.). Yet, at present, we remain uncertain about which pathogenic agents were the sources of the epizootics that expatriated these foundational species from the Caribbean (chapter 9—Raymundo et al.). While the root causes of these and other coral diseases are likely environmentally and genetically related, the mechanism by which most corals die has been hypothesized to be microbial in nature (chapter 4—Little et al.), and little work has been done on how ecosystem-level processes, outside of acute stress factors, may be related to such epizootics. These mysteries are likely to increase as invasive species spread new pathogens across the oceans (chapter 7—Pagenkopp Lohan et al.).

Importantly, given that disease rates are hypothesized to increase alongside increasing levels of environmental degradation (chapter 5—Burge and Hershberger; chapter 6—Bojko et al.; chapter 8—Harvell and Lamb and invasion (chapter 7—Pagenkopp Lohan et al.), understanding the linkages among hosts, their potential pathogens, and the necessary and sufficient environmental conditions to induce disease is critical if we are to mitigate their effects (chapter 9—Raymundo et al.). Yet, with the exception of non-commercially important species discussed here (chapter 2—McLaughlin et al.; chapter 3—Morton et al.), it remains to be seen if we can actually be successful at this for most marine hosts, as we are generally in the business of characterizing diseases after they have emerged, instead of actively conducting surveillance of the marine system (chapter 11—Frasca et al.; chapter 12—Ben-Horin et al.).

13.3 Marine epizootics: disease ecology without epidemiology

As discussed in this book, although we once relied on more simplistic models of disease (e.g., one disease and one pathogen, and those that conform to Koch's postulates), we now know that the origins of the majority of pathologies tend not to be straight forward (chapter 1—Bateman et al.; chapter 11—Frasca et al.; chapter 12—Ben-Horin et al.). Many of the pathologies that follow the one disease and one pathogen model are the systems we can track best (chapter 12—Ben-Horin et al.); the ones that defy this tantalizingly simplistic model tend to give us the biggest grief. As described in this collection of works, disease likely affects every macroscopic organism (and likely most single-celled organisms too) we have yet evaluated: disease can be the result of infection by parasites such as nematodes, protists, or strange metazoans such as myxozoans, or the result of single bacterial and viral pathogens (chapter 1—Bateman et al.; chapter 2—McLaughlin et al.).

However, more and more often we are finding that these diseases can be a myriad of infections that occur simultaneously or sequentially, the result of complicated ecological relationships among species and differences in pathogen virulence state. For example, black band disease of corals is caused not by one but by several metabolically interacting bacteria which together form a pathogenic cocktail (chapter 1—Bateman et al.; chapter 11—Frasca et al.). In marsh grasses, fungal diseases are driven by shifts in food-web dynamics that reduce the abundance of a keystone crab predator and the resulting increase in herbivorous snail-induced wounding and fecal deposition in the grass tissues (chapter 3—Morton et al.). The emergence or disappearance of disease also can often be the result of hidden factors such as the presence of toxin genes or bacterial cell suppression genes in prophage, respectively (chapter 4—Little et al.). *Vibrio cholera* is common in seawater, but only strains containing the CTX gene found on a phage gene cassette cause human disease. And in abalone, disease severity and infection load of the wasting disease inducing the Rickettsiales pathogen is modulated by the presence of a phage (chapter 5—Burge and Hershberger; chapter 8—Harvell and Lamb; chapter 12—Ben-Horin et al.).

13.4 Examples of marine disease surveillance in practice

There are many examples of how time series analyses allowed scientists to learn about marine disease ecology (chapter 3—Morton et al.; chapter 5—Burge and Hershberger; chapter 7—Pagenkopp Lohan et al.; chapter 8—Harvell and Lamb; chapter 12—Ben-Horin et al.). For commercially important fishery and aquaculture organisms, we know a significant amount about the implicated pathogens, mechanisms of disease, and the contextuality of outbreaks and mortality events (chapter 10—Behringer et al.). Yet the story is bleaker for emerging diseases where the host system has been underexplored microbiologically, genetically, or ecologically. And still progress is certainly being made. A recent example of how additional surveillance and holistic approaches can successfully track the nature of disease outbreaks comes from the remarkable and interdisciplinary efforts to understand the recent sea star wasting disease epizootic that has altered the ecology of temperate western coastal regions of the USA. This disease is responsible for decimating populations of the predatory sunflower star *Pycnopodia helianthoides* (chapter 3—Morton et al.; chapter 8—Harvell and Lamb). While both the implicated virus and many members of the bacterial microbiome are highly prevalent in healthy and diseased individuals, what seems to have facilitated high mortality rates in some areas was anomalous local temperatures (chapter 5—Burge and Hershberger; chapter 8—Harvell and Lamb). This important finding was only possible due to the remarkable longitudinal surveying efforts of citizen scientists that collectively gathered regional photographic, temperature, and depth information on the outbreak. Future efforts should use such approaches to harness the power of the local citizenry to gather data, or even samples, to track disease outbreaks in real time.

13.5 Moving forward by embracing tools to look at disease outbreaks in the past

In summary, we have seen many advances in the field of marine disease ecology, but we remained stymied because we often lack contextualized data and a better longitudinal record of how disease events proceed in our oceans and shore ecosystems. Moreover, for many lingering cold cases, we also need information about the past. However, such hind-casting surveillance tools are gradually coming online (chapter 11—Frasca et al.). Thus, as we amend our approaches to conduct better modern-day surveillance, we can also glimpse into the past to understand what may have occurred previously.

Such information is essential to understanding what contributes to disease resistance and resilience of marine host populations, two aspects of marine disease ecology we need to explore in order to make conservation and mitigation efforts more effective in the future.

Acknowledgments

This work was funded in part by NSF DOB grant #1442306 to R.V.T.

References

Arvin, A. 2005. Aging, immunity, and the varicella-zoster virus. *New England Journal of Medicine* 352 (22): 2266–67.

Loo, V.G., A.-M. Bourgault, L. Poirier, F. Lamothe, S. Michaud, N. Turgeon, B. Toye, et al. 2011. Host and pathogen factors for *Clostridium difficile* infection and colonization. *New England Journal of Medicine* 365 (18): 1693–703.

Rouphael, N.G., and D.S. Stephens. 2012. *Neisseria meningitidis*: biology, microbiology, and epidemiology. *Methods in Molecular Biology* 799: 1–20.

Index

Note: Tables, figures, and boxes are indicated by an italic *t*, *f*, and *b* following the page/paragraph number.